HANDBOOK OF OZONE TECHNOLOGY AND APPLICATIONS

volume one

Rip G. Rice and
Aharon Netzer, Editors

HANDBOOK OF OZONE TECHNOLOGY AND APPLICATIONS

volume one

230 Collingwood, P.O. Box 1425, Ann Arbor, Michigan 48106

Library of Congress Catalog Card Number 81-70869
ISBN 0-250-40324-2

Manufactured in the United States of America

Butterworths, Ltd., Borough Green, Sevenoaks,
Kent TN15 8PH, England

PREFACE

In the last 30 years only two treatises of note have been published in the United States dealing with ozone technology and its applications. The American Chemical Society (ACS) published *Ozone Chemistry and Technology,* Advances in Chemistry Series, No. 21 (American Chemical Society, Washington, DC, 1959). In 1972 F. L. Evans III edited a collection of papers presented at an ACS meeting held in Washington, DC, in late 1970. This book, *Ozone in Water and Wastewater Treatment* (Ann Arbor Science Publishers, Ann Arbor, MI, 1972), addressed that segment of ozone technology of most current interest at the time.

Since these two publications, a number of significant milestones in ozone technology and its applications have been passed, to wit:

1. The many uses of ozone in treating drinking water and swimming pool waters, which have been developed in Europe, are expanding into other continents of the world.
2. Ozone disinfection of municipal wastewater is being practiced full-scale in 19 U.S. sewage treatment plants.
3. Ozone treatment of various industrial wastewaters is being practiced commercially, sometimes for reuse purposes (e.g., metal finishing wastewaters, photoprocessing bleach baths).
4. Ozone treatment of cooling tower waters for air-conditioning systems has reached the commercial scale in the United States.
5. New types of ozone generators now are available that can produce ozone at significantly lower power requirements.
6. Catalytic ozonation, the simultaneous application of ozone along with ultraviolet radiation, ultrasonics or hydrogen peroxide, has been shown to increase significantly the rates of some otherwise slow oxidations with ozone alone.
7. Ozone treatment of municipal wastewater, followed by granular activated carbon adsorption, is being studied on large pilot-plant-scale in South Africa for potable water reuse purposes and in the United States for combined municipal-industrial wastewater treatment.
8. Medical therapeutic applications of ozone are being practiced more and more routinely by medical professionals in many countries.

In spite of these recent activities, which have speeded the evolution of ozone technology and its applications, no authoritative compendium dealing with each principal aspect of the field has been assembled to date. In an attempt to fill this void, the present editors have assembled the best available cadre of experts to describe fundamental principles of ozone technology and to discuss the parameters that determine its successful application. This undertaking will result in four volumes, each containing authoritative presentations of fundamental principles and practical experiences in the applications of ozone technology.

This first volume addresses the history of ozone technology, methods of generation, principles of mass transfer and contacting of ozone with aqueous systems, effects of ozone on materials of construction, analysis and control of ozone in aqueous systems, engineering aspects of ozonation systems and mechanisms of reactions of ozone with organic materials in aqueous media.

Other volumes in this set will address the following subjects:

Ozone Treatment of Drinking Water
Ozone Treatment of Municipal Wastewaters
Ozone Treatment of Industrial Water and Wastewater
Ozone Treatment of Swimming Pool Waters
Odor Control with Ozone
Ozone Oxidation of Organic Materials in Aqueous and Nonaqueous Solvents
Catalytic Ozonation
Atmospheric Ozone Technology
Medical Therapy with Ozone
Health and Safety Aspects of Ozone Technology

It is our hope that this compendium will aid those involved in the various aspects of ozone technology and its applications.

The editors are particularly grateful to the authors who have provided authoritative chapters for this series and to the many other professionals who have reviewed these contributions for accuracy and clarity of presentation.

Rip G. Rice
Aharon Netzer

Rice

Netzer

Rip G. Rice is Director of Environmental Systems for the Jacobs Engineering Group Advanced Systems Division and manages the Jacobs Washington, DC, office. Dr. Rice co-founded the International Ozone Institute (now the International Ozone Association) in 1973, and is President of the IOA Board of Directors for 1982–1983. He has authored or co-authored more than 25 papers dealing with various aspects of ozone technology, and is a consultant and invited lecturer on the application of ozone technologies to water and wastewater treatment. Dr. Rice has also edited or co-edited eight proceedings or monographs in the field of ozone technology and is co-author of a book entitled *Ozone for Industrial Water and Wastewater Treatment.*

Dr. Rice holds an Associate of Science degree from the University of Texas at Arlington, a BS (chemistry) from George Washington University and a PhD (organic chemistry) from the University of Maryland. He has 33 years of experience in various positions dealing with chemical research, engineering and environmental sciences with various agencies of the U.S. government and several private industrial firms.

Aharon Netzer is Professor of Water and Wastewater Engineering at the University of Texas at Dallas. He was previously associated with the Canada Center for Inland Waters and McMaster University. Dr. Netzer received his BSc, MSc and PhD degrees from The Hebrew University. His main research interests are water and wastewater engineering, advanced water and wastewater treatment, industrial wastewater treatment, and toxic substances treatment. Dr. Netzer has authored more than 60 publications in these areas. He is a member of several professional organizations and is on the Board of Directors of the International Ozone Association. Dr. Netzer organized and was General Chairman of several international and national conferences on ozone technology and its applications.

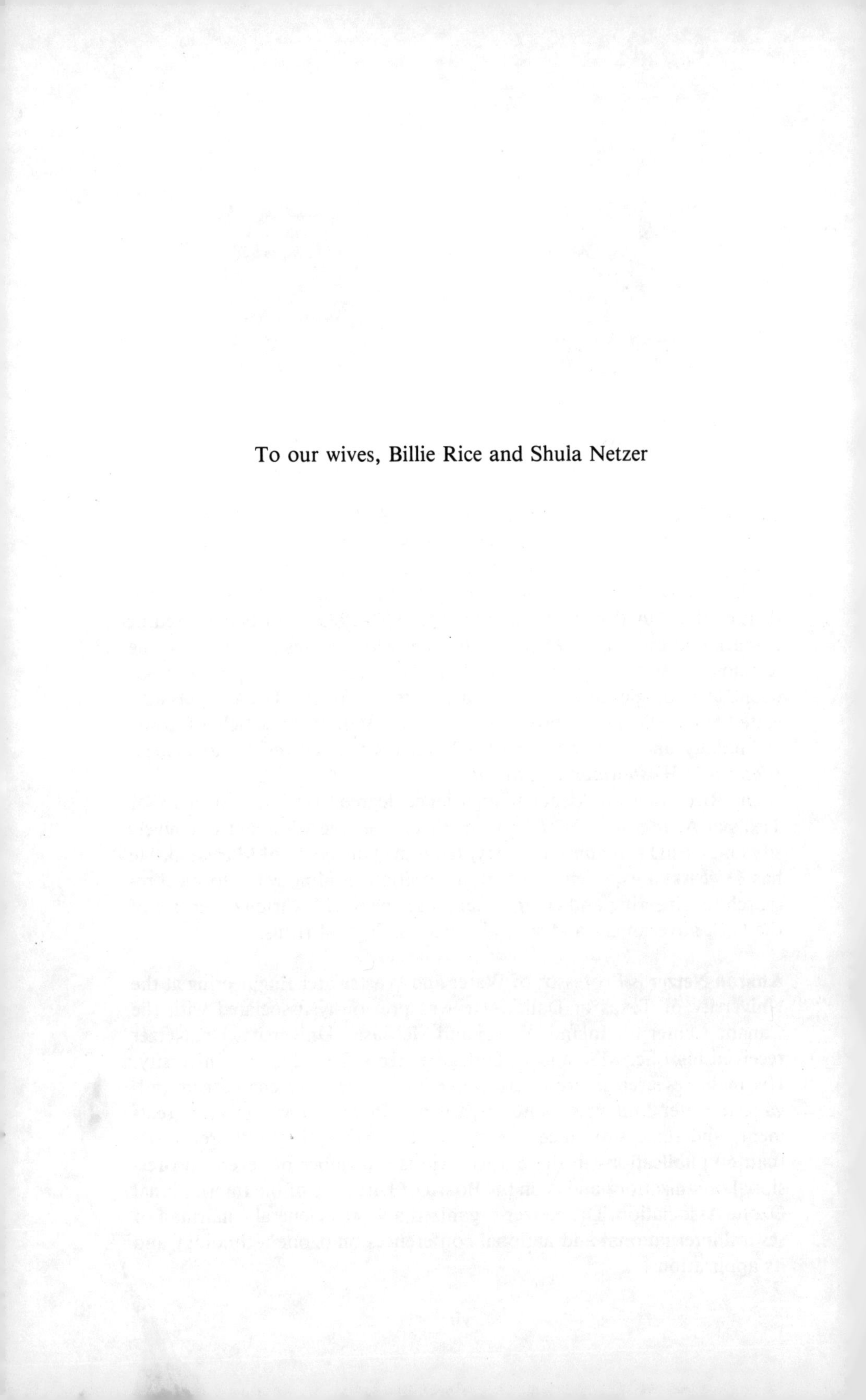

To our wives, Billie Rice and Shula Netzer

CONTENTS

Section 5: Engineering Aspects

Section 6: Mechanisms of Organic Oxidations

CHAPTER 1

HISTORICAL BACKGROUND, PROPERTIES AND APPLICATIONS

Archibald G. Hill

School of Chemical Engineering
Oklahoma State University
Stillwater, Oklahoma

Rip G. Rice

Jacobs Engineering Group Inc.
Washington, DC

Most of the general public today is aware of the term "ozone," primarily as a consequence of recent press releases concerning the threats to the environment that may result from destruction of the atmospheric ozone layer by chlorofluorocarbons and other pollutants. The discovery that ozone is a constituent of photochemical smog has led to additional opportunities for public familiarity through occasional media announcements of "ozone alerts" in areas subject to smog attacks. The fact that ozone may be produced synthetically for beneficial purposes is less well known. However, it was not very long after recognition that ozone was a new chemical oxidizing substance that means were developed to produce this compound in commercially significant quantities.

The purpose of this chapter is to trace the development of ozone technology, from its discovery through its early applications, particularly in the areas of water and wastewater treatment. Each of the major factors affecting the generation and application of ozone for a specific purpose and its major applications will be discussed in detail in subsequent chapters.

PROPERTIES AND STRUCTURE

Ozone (O_3) is an unstable gas having a pungent, characteristic odor. It is formed photochemically in the earth's stratosphere, but exists at ground levels only in low concentrations. At ordinary temperatures, ozone is a blue gas, but at the concentrations at which it is ordinarily produced, this color is not noticeable unless the gas is viewed through a considerable depth. Some of the major physical properties of ozone are listed in Table I [1,2].

At −112°C, ozone condenses to a dark blue liquid that explodes easily. Concentrated ozone-oxygen mixtures (above about 20% ozone) also are easily exploded, either in the liquid or vapor state. Such explosions may be initiated by small amounts of catalysts, organic matter, shocks, electrical sparks, or sudden changes in temperature or pressure. However, under the conditions whereby ozone is generated commercially, concentrations of ozone in oxygen above 10% cannot be obtained conveniently, and no instances of ozone explosions have been reported during its long history of use in treating drinking water.

Ozone gas is sparingly soluble in water, about 13 times more soluble than oxygen at standard temperature and pressure (STP) (see below). It decomposes back to oxygen, from which it is formed, rapidly in aqueous solution containing impurities, but more slowly in pure water or in the gaseous phase. The rate of ozone decomposition in water is affected greatly by the purity of water and the cleanliness of the glassware in which decomposition experiments are conducted. For example, Figure 1 [3] shows that the half-life of ozone in distilled or tap water (source unspecified) is about 20 min at 20°C. However, in double-distilled water, only 10% of the ozone decomposed over a period of 85 min. Longer half-lives have been measured at lower temperatures [4].

Structure

On the basis of the works of many investigators, the structure of the ozone molecule is described by Bailey [5] as a resonance hybrid of the four canonical forms represented by structure 1 in Figure 2. A simplified molecular orbital description of the molecule is illustrated by structure 2 in Figure 2.

EARLY HISTORY

The distinctive odor of ozone was associated with its first recorded observations and later inspired its discoverers with the basis for its name.

Table I. Properties of Pure Ozone [1,2]

Melting Point (°C)	-192.5 ± 0.4
Boiling Point (°C)	-111.9 ± 0.3
Critical Temperature (°C)	-12.1
Critical Pressure (atm)	54.6
Critical Volume (cm^3/mol)	111

Temperature (°C)	Liquid Density (g/cm^3)	Liquid Vapor Pressure (torr)
−183	1.574	0.11
−180	1.566	0.21
−170	1.535	1.41
−160	1.504	6.73
−150	1.473	24.8
−140	1.442	74.2
−130	1.410	190
−120	1.378	427
−110	1.347	865
−100	1.316	1605

Density of Solid Ozone at 77.4°K (g/cm^3)	1.728
Viscosity of Liquid (cP)	
At 77.6°K	4.17
At 90.2°K	1.56
Surface Tension (dyn-cm)	
At 77.2°K	43.8
At 90.2°K	38.4
Parachor[a] at 90.2°K	75.7
Dielectric Constant, liquid, at 90.2°K	4.79
Dipole Moment (Debye)	0.55
Magnetic Susceptibility (cgs units)	
Gas	0.002×10^{-6}
Liquid	0.150
Heat Capacity of Liquid from 90 to 150°K (Cp)	$0.425 + 0.0014(T - 90)$
Heat of Vaporization (kcal/mol)	
At −111.9°C	3410
At −183°C	3650

	Heat of Formation (kcal/mol)	Free Energy of Formation (kcal/mol)
Gas at 298.15°K	34.15	38.89
Liquid at 90.15°K	30.0	
Hypothetical Gas at 0°K	34.74	

[a] $M\gamma^{1/4}(D - d)$, where M = molecular weight, γ = surface tension, D = liquid density, d = vapor density.

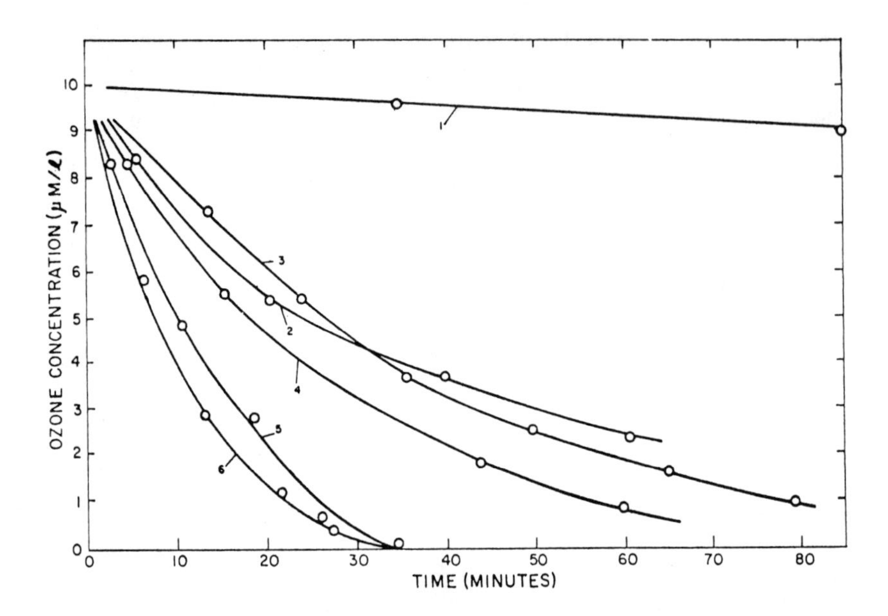

Figure 1. Ozone decomposition in different types of water at 20°C. 1 = double-distilled water; 2 = distilled water; 3 = tap water; 4 = groundwater of low hardness; 5 = filtered water from Lake Zurich; 6 = filtered water from the Bodensee [3].

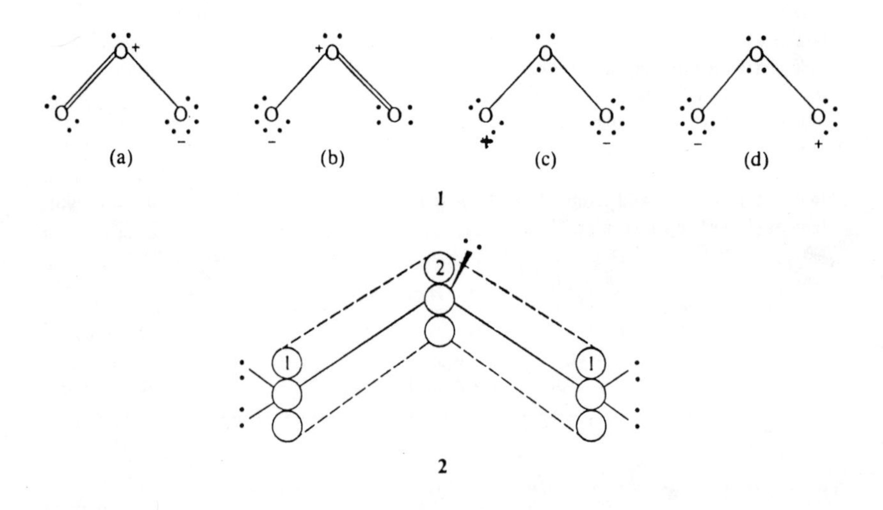

Figure 2. The ozone molecule [5].

Homer [6] noticed the smell that accompanies a thunderbolt and included his impressions in his *Iliad* and *Odyssey* [7]. Thus in Book XII, verse 417 of the *Odyssey,* Jupiter strikes a ship with a thunderbolt "quite full of sulphurous odor." In 1785 van Marum [8], a Dutch physicist experimenting with his powerful electrical machine, observed that air acquired a characteristic odor when subjected to the passage of a series of electrical sparks. Cruickshank [8] in 1801 observed the same odor in the gas formed at the anode during the electrolysis of water.

Schönbein announced the discovery of ozone in a memoir that he presented to the Academy of Munich in 1840 [9]. He had observed a peculiar odor during electrolysis and sparking experiments. He also recognized that this was the same odor observed after a flash of lightning. Schönbein concluded that this odor was due to a new substance, to which he gave the name "ozone," derived from the Greek word "ozein," meaning "to smell."

At first, Schönbein suggested that this odorous substance might be a new electronegative element belonging to the same class as chlorine and bromine. Van Marum had observed previously [10] that electrified air had the power to attack mercury. Schönbein found that ozone would attack potassium iodide in aqueous solution, liberating free iodine. He also observed that ozone could be destroyed on being passed through a heated glass tube. In his experiments conducted at Basle, Schönbein recorded [10] that ozone could be formed by passing moist air over a stick of phosphorus. He further claimed to have produced ozone by plunging a heated glass rod into a mixture of air and vapors of ether. Using an ozone test paper prepared by soaking in potassium iodide, he concluded that ozone occurs naturally in the ambient atmosphere.

In 1845 de la Rive and Marignac [11] obtained ozone by subjecting pure dry oxygen to the action of an electric spark. In 1848 Hunt [12] advanced the hypothesis that ozone is triatomic oxygen, drawing his conclusions from its then known properties. In 1860 Andrews and Tait [13] reported that oxygen gas, in being converted into ozone, diminishes in volume, but recovers its original volume when the ozone changes back into oxygen by the action of heat or otherwise. These workers found that small amounts of mercury or metallic silver had the power to destroy ozone.

In 1866 Soret [14] experimented with a mixture of oxygen and ozone obtained by electrolysis. He showed that when this mixture is brought into contact with oil of turpentine or oil of cinnamon, a reduction in volume takes place that is twice the amount of expansion the same mixture would sustain if the ozone were converted by heat into ordinary oxygen. Hence, Soret concluded that the density of ozone is 1.5 times that of oxygen gas. To check this result, Soret determined the rate at which ozone diffuses into air, and compared it with the rate for carbon monoxide, determined similarly. From the two rates, and on the basis of Graham's law, he calculated the ratio of the density of ozone to that of carbon dioxide and found it to be in satisfactory accordance with $O_3:CO_2 = 48:44$.

Ozone technology was advanced greatly by the development of ozone generating tubes, introduced by von Siemens in 1857 [15]. This type of ozone generator has subsequently served as a prototype for the majority of presently used electric discharge generators. Siemens' first ozone generator consisted essentially of two glass tubes, the outer coated externally and the inner coated internally with tin foil. Air feed gas was passed through the annular space. The metallic surfaces of the inner and outer tubes were connected to the terminals of an induction coil or electrical machine. Using such an arrangement, 3–8% of dry oxygen could be transformed into ozone. Modified versions of this apparatus were used by Brodie [1] and Bertholet [7], both of whom substituted electrolyte solutions for the metallic electrodes. This change allowed a certain amount of cooling to take place during ozone generation.

The thermochemical properties of ozone were investigated in 1868 by Hollman [16], who compared the heat released when various gases and vapors (H_2, C_2H_2 and others) were burned in a stream of pure oxygen, to the amount of heat generated when the same substances were burned in a stream of ozonated oxygen. He found that a greater quantity of heat was always released in the presence of ozone. Knowing the concentration of his ozone/oxygen mixture (1–2% by weight) and the incremental amount of heat generated in the presence of ozone, he was able to calculate the amount of heat released per gram of ozone transformed into ordinary oxygen. He determined a value of 355.5 cal/g ozone decomposed, which is equivalent to 17,064 calories (17.064 kcal)/g-mol. This is about half of the currently accepted value [1,2]. Later determinations by Berthelot in 1876 [7] gave 29.6 kcal, and Jahn [17] determined a value of 34.0 kcal in 1908. The discovery that ozone is an endothermic compound led to a number of attempts to design ozone generators based on thermal processes. These have failed due to the rapid decomposition of ozone at elevated temperatures.

Spontaneous decomposition of gaseous-phase ozone occurs over a period of hours at ambient temperatures. Similar decomposition takes place when ozone is dissolved in water, however at a faster rate, usually measurable in minutes. However, the quality of the water in which the stability of ozone is measured is very important. Thus, Figure 1 [3] shows the rates of decomposition of ozone in double-distilled water, distilled water, tap water, groundwater of low hardness and two filtered lake waters, all measured at 20°C. Although the half-life of ozone in distilled water is seen to be about 25 min, a statement that is to be found repeatedly throughout the ozone literature, it should be noted that in double-distilled water, ozone is only 10% decomposed after 80 min, even at 20°C. As water temperatures approach 0°C, ozone becomes even more stable [4].

Because of the variation in decomposition rates of ozone in aqueous solution, wide discrepancies for the solubility of ozone in water have been reported. Early values for the solubility of ozone in water often have been given in terms of the Bunsen coefficient (liters of gas in one liter of water). In 1873 Schöne [18] obtained a solubility value of 0.366 liter/liter water at 18°C; in 1874 Carius [19] found a value of 0.834 liter/liter water at 1°C. In 1894 Mailfert [20] determined the following values:

Temperature (°C)	Solubility (liter/liter Water)
0	0.64
11.8	0.5
15	0.456
19	0.381
27	0.27
40	0.112
55	0.031
60	0

These values are about 10–15 times the corresponding values for the solubility of oxygen.

Venosa and Opatken [21] pointed out that the fundamental relationship governing the solubility of ozone (or any gas) in a liquid is Henry's law. Expressed simply, Henry's law states that the weight of any gas that will dissolve in a given volume of liquid, at constant temperature, is directly proportional to the partial pressure that the gas exerts above the liquid [22]. Expressed as an equation:

$$Y = HX$$

where Y = partial pressure of the gas above the liquid (mm Hg)
X = concentration of the gas in the liquid at equilibrium with the gas above the liquid (mol gas/total mol gas + liquid)
H = Henry's law constant, which varies with temperature (mm Hg/mol fraction)

The terms in the above equation can be converted into units of concentration, such that:

- Y = concentration of gas above the liquid in equilibrium with the gas dissolved in the liquid (mg/l)
- X = concentration of gas in the liquid in equilibrium with the gas above the liquid (mg/l)
- H = [mg gas/l gas]/[mg gas/l liquid]

Thus Henry's law simply expresses the relationship between the concentration of gas above the liquid that must exist for a given concentration of gas to be dissolved in the liquid. The lower the value of H, the more soluble is the gas.

After converting Henry's law constants (taken from the International Critical Tables) to units of concentration, Venosa and Opatken [21] compared the solubilities of oxygen and ozone (generated at 1% in air) in water at temperatures of 0, 10, 20 and 30°C (Table II). They also pointed

Table II. Solubility of Ozone and Oxygen in Water According to Henry's Law [21]

	Temp (°C)	H (mg gas/l air per mg gas/l water)	Y (mg gas/ l air)	X (mg gas/ l water)
Oxygen (air)	0	20.4	299	14.6
	10	25.4	289	11.4
	20	29.9	279	9.3
	30	34.2	270	7.9
Ozone, 1.0 wt %				
	0	1.56	12.9	8.3
	10	1.86	12.5	6.7
	20	2.59	12.1	4.7
	30	3.80	11.7	3.1

out that the magnitude of Henry's constant is a function of temperature alone, not concentration. The values of H for oxygen in Table II are the same, whether the gaseous-phase oxygen concentration is 21% (air) or 100%. It is clear from the data of Table II that ozone is approximately 13 times more soluble than oxygen at STP (H = 20.4 for oxygen vs 1.56 for ozone) over the termperature range presented.

The observation of the stabilizing effect of low temperature on ozone led to an investigation of the ozone generation process under these conditions. Hautefeuille and Chappuis [23] passed an electric discharge through oxygen gas at very low temperatures. They obtained concentrations of 14.9% by weight of ozone at 0°C and 21.4% at −23°C. Subsequently these workers [24] succeeded in producing liquid ozone by applying a pressure of 125 atm to richly ozonated oxygen at −100°C. Liquid ozone was found to have a dark indigo-blue color. Gaseous ozone had to be compressed slowly with constant cooling, otherwise an explosion was likely to occur.

Values of the boiling point of ozone were determined by Olszewski [25,26] in 1887, who reported a range of −106 to −109°C and Troost [27] in 1898, who reported −119°C. The value for the boiling point of ozone accepted today is −111.9°C [1,2].

SEARCH FOR APPLICATIONS

The powerful oxidizing effect of ozone had been observed in many different reactions before the molecular formula for ozone was determined and also before the introduction of the Siemens ozone generator. Reactions of ozone with organic matter were studied by Schönbein [28–31], Baumert [32] and von Gorup-Besanez [33,34]. Wood, straw, cork, starch, humus, vegetable colors, natural rubber, fats, fatty acids, alcohol, albumin and blood all are acted on by this oxidizing reagent. The bleaching effect of ozone on indigo was used by Schönbein as a method of quantitative determination of ozone concentrations [28–30]. Andrews [14] reported that the bleaching properties of ozone were used commercially in sugar refining and linen bleaching. One of the first U.S. patents pertaining to ozone was issued in 1868 to de Gerbeth [35], and describes a process in which ozone was used to convert a coal-oil mixture into a product suitable for use in paints, varnish-making and oil-cloth. In 1870 a U.S. patent was issued to Fewson [36] for an ozone-producing apparatus through which noxious and sewer gases were passed and were deodorized.

A major reference on ozone technology was published in 1904 by de la Coux [37]. Approximately half of the 557 pages of this volume are devoted to the industrial applications of ozone. Up to that time, ozone had been used for the preservation of milk, meat products, gelatin, casein and albumin. Processes involving ozone had been developed for the purification and artificial aging of alcoholic beverages, including wine and spirits. (Although treatment of freshly prepared red wine with ozone can reduce the aging time of a Bordeaux- or Burgundy-type wine by several years, this artificial aging technique has been outlawed by French regulatory authorities [38].)

Ozone also had been applied as a selective disinfectant in brewing and cider manufacturing. Additional uses included the production of starch, oils, greases, dyes and soap. Additionally, ozone had been shown to be able to accelerate the aging and hardening of foods.

Later chapters in this book will provide details of the modern application of ozone for the treatment of drinking water, municipal and industrial wastewaters, and its uses in breweries, odor control, medical therapy and other purposes. In the balance of this chapter, however, we will trace the historical development of the application of ozone, primarily in the treatment of drinking water and municipal and industrial wastewaters, to provide a foundation for appreciating how ozone technology has evolved to this point.

POTABLE WATER TREATMENT

The earliest experiments on the use of ozone as a germicide were conducted in 1886 by de Meritens in France [37]. He showed that even dilute ozonated air could effect the sterilization of polluted water. A few years later, the bactericidal properties of ozone were reported by Frölich from tests conducted at an experimental pilot plant erected at Martinikenfeld by the firm of Siemens and Halske [37]. Baron Henry Tindal of Amsterdam developed a water disinfection process [39] that incorporated the nondielectric ozone generator patented in the United States by van der Sleen and Schneller in 1897 [40], which produced about 10 g/hr of ozone. Product gas from the ozone generator was compressed, then fed to the base of a tall bubble column equipped with perforated plates installed at intervals. The Tindal system was first operated in 1893 at Oudshoorn, the Netherlands, near Leiden, on a water flow of 3 m^3/hr. This process later was demonstrated at the Hygienic Exhibition held in Paris in 1896 and at the Brussels International Exhibition in 1897. Wessels [41] continued to develop the technique of ozonation at an experimental water treatment plant in Paris.

The German firm of Siemens and Halske constructed their first full-scale plants at Paderborn (1902, 60 m^3/hr) and Wiesbaden (1903, 250 m^3/hr), Germany [42]. At these plants, the waters were pretreated in roughing filters containing small pebbles. Absorption of ozone into the waters was carried out in packed towers using broken flint as the packing. These towers apparently did not operate very efficiently in terms of absorption. Vosmaer [39] reports that a large portion of the added ozone remained in the exit gases, and as a result, attempts were made to recycle the ozone-rich air. In addition, ferric hydroxide was found to cling to the flint and clog the water/gas passages. However, from a bacteriological point of view, these packed towers produced an effluent with a satisfactory bacterial count of 2–9 organisms/ml.

In 1898 Otto et al. tested a new prototype plate-type ozone generator at Lille, France [43]. As a result of the successful operation of this plant, a larger plant was constructed at Nice, France, to treat 13 m^3/min of water. This plant utilized the then new Otto plate-type ozone generators and an aspirated injector for the mixing of ozone with water. This Bon Voyage plant at Nice (Figure 3), completed in 1906, sometimes is referred to as "the birthplace of ozonation for drinking water treatment" because it holds the record for continuous operation for this purpose. The Bon Voyage plant operated from 1906 until 1970, during which time an additional two drinking water treatment plants incorporating ozonation were constructed in other locations in Nice as it and its suburbs grew. In 1970 the new and modern Super Rimiez plant was constructed on a mountaintop overlooking Nice [44] and the three older plants were shut down. The Bon Voyage building (Figure 3) is preserved in Nice as a museum exhibit [45].

After construction and operation of the Bon Voyage plant in Nice in 1906, full-scale water treatment plants incorporating ozone appeared in several countries. In 1916 Vosmaer [39] listed 49 treatment plants using ozonation, having a total treatment capacity of 221 m^3/min. Of these 49 plants, 26 were located in France. As of 1977, at least 1036 water treatment plants employing ozonation were known (Table III), more than half of which were located in France [44]. It is interesting that Switzerland, no larger than New Jersey, had 150 ozonation plants in operation as of 1977.

The first continuous process for the disinfection of drinking water in the United States may well have been ozonation. Vosmaer, a native of Haarlem, the Netherlands, was familiar with the first installation of ozone for treating drinking water by Tindal. In his book, Vosmaer [39] describes some plants that he installed in Philadelphia in 1900–1905. The largest was able to handle 3.1 m^3/min of rough-strained water from the Schuylkill River. Vosmaer's first ozone contacting experiments were conducted with a column made of glass sections 30.5 cm in diameter and 45.7 cm in length. Between these sections, perforated celluloid plates were inserted. The total column had a height of 10 meters. When water and ozone were applied flowing cocurrently from the bottom, he observed separation of the gas and liquid layers below each perforated plate. He decided that these plates were, in fact, inhibiting the transfer of ozone into the water, so he removed them and began applying the water at the top of the column, so that the ozone flowed upward, countercurrently to the direction of water flow. Using this method, Vosmaer found that he could achieve a high degree of ozone absorption.

Figure 3. The Bon Voyage Plant at Nice, France. This plant has used ozone for disinfection continuously since it was built in 1906. In 1970 a new, larger plant (Super Rimiez) was built, also using ozone, and the Bon Voyage was shut down. Today, a visitor to Super Rimiez can glimpse the Bon Voyage plant, which is now used as a museum.

Table III. Operational Plants Using Ozone, 1977 [44]

Country	Number of Plants
France	593
Switzerland	150
Germany	136
Austria	42
Canada	23[a]
England	18
Netherlands	12
Belgium	9
Poland	6
Spain	6
United States	5
Italy	5
Japan	4
Denmark	4
Russia	4
Norway	3
Sweden	3
Algeria	2
Syria	2
Bulgaria	2
Mexico	2
Finland	1
Hungary	1
Corsica	1
Ireland	1
Czechoslovakia	1
Singapore	1
Portugal	1
Morocco	1
Total	1039

[a] Includes expansions. Actual number of operating plants in Canada is 20, with three more under construction.

Higher-capacity columns of 61 and 91.5 cm diameter were later constructed of steel. These columns were tested on May 19, 1905 by Hale and Jackson of the New York City Department of Water Supply [39]. These tests showed 99.985 and 99.998% reductions in bacteria contents of the waters treated. The ozone dosage was about 11 mg/l from a gas containing a very low concentration of ozone, 1.3 mg/l. In addition, this water was found to have a high content of dissolved organics, 14.5–10.7 mg/l, as determined by the $KMnO_4$ demand. During ozonation, these organic contents decreased by 40%. Other reductions in water parameters measured included:

- turbidity: 80%
- color: 77%
- nitrites: 79.5%
- albuminoid ammonia: 11.9%

Military sanitarians were among those with a keen interest in the developing technology of disinfection (by all methods available). Annual reports of the U.S. War Department's Surgeon General [46] chronicle U.S. efforts to improve troop water supplies. In 1909 the application of ozone was investigated for use at Fort Niagara, NY, using an experimental apparatus installed at the Army Medical Museum. However, ozonation was not adopted by the military for drinking water disinfection, probably because of the many logistical advantages of hypochlorite, iodine and other more easily carried and applied water disinfectants.

Large-Scale Plants, 1910 to 1953

Economic difficulties and a shortage of electricity following World War I contributed to a reduced period of operation for two large European ozonation plants. The St. Petersburg (now Leningrad) installation, completed in 1910—the largest in the world at that time—had a capacity of 29 m^3/min. Operated until 1919, it utilized 126 Siemens and Halske ozone generators, five Otto emulsifier contactors and five sterilizing towers.

During 1914–1918, a 63-m^3/min ozone water treatment plant was installed at the St. Maur filtration plant in Paris [47]. Ten ozonation contact chambers were provided, each with four emulsifiers. The ozone dosage was around 2 mg/l from an ozone-air gas containing 2–3 mg/l of ozone. Due to the use of a provisional hypochlorination process, this ozonation plant was not operated until 1925. It then served for a period of seven years, to remove tastes imparted to the Marne River by industrial sources.

Despite the economic advantages of chlorine, applications of ozone for treating drinking water continued to increase. As of 1940, there were 114 ozonation systems in use treating public water supplies [48]:

- France: 90
- Italy: 14
- Belgium: 5

- England: 4
- Romania: 3
- United States: 2
- Russia: 1

In 1933 the Siemens Company introduced the use of dual silica gel dryers for treatment of the air to be fed to ozone generators. These were designed for continuous operation with the saturated adsorption cell being reactivated thermally in a closed circuit. Changeover of the air streams from the first (saturated) silica gel drying tower to the second (regenerated) at the time of reactivation was ensured by a solenoid valve, itself regulated by a timer equipped with adjustable electrical contacts. Otto ozone generation plants began using this method of air drying in 1938. With such adsorption equipment, dew points of the air feed gas could be attained in the range of about −20 to −50°C.

The Siemens Company also was the first to introduce high frequency (10,000 Hz) into ozone generators having tubular design with two concentric glass dielectrics. The outer glass tube was supported vertically in a tank through which water was circulated. Kozhinov [42] reported that this model could generate 8 mg/l of ozone at a power efficiency of 35 g O_3/kWh. Despite the advantage of high concentration, this unit was not very successful, due to its relative fragility and the requirement of a 50-to-10,000-Hz converter set.

Rideal [8] recorded in 1920 that "dry air" will produce an approximately 50% greater yield of ozone than undried air containing 0.93 kPa water vapor (5.5°C dew point). In more recent data provided by Sease [49], ozone production from air at −80°C dew point was found to be 50% greater than from air at a −5°C dew point. In view of these observations, it is clear that better air drying methods have been one key innovation in ozone generation technology up to the present time.

In 1933 contracts were awarded to expand the St. Maur, Paris facility to treat 208 m^3/min of water [47]. At this time, a refrigerative method of drying was preferred over calcium chloride for large volumes of air. When the first trial runs were conducted in 1941, the air-conditioning apparatus was found to be inadequate. The frost that deposited on the evaporator (of a direct expansion ammonia plant) was much less dense and had a lower thermal conductivity than had been anticipated. The planned operation of three ozone generating tube bundles in an alternating frost-defrost cycle proved to be unworkable. The plant was shut down and did not receive further attention until 1949 to 1950, because of war and postwar conditions. During this period, the drying system was altered to operate in two steps. The ammonia coils were used to cool a calcium chloride brine. In the first stage, brine cooled the air to 0°C, condensing liquid water. In the second stage, alternating bundles were cooled by brine to bring the dew point down to −12 to −13°C.

This plant began operation in 1953 and filtered water was treated with an average ozone dose of 1.1 mg/l. The total energy consumption was 39 W-hr/g ozone made available to the water. This figure includes 12 W-hr/g ozone produced for the compressors and 6 to 8 W-hr/g ozone for the air conditioner. The plant included 16 contact columns with a cross-sectional area of 14 m^2 and an effective contact depth of 6.3 m. At design flow, it was contemplated to have 12 units in operation, at a flow of 17.4 m^3/min each, with 4 similar units in reserve. Both water and ozone traveled cocurrently in an upward direction. Serving these columns were 16 banks of ozone generators, 16 stainless steel water ring compressors for ozonated air and 8 groups of frequency changers, 50–500 Hz. Half of the plant used Otto plate-type ozone generators and half used the van der Made generator. A bank of plate ozone generators, operated at 18,000–20,000 V, produced 1600 g of ozone per hour while consuming 35 kW of electricity. This corresponds to an ozone production rate of 45.7 g/kWh. The tubular ozone generators operated at 10,000 V produced the same amount of ozone, while consuming only 25 kW, yielding 64 g/kWh. These figures do not include power losses incurred by the frequency converter. The concentration of ozone in the ozone/air mixture was 2–3 mg/l. Following a shakedown period of several months, chlorination was discontinued in favor of ozonation. The St. Maur plant provided about one-third of the Paris potable water supply [47].

In 1949 the Belmont filtration plant in Philadelphia began operation of a 95-m^3/min ozonation process for removal of taste and odor [50]. Unlike the St. Maur facility, where ozone had been applied as the final treatment step, the point of ozone introduction in Philadelphia was prior to the filters and following a 24-hr presedimentation. Pilot-plant tests had indicated an average dosage requirement of 1.6 mg/l of ozone, with the maximum not exceeding 4 mg/l. The latter figure was adopted as the basis for the ozone generation capacity, which was specified to total 567 kg/day. The air-conditioning apparatus consisted of a combination of refrigerative and adsorptive drying, providing a dew point of −51°C. The employment of a new type of ozone generator, designed to be capable of withstanding internal pressure, eliminated the need for corrosion-resistant compressors. Rotary blowers increased the air pressure to 174 pKa prior to the drying section, improving the efficiency of moisture removal. Under these conditions it was possible to generate a high concentration (10–12 mg/l) of ozone in air at a power efficiency of 56 g/kWh. The total energy consumption per gram of ozone made available to the water was 25.5 W-hr/g ozone. This figure includes 4.4 W-hr ozone for the air blowers, 2.2 W-hr for the refrigerative and adsorptive dryers and 1.1 W-hr/g for the auxiliary equipment. It is apparent that the more effective air treatment equipment significantly reduced overall process power requirements.

Three contact chambers were provided at the Belmont plant, each 58 m^2 in area with a depth of 5.6 m. The theoretical retention time was 10 min at design flow. The ozone-air mixture entered through porous tubes at the bottom of the tanks. Due to the application of the water at the top in a countercurrent mode, and the high ozone concentration, efficiency of absorption reached 90–95%. At St. Maur, where gas concentration was low and the ozone/water flow was cocurrent, absorption varied from 60 to 80%, requiring the exhaust gases to be recycled. Ozonation at Philadelphia was discontinued in 1959 when the plant capacity was increased. Improved quality of the Schuylkill River water plus the availability of large basins providing 20 hours of contact, made the use of free residual chlorination more economical [50].

French Application of Ozone for Virus Removal

Systems supplying water to Paris and its suburbs constitute the largest concentration of ozonation plants in the world to date [51]. The combined production of ozone at the four largest plants totalled 8.7 ton/day in 1972. These four plants include Méry-sur-Oise (built in 1965, 208 m^3/min, 3.6 mg/l ozone dosage), Orly (1966, 208 m^3/min, 4 mg/l ozone dosage), Choisy-le-Roi (1968, 626 m^3/min, 3 mg/l ozone dosage) and Neuilly-sur-Marne (1972, 415 m^3/min, 4.8 mg/l ozone dosage). The ozone production lines include air preparation steps essentially similar to those used at the Welsbach installation, but utilizing modern equipment.

Enteric viruses, which infect the alimentary tract of humans, cause particular concern in water supply practice in many parts of the world. Included are polio-, coxsackie-, echo- and hepatitis viruses. French workers were among the first to measure the levels of concentration of these agents. Coin and co-workers [52,53] determined on laboratory scale that maintaining a residual dissolved ozone level of 0.4 mg/l over a period of 4 min will inactivate poliovirus types I, II and III to a level of 99.9%. As a result, ozone contacting systems for drinking water treatment in French water treatment plants were designed to attain this objective. A series of two, three or four ozone contacting chambers is provided. In the first chamber, the ozone demand of the water is satisfied and the threshold ozone residual of 0.4 mg/l is attained. In succeeding contact chambers, ozone lost through reaction and decomposition is replaced and the 0.4-mg/l dissolved ozone residual is maintained for the specified minimum amount of time (4 min).

In some of the newer ozonation plants using French designs, such as the one at Montreal, Canada [54], ozone is introduced initially into the second of three contacting chambers through porous diffusers to attain the 0.4-mg/l dissolved ozone residual. This level is maintained over the minimum 4 min by diffusion of additional ozone into the third chamber. Offgases from the second and third contact chambers are combined and drawn through nonozonated water entering the first tank by means of a submerged turbine contactor. In this manner, offgas ozone is "destroyed" while it helps to lower the ozone demand of the water as it enters the second contacting chamber.

Recent Developments

Biological Activated Carbon (BAC) Treatment

The uniqueness of following ozonation by filtration and granular activated carbon (GAC) adsorption steps was developed first in the Federal Republic of Germany, then quickly extended into France, Switzerland, Holland and Belgium [44]. Aerobic biological activity is promoted in the filters and GAC adsorbers which biologically convert some dissolved organics into CO_2 and water. Nitrification of ammonia to produce nitrate ions also can occur under these conditions. This biological removal of organics and ammonia can lower the loading of adsorbable organics onto the GAC medium, thus effectively extending its useful lifetime before reactivation is required. Thus costs for operating GAC adsorbers can be lowered.

Where possible, German water suppliers use unpolluted groundwater as the preferred source of drinking water [44]. In regions where increased demand has required the use of surface water supplies, more extensive treatment is utilized to produce drinking water of the same high quality as that of unpolluted groundwater. Such treatment results in a dissolved organic carbon (DOC) level below 2 mg/l; thus, only a small terminal dose of residual disinfectant (chlorine or chlorine dioxide), 0.3–0.5 mg/l, is necessary to provide disinfection and protection of the distribution systems.

During the 1960s and 1970s, several European waterworks began to use ozonation followed by GAC as additional treatment steps to offset increasing pollution of surface waters. At the Am Staad plant in Düsseldorf, Federal Republic of Germany, ozone was applied originally to river-sand-bank-filtered Rhine River water for iron and manganese oxidation, followed by filtration and granular activated carbon adsorption for taste and odor removal [55,56]. Shortly thereafter the Flehe and Holthausen plants in Düsseldorf installed the same treatment process. Experience has shown that this sequence of treatment steps removed more dissolved organics (and ammonia) than could be expected by oxidation and/or adsorption individually. German scientists discovered that trace organics and ammonia were removed by biological activity, apparently promoted in the dual-media and GAC-adsorption units that followed ozone oxidation.

Oxidation with ozone is known to convert many slowly or nonbiodegradable organic materials into biodegradable forms. Aeration, which occurs simultaneously with ozonation, increases the dissolved oxygen concentration of the water. The combination of increased biodegradable organics and dissolved oxygen in the ozonated water thus promotes aerobic biological activity in the filtration and adsorption media.

This process, termed biological activated carbon [57], has been found to be effective in removing ammonia (as well as DOC) and thus has replaced breakpoint chlorination in several European locations. Optimized biological activity in the GAC adsorption units has significantly extended the period of GAC operation before regeneration is necessary by factors of 4–6 times [58,59] in waters in which chlorinated organics are not present in high concentrations. Many chlorinated organics are not readily biodegradable nor are they readily oxidized by ozone. Therefore they can be adsorbed by GAC and will not be biodegraded by the microorganisms contained in the filter and adsorption media. Thus, BAC processing will show no advantages over GAC adsorption alone for removal of highly chlorinated organics from drinking water supplies.

Sontheimer [60] reported that the dosage of ozone used in combination with GAC must be carefully controlled. Adding too much ozone can reduce the adsorbability of the dissolved organics. He recommends a dosage of 0.2–0.5 mg O_3/mg nonbiodegradable organics. The nonbiodegradable organic concentration is estimated from the difference in performance of the GAC column when pretreated by oxygen only and by oxygen containing ozone.

The first French drinking water treatment plant to utilize the BAC process was started on full-scale in 1976 [61]. The Rouen-la-Chapelle plant had been treating well waters drawn adjacent to the Seine River by chlorination, later by chlorine dioxide. However, increasing pollution and a rising ammonia content (the Rouen plant is downstream of Paris) required a new treatment process. After two years of studying the sequence of ozonation, sand filtration, and GAC adsorption processes at Rouen in detailed pilot-plant studies [62,63], the new treatment plant came on-line in 1976. A report made in 1979 [64] indicated that the GAC contained in four 1-m-high filter beds still had given no signs of breakthrough after 3 years of continuous use. However, one of the GAC beds had been reactivated for the instruction of plant personnel and to determine the effects on postreactivation performance of the GAC when it was replaced in operation.

Multiple-Stage Ozonation

Early French treatment plants used ozonation as a single treatment step only, either as the terminal disinfectant or to be followed by low levels of chlorine or chlorine dioxide. Later German and French plants incorporated ozonation into the initial stages of water treatment for oxidation purposes (such as at Düsseldorf). Some of these plants utilized ozone in a second stage for disinfection (as at Rouen-la-Chapelle).

With a fuller understanding of the advantages of multiple application of small dosages of ozone at different points in the water treatment process, including promotion of biological activity, several French treatment plants now are installing three stages of ozone treatment. Schulhof [65] described how three of the large Paris suburbs plants (Méry-sur-Oise, Choisy-le-Roi and Neuilly-sur-Marne) are being modified in this regard (Figure 4). Raw water from each of the rivers is ozonated with an ozone dosage of less than 1 mg/l, then stored in a detention reservoir for 2–3 days. During this time, some of the dissolved organics and ammonia are converted into CO_2 and nitrate ions, respectively, by biological action. The primary purpose of this initial ozonation is to aerate the raw water and to oxidize some of the dissolved organic materials.

After the 2- to 3-day detention time, the water is again ozonated with a dosage of less than 1 mg/l, then chemically treated with poly(aluminum chloride) and powdered activated carbon. This second treatment with ozone assists the flocculation/sedimentation process by creating additional polar sites on the organic molecules for more efficient removal of dissolved organics and suspended solids. After sand filtration, the water is ozonated for primary disinfection (0.4 mg/l of dissolved ozone to be maintained under these conditions for 10 min) and also to further oxidize organic constituents and add additional dissolved oxygen to the water. Then GAC adsorption follows, with aerobic biological activity occurring in the GAC media for additional organics removal and nitrification of ammonia. Chlorination followed by dechlorination is the terminal treatment prior to sending the waters to their respective distribution systems.

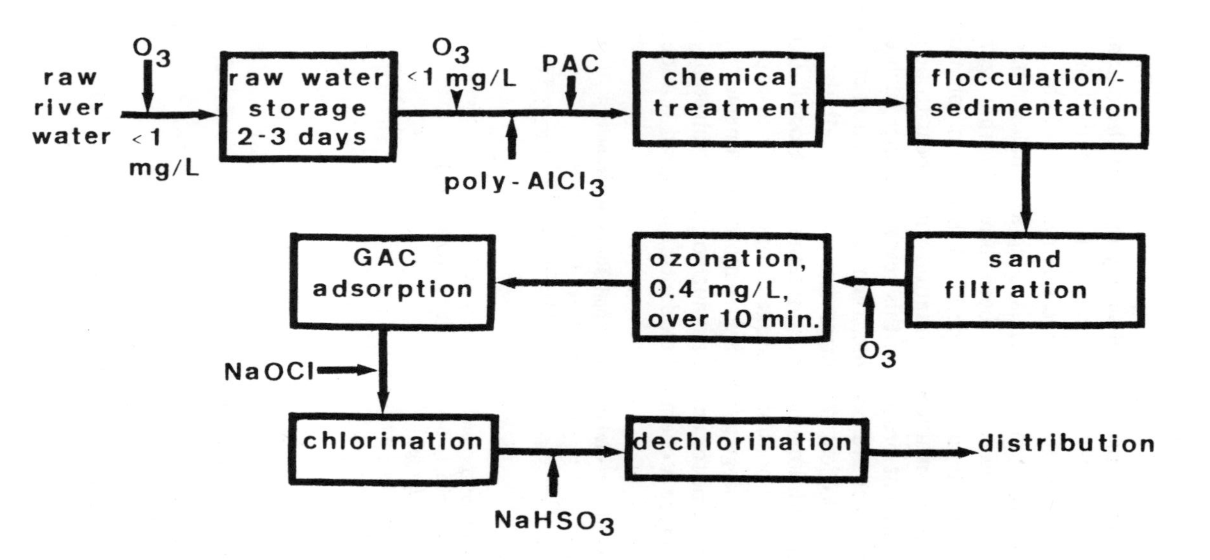

Figure 4. Treatment process ultimately planned for three Paris suburban plants [65].

Summary

Ozone is used in the treatment of potable water supplies for the purposes listed in Table IV [66]. Employing ozone as the primary oxidant before chlorination usually will satisfy most of the oxidant demand of the water being treated, thus lowering the subsequent demand for chlorine and minimizing the unwanted side reactions of chlorination to produce chlorinated organic materials, such as trihalomethanes (THM).

In some cases, however, it has been shown [67] that ozonation prior to chlorination actually increases the THM formation potential of dissolved organics. In such instances, ozonation will appear to be detrimental to the water treatment process in this respect. Under such circumstances, several approaches are possible. First, the ozonation step could be followed by a biological treatment step (slow sand filtration, reservoir detention) or additional chemical treatment to flocculate the now more polar organics, or possibly passage through GAC adsorption with maximized aerobic biological activity. A more fundamental water treatment approach would be to remove as much of the dissolved organic materials as possible before a chemical oxidant of any kind is added.

Table IV. Applications of Ozone in Drinking Water Treatment [66]

Application
Bacterial Disinfection
Viral Inactivation
Oxidation of Soluble Iron and/or Manganese
Decomplexing Organically Bound Manganese (Oxidation)
Color Removal (Oxidation)
Taste and Odor Removal (Oxidation)
Algae Removal (Oxidation)
Oxidation of Organics
Phenols
Detergents
Pesticides
Microflocculation of Dissolved Organics (Oxidation)
Oxidation of Inorganics
Cyanides
Sulfides
Nitrites
Turbidity or Suspended Solids Removal (Oxidation)
Pretreatment for Biological Processes (Oxidation)
On Sand
On Anthracite
On Granular Activated Carbon

MUNICIPAL WASTEWATER TREATMENT

In municipal wastewater treatment, ozone has been studied primarily for disinfection (following primary or secondary treatment), but also for lowering of biological (BOD) and chemical oxygen demand (COD) levels, oxidation of ammonia, and removal of color, nutrients and suspended solids. Most of the disinfection work has been pioneered in the United States since the early 1970s, where it has been developed to the point of full-scale commercial use [21]. A considerable amount of research has been conducted by the Japanese during the same period, especially for recycling of treated sewage effluents for nonpotable purposes in water-short regions of that country. More recently, French and German investigators have published results of pilot plant studies involving ozonation of municipal sewage. Israeli scientists have been studying the BAC process for treatment of sewage prior to ground recharge. In South Africa, ozonation has been shown to be cost-effective for extending the operating lifetimes of GAC columns used to remove dissolved organics from sewage being treated for potable reuse. The evolution of these and other studies will be reviewed briefly here.

United States

In 1956 the Armour Research Foundation reported results of the evaluation of ozone for sterilization of pathogenic organisms contained in sewage [68]. They found that *Bacillus anthracis, Clostridium botulinum* toxin, influenza virus and *Bacillus subtilis* var. *Niger* were completely removed by absorbed ozone dosages between 100 and 200 mg/l.

In 1965 workers at Louisiana Technological University [69] treated raw domestic sewage with absorbed ozone dosages of 6–11 mg/l accumulated over a 3-hr period. Under these conditions, the weight ratio of BOD_5 removed to ozone added was approximately 10:1. It is likely that at this slow rate of ozone addition, the concentration of microorganisms was not appreciably lowered. Therefore, oxidation of the organic constituents (as measured by BOD_5) could have occurred by both biological and chemical mechanisms.

In 1972 a U.S. patent was issued to the Union Carbide Corporation [70], which described laboratory results combining ozone disinfection with high-purity oxygen activated sludge treatment of municipal sewage. Clarified effluent was passed downflow through a vertical tray-type contactor against rising bubbles of ozone-containing oxygen. In one test, an

8-mg/l dose of 2.28% (by volume) ozone in oxygen lowered the coliform count from 42,000 to 45/100 ml. Such a process has many potential advantages. First, the oxygen requirements for the disinfection step are roughly equivalent to those of the activated sludge treatment of the clarified effluent studied (which contained levels of 10.5 and 31 mg/l of five-day BOD and COD, respectively), and the exhaust gases from the ozone contactor can be used for this purpose. Second, ozone generators can produce about twice as much ozone per unit of applied electrical energy when using pure oxygen as opposed to air feed. This enables installation of smaller ozone generation equipment. Third, oxygen is produced moisture-free, eliminating the need for air pretreatment equipment, which usually accounts for 25–30% of the capital costs in an air-feed ozone generation process.

Extensive studies of ozone disinfection of municipal sewage on pilot-plant scale were conducted during the late 1960s and early 1970s in Louisville, KY [71], Chicago, IL [72], Los Angeles, CA [73], and New York, NY [74]. In all of these studies, an ozone-induced froth was produced; provisions must be made to remove this material to avoid fouling of processing equipment. Results obtained for ozone disinfection varied in these studies, higher dosages of ozone being required in Los Angeles and Louisville because of the higher content of industrial wastewaters in the raw sewage at these locations.

In 1971 Airco, Inc. conducted an eight-month pilot plant study at the Blue Plains sewage treatment plant in Washington, DC, for the purpose of tertiary treatment with ozone (not for disinfection) [75]. This work was funded by the U.S. Environmental Protection Agency (EPA). Effluents from 9 different sewage treatment processes were subjected to ozonation in a series of 6 glass ozone contacting columns 61 cm in diameter and 6.1 m high. Ozone in oxygen was introduced into the base of each column. This study [75] established the apparent stoichiometry and kinetics for reactions of ozone with trace organics found in municipal sewage. For influent COD levels in the range of 30–80 mg/l, 50% removal could be achieved with weight ratios of dissolved ozone: COD of less than 3:1. However, for COD removals much above 20 mg/l, high dissolved ozone concentrations were required to complete the reactions in a reasonable period of time. With the type of contactor studied, less than 50% of the applied ozone was absorbed when dissolved ozone concentrations greater than 3–4 mg/l were maintained in the liquid.

As of this writing, there are 28 operational sewage treatment plants in the United States that include ozone disinfection [76]. An additional plant includes ozone for lowering suspended solids before ground recharge

[77]. Another plant uses ozonation during tertiary treatment before GAC adsorption to aid in removing organic components before ground recharge [78]. An additional 13 plants are under construction, and 9 more are being designed, all of which will incorporate ozonation, primarily for disinfection.

Most of the early sewage treatment plants incorporating ozonation are small, less than 19,000 m^3/day (5 mgd), and the ozone is generated from air. The larger and more recent sewage treatment plants are designed with oxygen activated sludge treatment, and thus have an available supply of cheap oxygen. This allows the ozone to be generated from pure oxygen rather than from air, with attendant savings in the energy cost of ozone generation and the ability to use smaller ozone generating equipment.

The largest operating plant today is located at Springfield, MO [79]. Here a 132,500-m^3/day (35-mgd) oxygen activated sludge plant has been in operation with ozone disinfection since 1978. However, in early 1982, the city of Indianapolis, IN, will start up two larger plants, one treating 473,000 m^3/day (125 mgd), the other treating 454,250 m^3/day (120 mgd), both using the oxygen activated sludge process with ozone disinfection [80]. When these plants are operational, the city of Indianapolis will be able to boast of having the largest sewage treatment plants in the world employing ozone disinfection.

In the United States, disinfection of sewage usually is defined as the attainment of a most probable number (MPN) of 200 fecal coliforms per 100 ml. This level of disinfection generally can be achieved in a biologically treated and filtered wastewater with absorbed ozone dosages of 4–8 mg/l. However, if the wastewaters contain significant amounts of ozone-demanding industrial wastes, the ozone dosages required to attain this level of disinfection can rise to 15 mg/l and even higher.

A more stringent disinfection level of 2.2 *total* coliforms per 100 ml has been set by the state of California for treated sewage to be discharged into areas intended for human consumption, human recreation or aquaculture. Conditions for attaining such stringent disinfection levels using ozone were determined recently by Stover and Jarnis [81] under an EPA-funded grant to the city of Marlborough, MA. Using filtered secondary treated effluents and a submerged turbine ozone contactor, the 2.2 total coliforms/100 ml objective was attained with absorbed ozone dosages of 35–40 mg/l. However, using filtered and nitrified secondary effluents, the objective could be attained with absorbed ozone dosages of only 15–20 mg/l. These results should be compared with the 4- to 8-mg/l absorbed ozone dosages required to attain the disinfection objective of 200 fecal coliforms/100 ml required in most other regions of the United States.

Japan

Somiya and co-workers [82–85] reported results of basic studies of the use of ozone in the tertiary treatment of sewage, beginning in 1973. A three-part study of ozone treatment of secondary effluents in countercurrent bubbling towers was published by Somiya and Tsumo [86–88] in 1975. These researchers studied the absorption characteristics of ozone by secondary effluent, ozonation of secondary effluents after dual-layer filtration and the action of ozone on suspended solids.

In 1973 Ikehata [89] described research studies being conducted in Japan on treatment of municipal sewage for reuse applications. Ozone was studied for the oxidation of manganese(II) compounds to form insoluble MnO_2; oxidation of cyanide compounds, ammonia, amino acids and alcohols and ABS resin; and inactivation of viruses. The successive treatment of sewage by biological treatment followed by alum flocculation, rapid sand filtration, ozonation and then GAC adsorption was found to produce water which met World Health Organization quality standards for drinking water. In addition, ozone treatment was also found to decolorize, deodorize, oxidize nitrite to nitrate, and oxidize and remove iron from the treated wastewaters.

These many years of research in Japan culminated in 1981 in the installation of ozonation treatment into onsite wastewater reclamation systems at seven commercial buildings and apartment complexes in Japan [90]. Reclaimed wastewaters at these sites are used for toilet flushing, automobile washing water, and ornamental streams and ponds.

France

Gomella [91] described pilot-plant studies conducted on the ozonation of sewage from the Paris district at Colombes in 1976. Subsequent pilot-plant studies in this field include those reported by Gaissa et al. [92] for disinfection of effluent in the city of Nice, and several studies conducted by Légeron and co-workers [93–95]. Richard and Conan [96] showed that the addition of polyelectrolyte to sewage can greatly change the operative interfacial forces during transfer of ozone from the gas bubbles into sewage, and thus lower the amount of ozone needed to attain a given level of disinfection.

Germany

Wölfel and Sontheimer [97] described laboratory experiments conducted in 1976 and 1977 in which effluent from the West Berlin sewage treatment plant was ozonated to make the organic components more biodegradable before injection of the effluent through sand banks prior to passing into the receiving river. In 1981 Sarfert and Altmann [98] reported pilot-plant studies in which the biologically treated effluent from the Berlin sewage treatment plant was treated with ozone for the same purpose. This pilot plant began operating in February 1981, and will be operated for an extended period of time to develop performance data sufficient to convert the full-scale plant to this new process. The effluent is flocculated by ferric chloride, then ozonated with 1 mg/l of ozone, double-layer filtered and injected underground into an adsorption well. At this low level of ozonation, the nitrification processes that occur in the filters are not affected.

Israel

Wachs et al. [99] described experiments conducted in 1976 and 1977 in which sewage treatment plant effluent was treated with lime at high pH, then was ozonated and passed through GAC media. The combined tertiary treatment process allowed biorefractory organic materials to be converted into more readily biodegraded materials, which then were degraded in the filters and GAC adsorbers placed downstream of ozonation. Total organic carbon (TOC) levels thus could be reduced to below 2 mg/l.

In 1980 Rom et al. [100] reported on more detailed pilot plant studies of this same treatment process.

South Africa

The city of Windhoek, South West Africa, has been processing municipal sewage for addition to its potable water supply for a number of years. In this process, breakpoint chlorination is used before GAC adsorption for removal of ammonia and dissolved organics. Recently, van Leeuwen and Prinsloo [101,102] discussed this process in papers describing experiments conducted at the Stander Water Reclamation Plant in Pretoria, South Africa. The objective of these studies is to improve on the process being utilized at Windhoek and to test the substitution of ozonation followed by a biological treatment step for breakpoint chlorination, currently used to remove ammonia, and GAC adsorption for removal of dissolved organic materials. Such substitution has been accomplished successfully at the 3785-m^3/day (1-mgd) Stander Water Reclamation Plant. One advantage of the substitution is that the GAC medium following ozonation did not have to be regenerated during the one-year study reported. In contrast, the GAC medium following breakpoint chlorination had to be reactivated every 90 days. The savings in GAC regeneration produced by substituting preozonation for breakpoint chlorination were found to be more than sufficient to pay for the costs of ozone treatment. This program at the Stander Water Reclamation Plant will be described in greater detail in a later chapter.

INDUSTRIAL WASTEWATER TREATMENT

In this area, ozone has been developed for oxidation of cyanide in electroplating wastewaters [103–105] decolorization of dyestuffs in textile wastewater in Japan [106], removal of phenolic compounds from refinery wastewaters in Canada [107], disinfection of seawater used for depuration of shellfish in France, Spain, Greece and other countries [108], recovery and reuse of spent iron cyanide photoprocessing bleach waters [109] and recycle and reuse of auto washing wastewaters in Vienna, Austria [110]. Many of these applications for ozonation will be described in detail in later chapters.

OTHER APPLICATIONS

The combination of ozonation conducted simultaneously with exposure of wastewaters to ultraviolet radiation has been developed for a variety of purposes [111,112]. One of these is the destruction of polychlorinated biphenyls (PCB), which has been determined by EPA to be the best available technology for this purpose [113]. Bleaching of paper pulps has been developed to the commercial demonstration stage in Norway [114]. Swimming pool waters are treated with ozone in a number of European countries [115–117]. A number of air-conditioning cooling waters are treated with ozone in the United States [118,119]. Many applications for ozone as a deodorizing agent are known in sewage treatment [120–122], fish processing plants in Japan [123–125], industrial processing plants, and restaurants [126].

For a number of years, German, Austrian, Swiss, Romanian, French and Italian medical and dental personnel have been using ozone for a variety of therapeutic applications. In dentistry, solutions of ozone in water are employed to maintain sterility in the mouth during oral surgery and implantations [127]. In medical therapy, ozone is used for treating cancer in conjunction with radiation therapy [128], fistules, osteolides and stasis ulcers [129], gangrene and atonic ulcers [130,131], liver ailments [132,133], circulatory disturbances [134–135], in gynecology [136,137], and in hematogenic oxidation therapy (injection of ozone/oxygen mixtures into the bloodstream) [138,139].

Finally, since the early 1950s, Emery Industries in Cincinnati, OH, has been the largest single user of ozone for the manufacture of azelaic and pelargonic acids by the ozonolysis of oleic acid [140]. The reader interested in details of the reactions of ozone with olefins in the nonaqueous liquid and in the gas phases is referred to Bailey [5] and later chapters in this series.

There is little doubt that with a better understanding of the properties and methods of application of ozone, these and other beneficial uses for this versatile oxidizing agent will continue to be developed.

REFERENCES

1. Manley, T. C., and S. J. Niegowski. "Ozone," in *Encyclopedia of Chemical Technology, Vol. 14,* 2nd ed. (New York: John Wiley & Sons, Inc., 1967), pp. 410–432.
2. Nebel, C. "Ozone," in *Encyclopedia of Chemical Technology, Vol. 14,* 3rd ed. (New York, John Wiley & Sons, Inc., 1981), pp. 683–713.
3. Rosenthal, H. "Selected Bibliography on Ozone, Its Biological Effects and Technical Applications," Fisheries Research Board of Canada Technical Report No. 456, Fisheries and Marine Service, Pacific Biological Station, Nanaimo, BC (1974).
4. Sease, W. S. "Ozone Mass Transfer and Contact Systems," in *Proceedings of the Second International Symposium on Ozone Technology,* R. G. Rice, P. Pichet and M.-A. Vincent, Eds. (Vienna, VA: International Ozone Association, 1976), pp. 1–14.
5. Bailey, P. S. *Ozonation in Organic Chemistry, Vol. I, Olefinic Compounds* (New York: Academic Press, Inc., 1978), p. 8.
6. Homer. *Iliad* 8:135, 14:415; *Odyssey* 12:417, 14:307; *The Complete Works of Homer,* Modern Library Edition (New York: Random House, 1950).
7. Mohr, F. *Ann. Phys.* (Leipzig) 91(2):625 (1854).
8. Rideal, E. K. *Ozone* (London: Constable & Co., Ltd., 1920).
9. Schönbein, C. F. *Comptes rendus Hebd. Seances Acad. Sci.* 10:706 (1840).
10. Schönbein, C. F. "Notice of Christian Frederic Schönbein, The Discoverer of Ozone," Annual Report of the Board of Regents of the Smithsonian Institution, 1868 (Washington, DC: U.S. Government Printing Office, 1869), pp. 185–192.
11. de la Rive and Marignac. *Comptes rendus Acad. Sci.* 20:1291 (1845).
12. Hunt, T. J. *Am. J. Sci.* 6:171 (1848).
13. Andrews, T., and P. G. Tait. "On the Volumetric Relations of Ozone, and the Action of the Electrical Discharge on Oxygen and Other Gases," *Philosoph. Trans.* (1860), pp. 113–132.

14. Andrews, T. "Address on Ozone," presented before the Royal Society of Edinburgh, December 22, 1873, in *Stokes Collection on Inorganic Chemistry,* Johns Hopkins University Library, Baltimore, MD.
15. von Siemens, W. *Poggendorff's Ann.* 102:120 (1857).
16. Hollman, P. J. *Memoire sur l'Equivalent Calorifique de l'Ozone,* Library of C. van der Post, Jr., Utrecht, The Netherlands (1868).
17. Jahn. *Zeit. Anorg. Chem.* 60:357 (1908); 68:250 (1910).
18. Schöne. *Ber.* 6:1224 (1873).
19. Carius. *Ann.* 174:30 (1874).
20. Mailfert. *Compt. rend.* 119:951 (1894).
21. Venosa, A. D., and E. J. Opatken. "Ozone Disinfection—State of the Art," paper presented at the 52nd Annual Water Pollution Control Federation Conference, Pre-Conference Workshop on Wastewater Disinfection, Houston, TX, October 7, 1979.
22. Sawyer, C. N. and P. L. McCarty. *Chemistry for Sanitary Engineers,* 3rd ed. (New York: McGraw-Hill Book Company, 1978).
23. *Encyclopedia Britannica,* Vol. XVIII (ORN-PHT), 9th ed. (New York: Charles Scribner's Sons, 1885), p. 113.
24. Hautefeuille and Chappuis. *Compt. rend.* 94:1249 (1882).
25. Olszewski. *Monatsh.* 8:109 (1887).
26. Olszewski. *Ann. Physik.* 3:31 (1887).
27. Troost. *Compt. rend.* 126:1751 (1898).
28. Schönbein, C. F. *J. Prakt. Chem.* [1] 34:492 (1845).
29. Schönbein, C. F. *J. Prakt. Chem.* 52:135 (1851).
30. Schönbein, C. F. *J. Prakt. Chem.* 105:198 (1868).
31. Schönbein, C. F. *J. Prakt. Chem.* [1] 105:232 (1868).
32. Wetherill, C. M. "Ozone and Antozone," Annual Report of the Board of Regents of the Smithsonian Institution (Washington, DC: U.S. Government Printing Office, 1865), pp. 167–177.
33. von Gorup-Besanez, E. *Justus Liebig's Ann. Chem.* 110:86 (1859).
34. von Gorup-Besanez, E. *Justus Liebig's Ann. Chem.* 125:207 (1863).
35. de Gerbeth, F. L. U.S. Patent 81,071 (1868).
36. Fewson, H. U.S. Patent 387,286 (1888).
37. de la Coux, H. *l'Ozone et ses Applications Industrielles* (Paris: Vve. Ch. Dunod, 1904), p. 49.
38. Le Pauloué, J., Soc. Trailigaz, Garges-lès-Gonesse, France. Personal communication (1977).
39. Vosmaer, A. *Ozone, Its Manufacture, Properties and Uses* (New York: D. Van Nostrand Co., 1916).
40. van der Sleen, N. and A. Schneller. "Apparatus for Causing Chemical Changes in Gases," U.S. Patent 587,770 (1897).
41. "The Production and Utilization of Ozone With Especial Reference to Water Purification," *Eng. News* 63(17):488–496 (1910).
42. Kozhinov, V. F. *Equipment for the Ozonation of Water* (Moscow: Publishing House on the Literature of Construction, 1968).
43. Grubbs, S. B. "Sterilization of Water by Chloric Peroxide and Ozone," U.S. Marine Hospital Service Public Health Reports 15(144) (1900).
44. Miller, G. W., R. G. Rice, C. M. Robson, R. L. Scullin, W. Kühn and H. Wolf. "An Assessment of Ozone and Chlorine Dioxide Technologies

for Treatment of Municipal Water Supplies," U.S. EPA Report EPA-600/2-78-147 (Washington, DC: U.S. Government Printing Office, 1978).

45. Tschaeglé-Appert, P., Cie. Génerale des Eaux, Nice, France. Personal communication (1977).
46. *Reports of the Surgeon General,* War Dept. of the United States of America (1910, 1913, 1914).
47. Guinvarc'h, P. "Three Years of Ozone Sterilization of Water in Paris," in *Ozone Chemistry and Technology,* Advances in Chemistry Series No. 21 (Washington, DC: American Chemical Society, 1959), pp. 416–429.
48. Baker, M. N. *The Quest for Pure Water* (New York: The American Water Works Association, 1949).
49. Sease, W. S. "Supplementary Material, Short Course on Ozone Technology," The Center For Professional Advancement, Somerville, NJ (1975).
50. Bean, E. L. "Ozone Production and Costs," in *Ozone Chemistry and Technology,* Advances in Chemistry Series, No. 21 (Washington, DC: American Chemical Society, 1959), pp. 430–442.
51. Gomella, C. "Ozone Practices in France," *J. Am. Water Works Assoc.* 64(1):39–45 (1972).
52. Coin, L., C. Hannoun and C. Gomella. "Inactivation of Poliomyelitis Virus by Ozone in the Presence of Water," *La Presse Médicale* 72(37): 2153–2156 (1964).
53. Coin, L., C. Gomella, C. Hannoun, and J. L. Trimoreau. "Ozone Inactivation of Poliomyelitis Virus in Water," *La Presse Médicale* 75(38): 1883–1884 (1967).
54. Bouchard, J. C. "l'Ozone Depuis Vingt Ans Dans le Traitement des Eaux Potables," in *Proceedings of the Second International Symposium on Ozone Technology,* R. G. Rice, P. Pichet and M.-A. Vincent, Eds. (Vienna, VA: International Ozone Association, 1976), pp. 705–714.
55. Hopf, W. "Treatment of Water with Ozone and Activated Carbon (Düsseldorf Process), Part I," *Wasser-Abwasser* 111(2):83–92 (1970).
56. Hopf, W. "Treatment of Water with Ozone and Activated Carbon (Düsseldorf Process), Part II," *Wasser-Abwasser* 111(3):156–164 (1970).
57. Rice, R. G., G. W. Miller, C. M. Robson and W. Kühn. "A Review of the Status of Preozonation of Granular Activated Carbon for Removal of Dissolved Organics from Water and Wastewater," in *Carbon Adsorption Handbook,* P. N. Cheremisinoff and F. Ellerbusch, Eds. (Ann Arbor, MI: Ann Arbor Science Publishers, Inc., 1978), pp. 485–538.
58. Sontheimer, H. "Applying Oxidation and Adsorption Techniques: A Summary of Progress," *J. Am. Water Works Assoc.* 71(11):612–617 (1979).
59. Schalekamp, M. "The Use of GAC Filtration to Ensure Quality in Drinking Water From Surface Sources," *J. Am. Water Works Assoc.* 71(11):638–647 (1979).
60. Sontheimer, H. "Use of Biological Activated Carbon in German Drinking Water Treatment Plants," paper presented at the Seminar on Current Status of Wastewater Treatment and Disinfection with Ozone, International Ozone Association, Cincinnati, OH, September 15, 1977.

61. Rice, R. G., C. Gomella, and G. W. Miller. "Rouen, France Water Treatment Plant: Good Organics and Ammonia Removal With No Need to Regenerate Carbon Beds," *Civil Eng.* (May 1978), pp. 76–82.
62. Gomella, C., and D. Versanne. "le Role de l'Ozone dans la Nitrification Bactérienne de l'Azote Ammoniacal—Cas de l'Usine de la Chapelle Banlieue Sud de Rouen (Seine Maritime), France," paper presented at 3rd International Congress on Ozone Technology, International Ozone Assoc., Paris, France, May 1977.
63. Gomella, C., and D. Versanne. "Role de l'Ozone dans la Nitrification Bactérienne de l'Azote Ammoniacal: Cas de l'Usine de la Chapelle à St.-Etienne Rouvray (Seine Maritime)," *Tech. San. Municipale* (1977), pp. 78–81.
64. Schulhof, P. "French Experiences in the Use of Activated Carbon For Water Treatment," paper presented at the NATO/CCMS Conference on Adsorption Techniques in Drinking Water Treatment, U.S. EPA, Reston, VA (April 30–May 2, 1979).
65. Schulhof, P. "Water Supply in the Paris Suburbs: Changing Treatment for Changing Demands," *J. Am. Water Works Assoc.* 72(8):428 (1980).
66. Rice, R. G., C. M. Robson, G. W. Miller, and A. G. Hill. "Uses of Ozone in Drinking Water Treatment," *J. Am. Water Works Assoc.* 73(1):44–57 (1981).
67. Umphries, M. D., R. R. Trussell, A. R. Trussell, L. Y. C. Leong, and C. H. Tate. "The Effects of Preozonation on the Formation of Trihalomethanes," *OZONews* 6(3), Part 2 (March 1979).
68. Miller, S., B. Borkhardt and R. Ehrlich. "Disinfection and Sterilization of Sewage by Ozone," in *Ozone Chemistry and Technology,* Advances in Chemistry Series (Washington, DC: American Chemical Society, 1959), pp. 381–387.
69. Marsh, G. R., and G. H. Panula. "Ozonization in the BOD Reduction of Raw Domestic Sewage," *Water Sew. Works* (October 1965), pp. 372–377.
70. McWhirter, J. R., and E. K. Robinson. U.S. Patent 3,660,277 (1972).
71. Nebel, C., R. D. Gottschling, R. L. Hutchinson, T. J. McBride, D. M. Taylor, J. L. Pavoni, M. E. Tittlebaum, H. E. Spencer, and M. Fleischman. "Ozone Disinfection of Industrial-Municipal Secondary Effluents," *J. Water Poll. Control Fed.* 45(12):2493–2507 (1973).
72. Zenz, D. "Ozonation Pilot Plant Studies at Chicago," paper presented at Seminar on Ozonation in Sewage Treatment, Univ. of Wisconsin at Milwaukee, November 9–10, 1971.
73. Ghan, H. B., C.-L. Chen, R. P. Miele, and I. J. Kugelman. "Wastewater Disinfection with Ozone Works Best with a Clean Effluent and Multiple, Low-Dosage Injection Points," *Bull. Calif. Water Poll. Control Assoc.* 12(2):47–52 (1975).
74. Bollyky, L. J. and B. Siegel. "Ozone Disinfection of Secondary Effluent," *Water Sew. Works* (April 1977), pp. 90–92.
75. Wynn, C. S., B. S. Kirk and R. McNabney. "Pilot Plant for Tertiary Treatment of Wastewater With Ozone," U.S. EPA Report EPA-R2-73-146 (Washington, DC: U.S. Government Printing Office, 1973).
76. Rice, R. G., L. M. Evison and C. M. Robson. "Ozone Disinfection of Municipal Wastewater—Current State-of-the-Art," *Ozone Sci. Eng.* (submitted).

77. Trussell, R., T. Nowak, C. Tate, S. Lo and F. Ismail. "Ozone Pretreatment for Coagulation/Filtration of Secondary Effluents," in *Proceedings of the Second International Symposium on Ozone Technology,* R. G. Rice, P. Pichet and M.-A. Vincent, Eds. (Vienna, VA: International Ozone Association, 1976), pp. 586–610.
78. Jenks, J. H. and B. L. Harrison. "Multi-feature Reclamation Project Accomplishes Multi-objectives," *Water Wastes Eng.* (November 1977).
79. Shifrin, W. G. and P. F. Johnson. "Ozonation Takes a Giant Step," *Water Wastes Eng.* (March 1978), pp. 50–53.
80. Robson, C. M., and D. Wells. "Combined Municipal/Industrial Wastewater Treatment at Indianapolis, IN," paper presented at the Conference on Combined Municipal/Industrial Wastewater Treatment, U.S. EPA, Univ. of Texas at Dallas, March 25–27, 1980.
81. Stover, E. L. and R. N. Jarnis. "Engineering and Economic Aspects of Wastewater Disinfection with Ozone under Stringent Bacteriological Standards," *Ozone Sci. Eng.* 2(2):159–176 (1980).
82. Somiya, I. "Ozone Treatment in Tertiary Treatment," *J. Jap. Sew. Works Assoc.* 10(109):1–14 (1973).
83. Somiya, I. "Quality Changes of Secondary Effluent Treated by Ozone and Chlorine," in *Forum on Ozone Disinfection,* E. G. Fochtman, R. G. Rice and M. E. Browning, Eds. (Vienna, VA: International Ozone Association, 1977), pp. 402–421.
84. Somiya, I., H. Yamada and T. Goda. "The Ozonation of Nitrogenous Compounds in Water," paper presented at Symposium on Advanced Ozone Technology, International Ozone Association, Toronto, Ontario, November 16–18, 1977.
85. Somiya, I., and T. Tokuda. "Model Identification of Ozone and Activated Carbon Treatment Processes Based on a Population Balance Model," paper presented at Symposium on Advanced Ozone Technology, International Ozone Association, Toronto, Ontario, November 16–18, 1977.
86. Somiya, I., and H. Tsumo. "Research on Ozone Treatment of Secondary Effluent by Counter-current Bubbling Tower (1)—Absorption Characteristics of Ozone," *Water Purif. Liquid Wastes Treat.* 16(7):647–658 (1975).
87. Somiya, I., and H. Tsumo. "Research on Ozone Treatment of Secondary Effluent by Counter-current Bubbling Tower (2)—Performance of Ozonation on Filtrate by Double Layer Filter," *Water Purif. Liquid Wastes Treat.* 16(8):725–738 (1975).
88. Somiya, I., and H. Tsumo. "Research on Ozone Treatment of Secondary Effluent by Counter-current Bubbling Tower (3)—Performance of Ozonation on Suspended Solids," *Water Purif. Liquid Wastes Treat.* 16(9):859–866 (1975).
89. Ikehata, A. "Treatment of Municipal Wastewater by the Use of Ozone to Yield High Quality Water," in *Proceedings of the First International Symposium on Ozone for Water and Wastewater Treatment,* R. G. Rice and M. E. Browning, Eds. (Vienna, VA: International Ozone Association, 1975), pp. 227–231.
90. Asano, T., Y. Nagasawa, N. Hayakawa, and T. Tamaru. "On-site Wastewater Reclamation and Reuse Systems in Commercial Buildings and Apartment Complexes," paper presented at Water Reuse

Symposium II, U.S. Dept. of the Interior, Office of Water Research and Technology, Washington, DC, August 23–28, 1981.

91. Gomella, C. "Contribution to the Study of Treated Sewage Disinfection by Ozone," in *Forum on Ozone Disinfection,* E. G. Fochtman, R. G. Rice and M. E. Browning, Eds. (Vienna, VA: International Ozone Association, 1977), pp. 205–215.
92. Gaissa, J., B. Hughes, J. L. Plantat, M. Plisser, M. Roullet, J. P. Torres, J. P. Bouchez, and M. Paulet. "Contribution to the Study of Wastewater Disinfection by Ozonation," paper presented at Third International Congress on Ozone Technology, International Ozone Association, Paris, France, May 1977.
93. Lebesgue, Y., and J.-P. Légeron. "Disinfection of Wastewater with Ozone," paper presented at Ozone Technology Symposium, International Ozone Association, Los Angeles, CA, (May 23–25, 1978).
94. Légeron, J.-P. "Comparative Study of Ozonation Conditions in Wastewater Tertiary Treatment," *Ozone Sci. Eng.* 2(2):123–138 (1980).
95. Légeron, J.-P. "Contribution to the Optimization of Ozone Disinfection of Domestic Wastewater," paper presented at Fifth World Ozone Congress, International Ozone Association, Berlin, Federal Republic of Germany, March 31–April 3, 1981.
96. Richard, Y., and M. Conan. "Ozone Disinfection and Wastewater Treatment: Importance of Interface Action," *Ozone Sci. Eng.* 2(2):139–158 (1980).
97. Wölfel, P., and H. Sontheimer. "Improvement of the Biological Degradation of Wastewaters by an Ozone Treatment," presented at Third International Congress on Ozone Technology, International Ozone Association, Paris, France, May 1977.
98. Sarfert, F., and J.-J. Altmann. "Use of Ozone for Wastewater Treatment in the Experimental Installations of WAR-Berlin," paper presented at Fifth Ozone World Congress, International Ozone Association, Berlin, Federal Republic of Germany, March 31–April 3, 1981.
99. Wachs, A. M., N. Narkis, M. Schneider, and P. Wasserstrom. "Use of Ozone in Water Renovation. Part 1—Removal of Organic Matter from Effluents by Lime Treatment, Ozonation and Biologically Extended Carbon Adsorption," presented at Third International Congress on Ozone Technology, International Ozone Association, Paris, France, May 1977.
100. Rom, D., A. M. Wachs, and M. Rotel. "Pilot Plant Studies of Water Renovation in a System Combining Ozonation with Activated Carbon Treatment," presented at 53rd Annual Water Pollution Control Federation Conference, Las Vegas, NV, October 1, 1980.
101. van Leeuwen, J., and J. Prinsloo. "The Effect of Various Oxidants on the Performance of Activated Carbon Used in Water Reclamation," presented at Fifth Ozone World Congress, International Ozone Association, Berlin, Federal Republic of Germany, March 31–April 3, 1981.
102. van Leeuwen, J., and J. Prinsloo. "Ozonation at the Stander Water Reclamation Plant," *Water SA* 6(2):96–102 (1980).
103. Walker, C. A., and W. Zabban. "Disposal of Plating Room Wastes," *Plating* (July 1953), pp. 777–780.
104. Selm, R. P. "Ozone Oxidation of Aqueous Cyanide Solution in Stirred Batch Reactors and Packed Towers," in *Ozone Chemistry and Tech-*

nology, Advances in Chemistry Series No. 21 (Washington, DC: American Chemical Society 1959), pp. 66–77.
105. Bollyky, L. J. "Ozone Treatment of Cyanide-Bearing Plating Waste," U.S. EPA Report EPA-600/2-77-104 (Washington, DC: U.S. Government Printing Office, 1977).
106. Matsuoka, H. "Ozone Treatment of Industrial Wastewater," *PPM* 4(10):57–69 (1973).
107. McPhee, W. T., and A. R. Smith. "From Refinery Wastes to Pure Water," Engineering Experimental Station, Engineering Bulletin 109, Purdue University, West Lafayette, IN (1962), pp. 311–326.
108. Rice, R. G., and M. E. Browning. "Ozone for Industrial Water and Wastewater Treatment—A Literature Survey," U.S. EPA Report EPA-600/2-80-060 (Washington, DC: U.S. Government Printing Office, 1980).
109. Lorenzo, G. A., and T. N. Hendrickson. "Ozone in the Photoprocessing Industry," *Ozone Sci. Eng.* 1(3):235–248 (1979).
110. Baer, F. H. "Ozone Step Allows Recycle of Organic-Fouled Water," *Chem. Eng.* (August 1970), p. 42.
111. Prengle, H. W., Jr. "Evolution of the Ozone/UV Process for Wastewater Treatment," presented at Seminar on Current Status of Ozonation for Wastewater Treatment and Disinfection with Ozone, International Ozone Association, Cincinnati, OH, September 15, 1977.
112. Leitis, E., J. D. Zeff and M. M. Smith. "Chemistry and Application of Ozone and Ozone/UV Light For Water Reuse," U.S. Department of the Interior, Office of Water Research and Technology, Report No. OWRT/RU-80/14 (Washington, DC: U.S. Government Printing Office, 1981).
113. U.S. EPA. "Polychlorinated Biphenyls (PCBs), Toxic Pollutant Effluent Standards," *Federal Register* 42(22):6532 (1977).
114. Adabie-Maumert, F. A., B. Fritzvold, and N. Soteland. "The Norwegian Semi-industrial Pilot Plant for Processing of Paper Pulp by Ozone," paper presented at Third International Congress on Ozone Technology, International Ozone Association, Paris, France, May 1977.
115. Orgler, K., and J.-P. Légeron. "Contribution of Ozonation to the Treatment of Swimming Pool Waters," paper presented at Third International Congress on Ozone Technology, International Ozone Association, Paris, France, May 1977.
116. Gagnaux, A. "Ozone Treatment of Swimming Pool Water—Development and Prospects in Switzerland," paper presented at Symposium on Ozone Technology, International Ozone Association, Los Angeles, CA, May 1978.
117. Eichelsdörfer, D. "Use of Ozone For the Treatment of Swimming Pool Water," in *Wasser Berlin '77* (Berlin, Federal Republic of Germany: AMK Berlin, 1977), pp. 323–340.
118. Merrill, D. T., and J. A. Drago. "Evaluation of Ozone Treatment in Air-Conditioning Cooling Towers," in *Condenser Biofouling Control,* J. F. Garey, R. M. Jorden, A. H. Aitken, D. T. Burton and R. H. Gray, Eds. (Ann Arbor, MI: Ann Arbor Science Publishers, Inc., 1980), pp. 487–496.
119. Hill, J., P. J. Kelleher, R. L. Kreiling and J. S. McBride. "Ozone Cooling Water Treatment in Hotels and Large Commercial Complexes," paper presented at Special Seminar on the Modern Use of Ozone Tech-

nology in Treating Cooling Tower Water, International Ozone Association, Cincinnati, OH, June 26, 1981.

120. Diaper, E. W. J. "Ozonation Systems For Odor Control," in *Ozone: Analytical Aspects and Odor Control,* R. G. Rice and M. E. Browning, Eds. (Vienna, VA: International Ozone Association, 1976), pp. 64–70.

121. McCabe, B. R. "Ozone for Odour Control in Sewage Treatment and Industrial Processes," in *Ozone: Analytical Aspects and Odor Control,* R. G. Rice and M. E. Browning, Eds. (Vienna, VA: International Ozone Association, 1976), pp. 82–89.

122. Perkins, N. J. "Odour Control at the Main Sewage Treatment Plant—Ashbridges Bay, Toronto, Ontario, Canada," in *Ozone: Analytical Aspects and Odor Control,* R. G. Rice and M. E. Browning, Eds. (Vienna, VA: International Ozone Association, 1976), pp. 90–97.

123. Kutsuma, J., T. Ohira, K. Koyama, and T. Ishiguro. "Deodorization of Desiccator Flue Gas at Fish Meal Plants," *J. Jap. Soc. Air Poll.* 4(1):72 (1969).

124. Kuzuma, J., T. Sakurai, I. Koyama, T. Ishiguro, and T. Odaira. "Research on Odor Emitted from Fish Meal Plants," Annual Report, Tokyo Metropolitan Research Institute of Environmental Protection, (1970), pp. 140–155.

125. Odaira, T., J. Kazumi, K. Koyama and T. Ishiguro. "Control of Fish Meal Plant Odors in Dryer Exhaust," *J. Jap. Soc. Air Poll.* 4(1):72 (1969).

126. Nebel, C., and N. Forde. "Principles of Deodorization with Ozone," in *Ozone: Analytical Aspects and Odor Control,* R. G. Rice and M. E. Browning, Eds. (Vienna, VA: International Ozone Association, 1976), pp. 52–63.

127. Kramer, F. "Range of Application of Ozone in Dentistry," paper presented at Fifth Ozone World Congress, International Ozone Association, Berlin, Federal Republic of Germany, March 31–April 3, 1981.

128. Werkmeister, H. "Ozone Utilization in Radiation Therapy," presented at 3rd Intl. Congress on Ozone Technology, Paris, France (May, 1977). International Ozone Association, Vienna, VA.

129. Werkmeister, H. "Low Pressure Ozone Gassing for Decubitis, Chronic Ulcerations, Fistulas and Radiation Effects," paper presented at Ozone Therapy Congress, Medical Society for Ozone Therapy, Baden-Baden, Federal Republic of Germany, October 30–November 1, 1979.

130. Landauer, W. "Ozone Gassing of Extremeties with Simultaneous Intravenous Application of an Ozone-Blood Foaming for Diabetic Gangrene with Expanded Nekro-Compounds in the Proper Canal," *Erfahrungsheilkunde* 28(10):802–804 (1979).

131. Tabakova, M. G. "Intervascular Ozone Therapy for Obliterating Gangrenous Illnesses, for Ulcera cruris atonicum, Mal perforans, Elephantiasis and Other Diseases," *Erfahrungsheilkunde* 21:440–441 (1971).

132. Binder, T. "Ozone Therapy in the Practice of the Tropical Physician with Particular Consideration of Liver Diseases," *Erfahrungsheilkunde* 29(12):978–981 (1980).

133. Hildemann, H. "Geriatric Treatment With HOT (Hematogenic Oxidation Therapy). Post-treatment of Liver Patients," *Erfahrungsheilkunde* 27(7):436–438 (1978).

134. Rokitansky, O., J. Washüttl, I. Steiner, and A. Rokitansky. "Blood Chemistry Studies of Peripheral Arterial Blood Supply Disturbances During Ozone Therapy," *Erfahrungsheilkunde* 28(10):804–810 (1979).
135. Wolff, H. "The Treatment of Peripheral Blood Supply Disturbances with Ozone," *Erfahrungsheilkunde* 23:181–184 (1974).
136. Salzer, H., S. Szalay, I. Steiner and J. Washüttl. "Changes in Several Biochemical Parameters During Ozone Therapy with Patients Having Gynaecological Malignomas," *Erfahrungsheilkunde* 28(10):770–775; 811–814 (1979).
137. Hernuss, P. "Ozone and Gynaecological Radiotherapy," *Strahlentherapie* (November 1975), pp. 493–499.
138. Hildemann, H. "HOT (Hematogenic Oxidation Therapy) Treatment for Arterial and Peripheral Blood Supply Disturbances—Objectives and Experiences," *Erfahrungsheilkunde* 25(11):480–483 (1976).
139. Doerfler, J. "Hematogenic Oxidation Therapy (HOT)—F.X. Mayer-Kur-Guided Symbiosis. Experiences in an Internistic Practice," *Erfahrungsheilkunde* 27(8):503–511 (1978).
140. *Chem. Eng.* 59(9):246 (1952).
141. *Am. Inkmaker* 45(10):36 (1967).

SECTION 1

GENERATION

CHAPTER 2

OZONE GENERATION BY CORONA DISCHARGE

James J. Carlins and Richard G. Clark
Union Carbide Corporation
Linde Division
Tonawanda, New York

Ozone, the triatomic allotrope of oxygen, has the formula O_3 and generally exists as a relatively unstable, reactive gas. Small, atmospheric quantities of ozone are produced by several natural and man-made sources. These include: lightning, the action of sunlight on smog components, faulty light switches and motor brushes, power transmission lines, copying machines; nuclear radiation, and ultraviolet (UV) light. Commercially, large quantities of ozone are produced in specially engineered corona discharges. This chapter presents the theory and practice of commercial ozone generation by this method. The discussion includes the electrical characteristics of a corona discharge, factors affecting ozone production and descriptions of generic approaches commonly taken by designers of commercial ozone generators.

The corona discharge method involves the passage of a dry oxygen-bearing gas through a corona discharge. Common feed gas streams are: oxygen, air and recycle streams containing oxygen, nitrogen, argon, carbon dioxide and perhaps other diluents.

CHEMISTRY

Although there are a number of mechanisms that may contribute to ozone formation in a corona, one particular reaction path is considered dominant [1]. The reaction is initiated when free, energetic electrons in the corona dissociate oxygen molecules:

$$e^{-1} + O_2 \rightarrow 2O + e^{-1} \quad (1)$$

Following this, ozone is formed by a three-body collision reaction:

$$O + O_2 + M \rightarrow O_3 + M \quad (2)$$

where M is any other molecule in the gas. At the same time, however, atomic oxygen and electrons also react with ozone to form oxygen:

$$O + O_3 \rightarrow 2O_2 \quad (3)$$

$$e^{-1} + O_3 \rightarrow O_2 + O + e^{-1} \quad (4)$$

In addition, the gas in the corona is at a high temperature, which promotes ozone decomposition reactions. Therefore, the net ozone yield or outlet composition is the sum of all of the reactions that form and decompose ozone. This net rate depends on many factors, including the oxygen content and temperature of the feed gas, contaminants in the feed gas, the ozone concentration achieved, the power density in the corona, the coolant temperature and flow, and the effectiveness of the cooling system. All of these influence the design of practical, economically attractive ozone generators and systems.

ELECTRICAL CHARACTERISTICS

A corona is characterized by a low-current electrical discharge across a gas-filled gap at a voltage gradient on the order of the sparking (electrical breakdown) potential of the gap. During breakdown, the gas becomes partially ionized and a characteristic, diffused bluish glow results. Kilovolt voltages and milliampere-to-ampere currents are typical within a corona. On the opposite end of the spectrum, an arc discharge is characterized by a high current density, causing a highly ionized gas and a low voltage gradient in the gap. In general, an arc is a localized discharge producing high temperatures and a bright white light [2].

Figure 1 depicts a typical corona cell consisting of two metallic electrodes separated by a gas-filled gap and a dielectric material. An oxygen-bearing gas flows through the discharge gap while high voltage is applied to the electrodes. Most of the electrical energy input to the corona is dis-

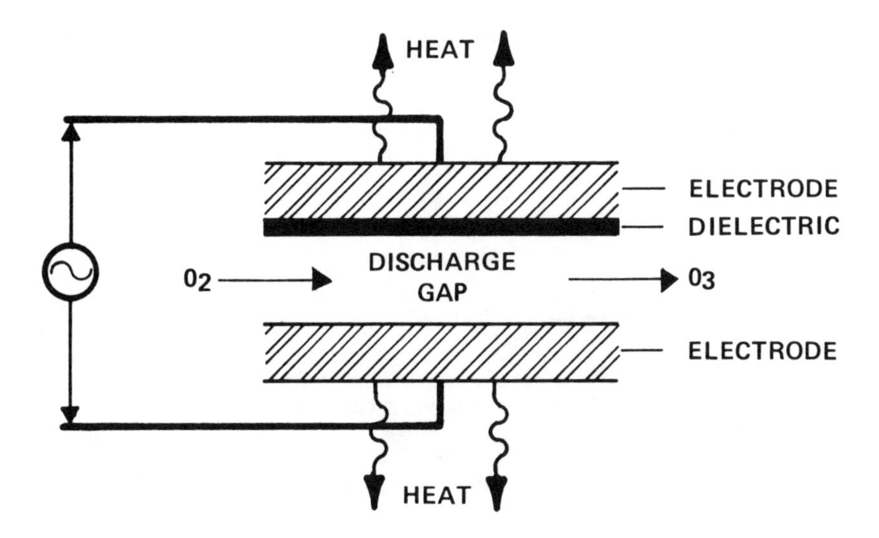

Figure 1. Typical corona cell configuration.

sipated primarily as heat with smaller portions going to light, sound and chemical reactions.

Electrically, a corona cell presents a capacitive load to the power supply due to both the gas-filled gap and the dielectric materials present. Ozone is produced in the corona as a direct result of power dissipation in the corona. Therefore, a fitting starting point for understanding ozone generation is the consideration of the electrical characteristics of a corona cell. The relations defining corona power consumption are easily derivable. It is useful to define the following fundamental parameters:

V_d = dielectric potential (V)
V_g = gap potential (V)
V_o = driving potential (V peak)
V_s = gap sparking potential (V peak)
V_{cs} = corona start potential (V peak)
f = driving voltage frequency (H)
P = corona power (W)
C_d = dielectric capacitance (f)
C_g = gap capacitance (f)
C_t = total cell capacitance (f)
V_e = gap corona extinction potential (V)

Commercial ozone generators use two basic corona cell geometries:

concentric tubes and parallel flat plates. The cell capacitance formulas are, for the concentric tube:

$$C = \frac{2 \pi \epsilon \epsilon_0 L}{ln(b/a)} \tag{5A}$$

for the parallel flat plate:

$$C = \frac{\epsilon \epsilon_0 A}{t} \tag{5B}$$

where b = radius of the outer concentric electrode
a = radius of the inner concentric electrode
L = length of the concentric tubes
t = flat plate electrode separation
ϵ = relative dielectric constant (5–7 for glass, ceramic, etc., 1 for air, oxygen, etc.)
ϵ_0 = absolute dielectric constant = 8.854×10^{-12} f/m
A = area of capacitor surface

Typically, the dielectric material chosen has a high dielectric strength (V/mm) and a high dielectric constant, such as glass or ceramic.

The driving potential of the cell can be expressed mathematically as $V = V_o \sin(wt)$, where $w = 2\pi f$. This is represented graphically in Figure 2. During the initial voltage increase, from t_o to t_1, the voltage gradient in the gas space is less than that required to ionize the gas. Here the cell behaves as two capacitances in series due to the dielectric and the air gap. These act as a capacitive voltage divider, which demands that the charges be equal on two series-connected capacitors. Therefore,

$$Q_d = Q_g$$

$$C_d V_d = C_g V_g$$

$$\frac{V_g}{V_d} = \frac{C_d}{C_g} \tag{6}$$

The dielectric constant of the gas is roughly one-sixth that of the dielectric material. Therefore, the capacitance of the dielectric material is considerably greater than that for the gas gap. From Equation 6, then, the voltage drop across the gap is considerably larger than that across the dielectric prior to gas breakdown.

At point wt, the driving voltage has reached a value V_{cs}, such that the voltage across the gas space has reached its sparking potential V_s and

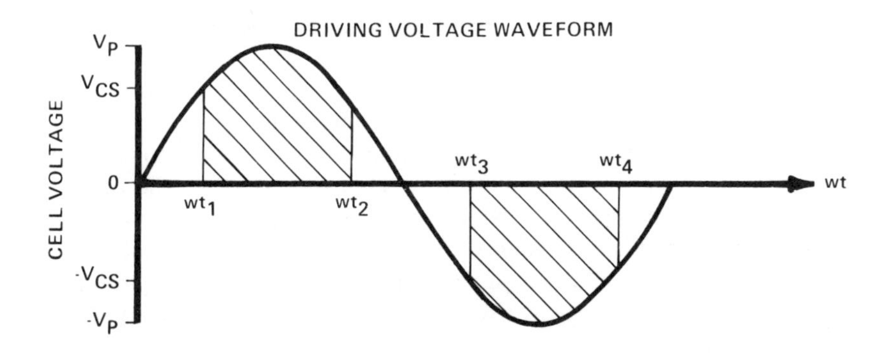

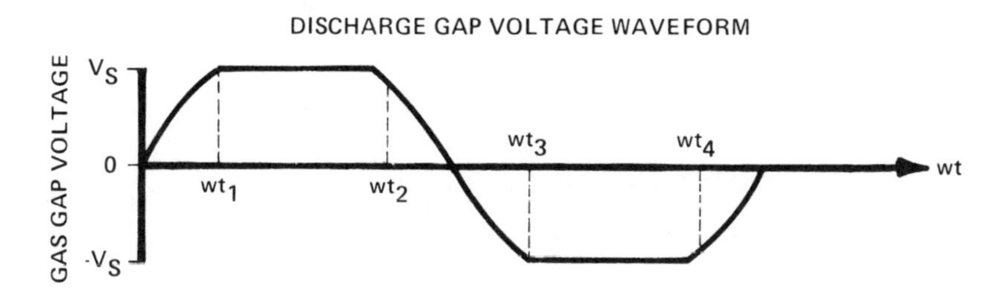

Figure 2. Driving voltage and discharge gap waveforms.

breaks down into a corona. Cobine [3] reports that the sparking potentials for air and oxygen are:

$$V_s = 29.64\ pt_g + 1350\ \text{(air)} \tag{7}$$

$$V_s = 26.55\ pt_g + 1480\ \text{(oxygen)} \tag{8}$$

where p = absolute gas pressure (kP)
t_g = gap length (thickness) (mm)

Once again, using the capacitive voltage divider calculation, a relationship between the corona start potential V_{cs} and V_s can be obtained:

$$C_t V_{cs} = \frac{C_g C_d}{C_g + C_d} V_{cs} = C_g V_s$$

$$V_{cs} = \frac{C_d + C_g}{C_d} V_s \tag{9}$$

Because the gas in the gap becomes conductive with an impressed voltage of V_{cs}, the voltage drop across the gap itself remains essentially constant at V_s with further increases in V_{cs}. Rosenthal and Davis [4] successfully modeled a conducting corona discharge gap as a zener diode that characteristically maintains a constant voltage drop during conduction. This wave shape is shown in Figure 2. Note that the voltage at the corona extinction (wt_2 & wt_4) is less than the sparking potential V_s. This phenomenon is due to ionization of the gas in the corona [5].

Viewed another way, as the gap goes into corona at point wt_1, the cell changes in character from two series capacitors to a capacitor in series with a voltage-clamped conductor (e.g., a zener diode). Therefore, the voltage drop across the dielectric must first increase more rapidly and then fall with the driving potential during the conduction period.

The characteristic shape of the dielectric potential during conduction can be expressed mathematically as:

$$V_d = V_o \sin(wt) - V_s \tag{10}$$

The charge stored on the dielectric is then:

$$\begin{aligned} Q_d &= C_d V_d \\ &= C_d [V_o \sin(wt) - V_s] \end{aligned} \tag{11}$$

The displacement current through the dielectric material is:

$$\begin{aligned} I_d &= dQ/dt \\ &= C_d w V_o \cos(wt) \end{aligned} \tag{12}$$

Since the currents through the gap and dielectric are equal, because they are in series, the instantaneous corona power draw is:

$$\begin{aligned} P_i &= V_s I_d \\ &= w C_d v_s V_o \cos(wt) \end{aligned} \tag{13}$$

The average power can be obtained by integrating the instantaneous power over the duration of the corona and accounting for the number of corona discharges per second (2f).

$$P = C_d V_s w V_o \left[\int_{wt_1}^{\pi/2} \cos(wt)\, d(wt) \right] 4/2\pi$$

$$= 4C_d V_s f V_o [\sin(wt)_{wt_1}^{\pi/2}]$$

$$= 4C_d V_s f V_o [1 - V_{cs}/V_o]$$

$$= 4C_d V_s f (V_o - V_{cs})$$

$$= 4C_d V_s f \left[V_o - \left(\frac{C_d + C_g}{C_d} \right) V_s \right] \tag{14}$$

This expression has been verified experimentally by Reynolds [6].

An understanding of the power equation (Equation 14) reveals the theoretical basis for the various practical approaches commonly followed by designers of corona cells. For a given geometry, gap thickness and gas pressure, the power drawn by the corona can be increased (1) by operating at higher frequencies; (2) by employing thinner dielectrics with higher dielectric constants; or (3) by operating at higher peak driving voltages. Many generators in operation today are powered at 60 Hz, although medium-frequency (400–600 Hz) and high-frequency (nominally 2000 Hz) equipment is available. Dielectric materials commonly are glass or ceramic, with thicknesses ranging from about 0.5 to 3 mm. Driving peak voltages commonly range from about 8 to 30 kV_p.

From a practical standpoint, however, minimum dielectric thicknesses usually are dictated by structural considerations, material quality (freedom from imperfections) and dielectric strength. High peak voltages tend to enhance dielectric failures. Therefore, lower peak voltages are recommended for longer dielectric life. For a given dielectric cooling system, extremely high power densities (W/m^2) also increase dielectric failures, due to dielectric heating. This imposes practical limits on power density and stringent requirements for an effective cooling system for highest dielectric reliability. Therefore, the corona cell engineer must strike a practical compromise between high power draw, cooling efficiency and dielectric reliability and maintenance.

The above comments apply to the interrelationships between dielectric voltage stress, corona power consumption, dielectric cooling and dielectric reliability. When ozone production due to corona power consumption is discussed (below), it will be demonstrated that efficient cooling of the corona discharge itself is also extremely important. This factor is not reflected in the power draw equation (Equation 14), but cannot be overlooked in corona cell design. The requirement for efficient corona cooling generally leads the designer to consider thin discharge gaps cooled from both sides.

Rosenthal and Davis [4] developed a valuable tool for investigating the operation of a corona cell. By creating an oscilloscope plot of voltage

V(t) vs charge Q(t), characteristic information concerning cell capacitances and critical voltages can be obtained. Figure 3 depicts the resultant idealized parallelogram. The two slopes of the parallelogram represent reciprocal capacitances for the conducting C_d and nonconducting C_T portions of the corona cycle. With these values, the gap capacitance C_g then can be calculated using:

$$C_T = \frac{C_g C_d}{C_g + C_d} \tag{15}$$

The straight lines indicate constant dynamic capacitances. The peak driving voltage V_o and sparking potential V_s are also readily obtainable as

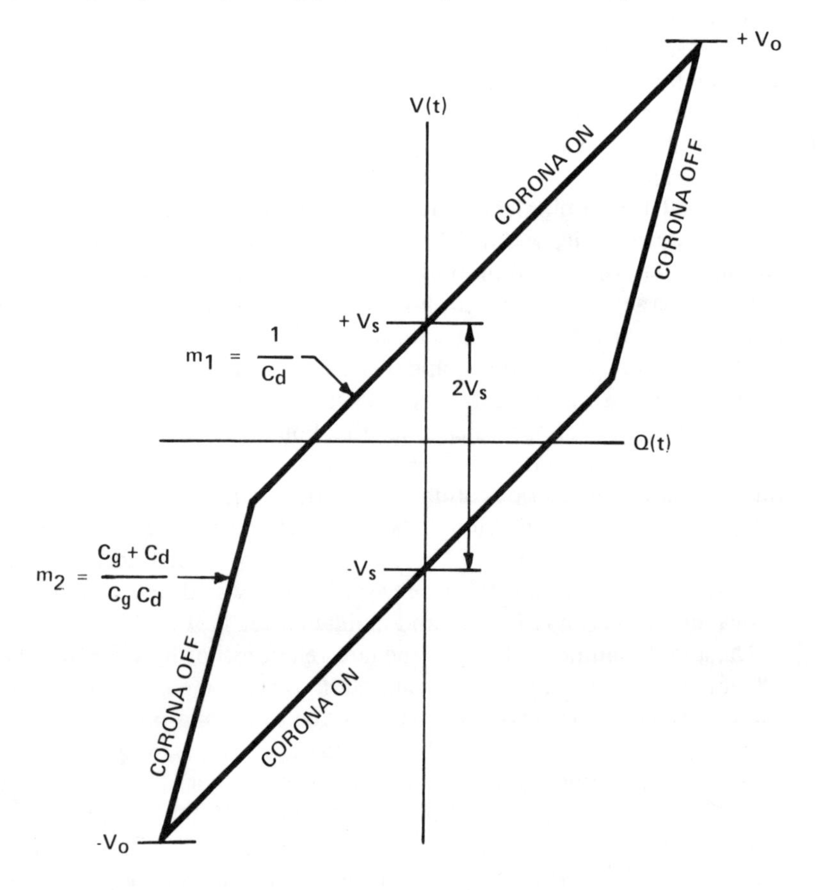

Figure 3. Idealized Q-V plot for a corona cell [4].

indicated. The energy dissipated per cycle can be determined by computing the area enclosed by the parallelogram.

The experimental setup to create a Q-V plot on an oscilloscope is shown in Figure 4. The voltage can be viewed using a 1000X voltage attenuation probe or a capacitive divider. The charge is determined by monitoring the voltage across an integrating capacitor C_I. This capacitance must be chosen much larger than C_d, so that it does not interfere with normal cell operation. Figure 5 represents an actual photo of a Q-V plot utilizing a typical corona cell with ceramic dielectric, an arbitrary air gap thickness and a high-frequency power supply. The variation in charge for the "corona off" portion of the parallelogram indicates the magnitude of voltage ripple in the power supply.

In summary, it is important that the factors giving rise to high corona power dissipation are thoroughly understood. However, these relations must not be followed blindly merely to maximize corona power draw. In the last analysis, high dielectric reliability and high ozone yield from the corona are the "bottom line" criteria for the designer. For these reasons, the thrust of the state-of-the-art developments currently in progress throughout the industry are toward higher quality and thinner dielectrics; thinner corona gaps; higher operating frequencies; lower driving voltages; higher corona power levels per unit electrode area; and improved cooling of both the dielectric material and the corona discharge. It is likely that significant generator improvements will continue to appear as the technology advances.

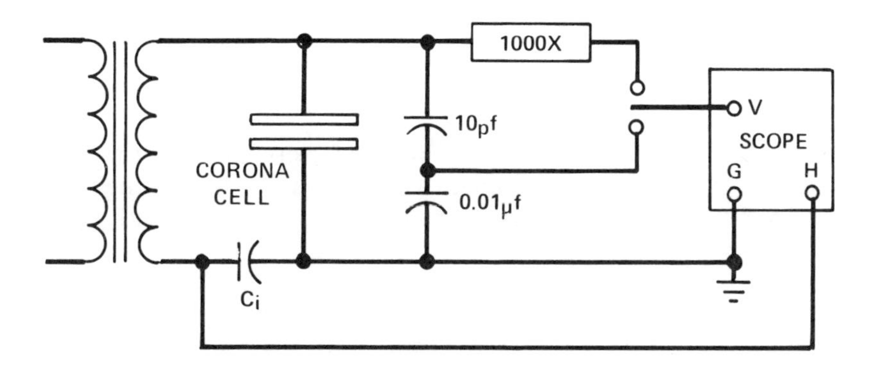

Figure 4. Experimental setup for Q-V plot [4].

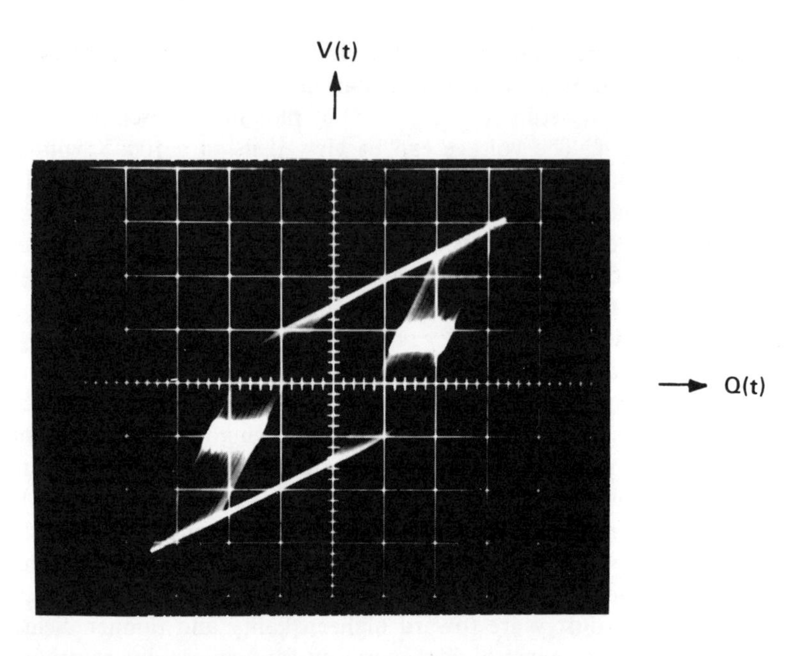

VERTICAL DEFLECTION, V(t) = 1130 VOLTS/DIV.
INTEGRATING CAPACITOR, C_i = 5.0μ FARAD
HORIZONTAL DEFLECTION = 0.5 VOLT/DIV.
Q(t) = (0.5V/d) (5.0μF) = 2.5μ COULOMB/DIV.
OPERATING FREQUENCY, f = 660 HERTZ

Figure 5. Actual Q-V plot for a corona cell.

OZONE YIELD

The previous section dealt with some of the factors relating to the electrical power consumed by a corona. This section discusses ozone production in the corona and some of the major factors that affect it. The discussion will deal primarily with the practicalities of ozone generation rather than with a theoretical understanding. To simplify the present discussion, power is taken to be corona power. It includes all power supply losses involved in the steady-state operation of the corona generator. Specifically excluded, to be discussed later, are the power or other costs associated with cooling of the corona and dielectrics. Corona power is usually measured by means of a calibrated wattmeter or watt-hour meter connected ahead of the generator. In this way, one is assured that all electrical losses within the generator are accounted for. The resulting energy consumption is directly related to the utility charges for corona operation.

Perhaps the most useful means for understanding the corona generation of ozone is to consider how specific energy input relates to ozone concentration. The unit for specific energy is kWh/kg. This can be understood either as the amount of energy that produces a kilogram of ozone or as the corona power divided by the ozone production rate. Clearly it is reciprocally related to energy efficiency. Concentrations are often expressed either in percent by weight or in milligrams of ozone per liter of gas.

Figure 6 illustrates the range of representative values of the corona specific energy as a function of ozone concentration for dry oxygen feed. The values are considered representative in that they typify the spectrum of large-scale generators in commercial use at this time. One must survey the major generator suppliers to determine the actual performance each can offer for any particular set of requirements.

The major conclusion to be drawn from Figure 6 is that the energy efficiency decreases rapidly as one tries to make ever higher ozone concentrations. This, of course, is the direct result of the reverse reaction, represented in Equations 3 and 4, in which the corona makes oxygen out of ozone. The reverse reactions are favored not only by higher ozone concentrations, but also by higher temperatures, because ozone is thermally unstable. The half-life of ozone decreases significantly as the temperature increases. Therefore, as will be illustrated later, heat removal from the corona strongly influences the net specific energy used, which includes both the forward and reverse reactions. Commercial ozone systems are commonly designed to operate in the range of 2–3 wt % of ozone from oxygen feed.

There are several approaches to varying the rate of ozone production. One can, for example, maintain the corona power constant and vary the feed gas flow through the discharge. The relative yield will vary with a characteristic represented by Figure 7. For very low flows, the ozone concentration is very high. Therefore, small quantities of ozone are produced because the specific energy is very high. As the gas flowrate increases, the ozone yield approaches a limiting value because the specific energy is nearly constant for very low ozone concentrations.

Alternatively, one can maintain a constant gas flow through the corona and vary the corona power level to achieve different ozone yields. Characteristics similar to those in Figure 8 are common.

Note that the ozone yield increases more slowly at higher power levels. Depending on the efficiency of the corona cooling system, it is possible

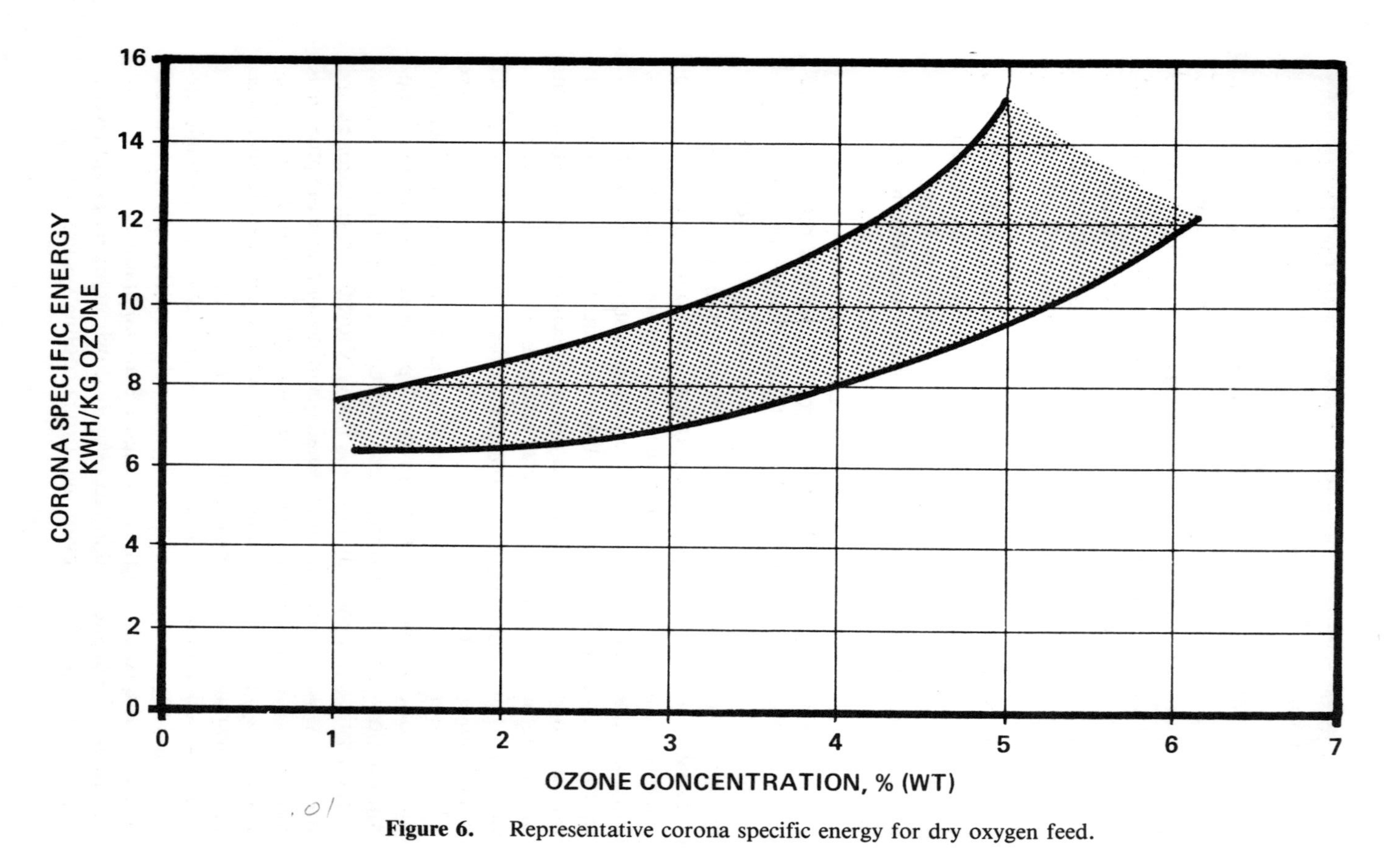

Figure 6. Representative corona specific energy for dry oxygen feed.

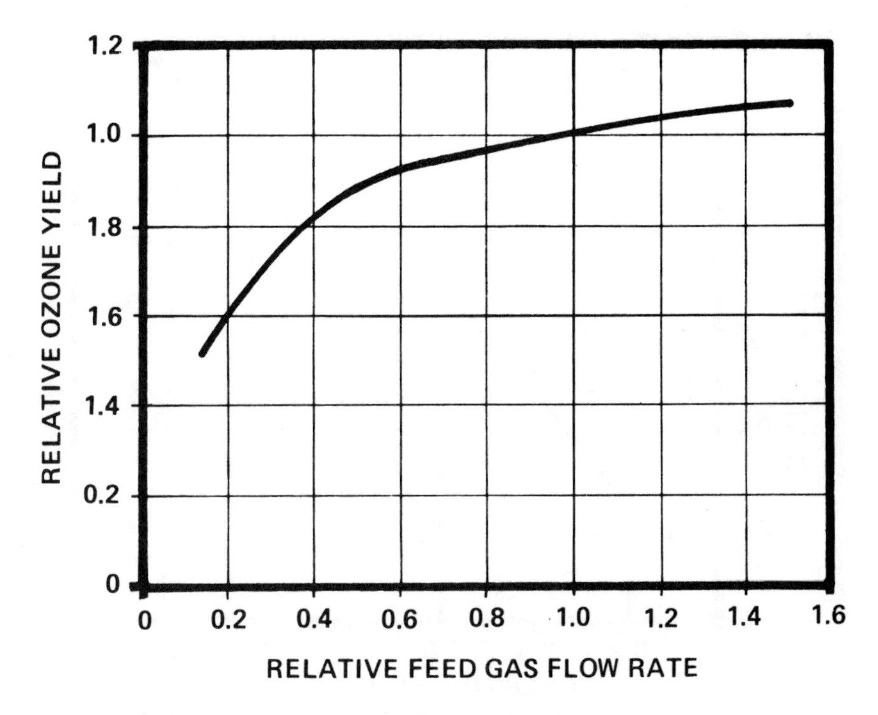

Figure 7. Ozone yield vs gas flowrate for constant corona power.

to increase the corona power (heat generation) to the point that no further ozone yield is achieved.

In Figure 9, representative energy values are shown for the production of ozone with dry air as the feed gas. Again, a range of values is given to represent the major types of commercial generators available at this time. As suggested by Figure 9, it is generally not economically attractive to generate high ozone concentrations with air feed. Commercial air-based systems are commonly designed to operate at about 1–2 wt % concentrations.

Because the oxygen content of the air is about 21%, one might expect the specific energy for ozone production from air to be about five times less than that for production from oxygen. In fact, only about twice as much energy is required to produce the concentrations of practical interest. Therefore, it seems that nitrogen is not an inert diluent but must somehow contribute to ozone formation. Popovich et al. [7] suggest that activated nitrogen atoms collide with and dissociate oxygen molecules in the corona. As is suggested by Equation 2, any mechanism that increases the atomic oxygen concentration in the corona will enhance ozone production.

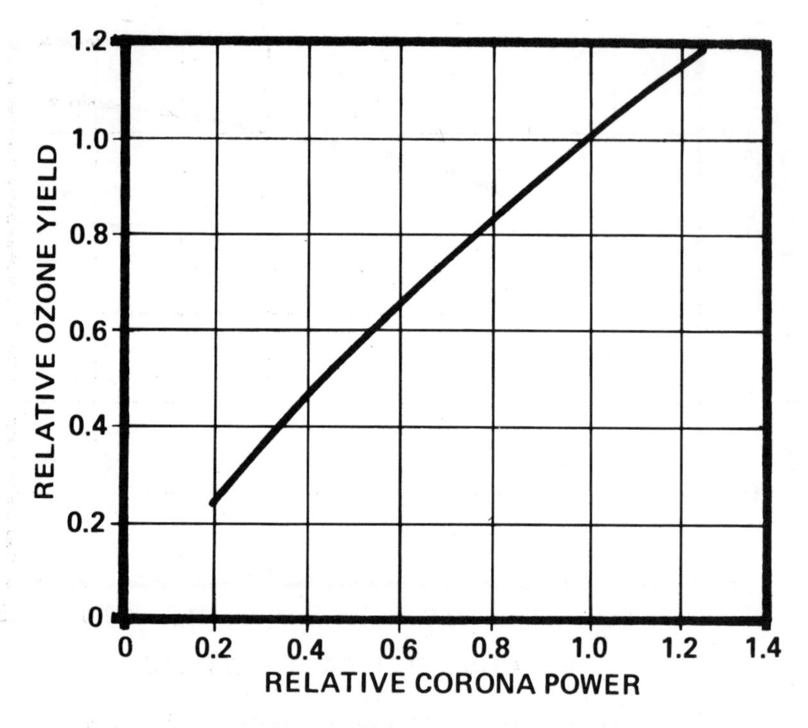

Figure 8. Ozone yield vs corona power for constant feed gas flowrate.

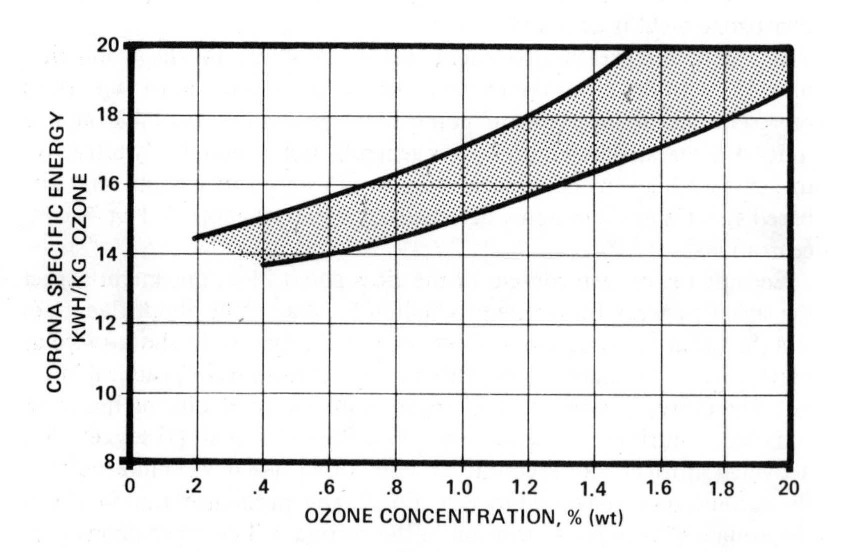

Figure 9. Representative corona specific energy for dry air feed.

The designers of ozone generators and ozonation systems must not forget that ozone generation is an extremely inefficient process. This can easily be illustrated. The theoretical heat of formation of ozone is reported to be 0.835 kWh/kg [8]. If one assumes that the "real-world" specific energy to make ozone from oxygen at 2 wt % is 7 kWh/kg, then it is evident that 88% of the electrical energy supplied to the corona is wasted and ultimately must be rejected from the generator. For a specific energy of 15.5 kWh/kg using air feed, 95% of the corona power must be rejected. Generator and systems designers must be prepared to deal with essentially all of the applied corona energy as waste heat. The actual total costs of cooling must be evaluated for each generator type before a true economic evaluation can be made.

From the viewpoint of the generator designer, the corona occurs as a gas-phase reactor with internal heat generation. Because air and other such gases are poor heat conductors, the gases in the corona can reach temperatures high enough that the thermal decomposition of ozone becomes quite significant, thereby reducing the net ozone yield. Some even suggest that the reverse reaction (ozone decomposition) dominates above 40°C [9]. For this reason, generators are usually designed with fairly thin gaps to facilitate heat removal from the corona. Gap thicknesses of 1-3 mm are typical.

Because of the influence of heat removal on ozone yield, the ozone yield from essentially all generators is sensitive to changes in the temperature of the coolant used. For air-cooled generators, the coolant is usually ambient air. For water-cooled or water/oil-cooled units, the excess heat is ultimately rejected to water. Figure 10 [10] presents the relative influence of cooling water temperature on the ozone yield of a typical water-cooled generator. More specific data should be obtained from the various generator manufacturers for each equipment type.

In addition to coolant temperature, coolant flowrate is often important, too. It is not uncommon for suppliers of water-cooled generators to recommend 2500–4000 liter H_2O/kg ozone (300–480 gal/lb) of 15–20°C cooling water, depending on the power density in use. Operating values as high as 16,700 liter/kg (2000 gal/lb) have been reported in the literature [11]. Increased coolant flowrates tend to offset both yield deterioration due to high pwer densities (gas temperature) and dielectric failure due to increased dielectric temperature.

Another important design consideration is that a relatively uniform gas thickness must be maintained in the corona gap. Recall that the power draw is proportional to $V_s(V_o - V_{cs})$, and that V_s and V_{cs} are both functions of gap thickness. If the gap thickness is relatively nonuniform, narrower portions will tend to draw more power per unit area and therefore establish a hotter corona. This results in less efficient ozone

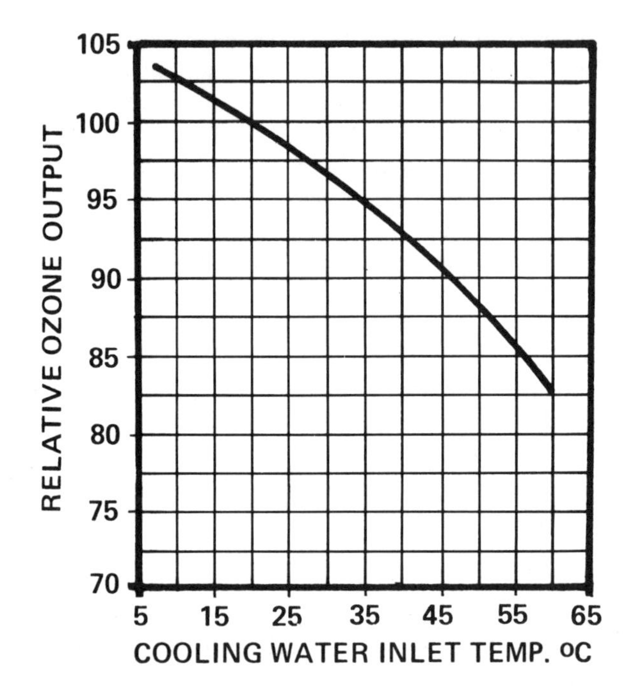

Figure 10. Typical effect of cooling water temperature on ozone production [10].

production (kWh/kg). Further, if the narrowing of the gap is severe enough to cause gas flow restriction in that region, the localized ozone concentration and the specific energy increase even further. Therefore, the generator designer must establish a uniform gap thickness to achieve the highest ozone generation efficiency.

Diluents in the feed gas have a strong, generally detrimental influence on the ozone yield in a corona. The most common diluent that causes trouble is water vapor. Trace amounts of water can substantially reduce the ozone output from an ozone generator. Figure 11 presents the relative yield of ozone as a function of the dew point of the feed gas [12]. These data generally lead ozone system designers to dry the incoming feed gas to at least −40°C, and usually to below −60°C dew point. The increase in ozone yield usually more than compensates for the additional cost of drying the feed gas to such low dew points. These comments apply, of course, to feed gases that contain water vapor, such as ambient air or gases being recycled back to the generator from a contact reactor. Pure oxygen gas being fed to the generator from either a liquid oxygen

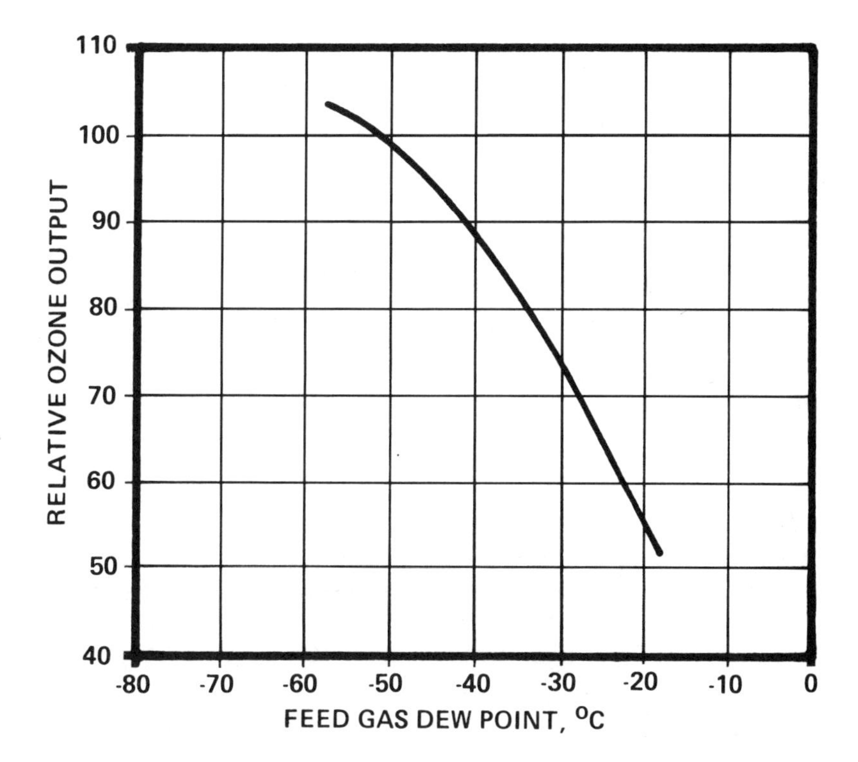

Figure 11. Effect of water vapor on ozone production [12].

facility or a pressure-swing plant already has a dew point well below −60°C and requires no further processing.

Water vapor in the feed gas can cause problems other than reduced ozone yield. If the feed is air or nitrogen-containing recycle gas, oxides of nitrogen formed in the corona continue to react with the water to form nitric acid. The nitric acid can cause damage to both generator components (for example, welds exposed to the gas stream) and to downstream equipment not designed to handle wet gas and nitric acid. These problems also can be avoided by designing an effective, reliable drying system ahead of the ozone generator.

For many systems, oxygen is the most attractive generator feed gas because of the high ozone yields and high concentrations possible with it. With oxygen, however, economics usually dictate that the offgases from the ozone contactor be recycled back through the system rather than vented, because of the high cost of the oxygen itself. Consider the ozone disinfection of secondary effluent as a typical example. The contactor offgases will include not only ozone and water vapor, but also nitrogen, argon and carbon dioxide, which are scrubbed from the effluent. In a closed-loop recycle system, the recycling gas must be dried before redelivery to the ozone generator, for high energy efficiency and system reliability. If the recycle compressor, dryer and piping components are not ozone-compatible, an in-line ozone destruction system must be used after the contactor. To keep the nitrogen concentration from continuing to build up in the recycle loop, with a resultant dramatic decrease in ozone yield, some of the recycling gas is vented to the atmosphere. The vent flow is adjusted to control the nitrogen concentration at the desired value. The argon and carbon dioxide levels also remain constant with constant venting. The vent gas also contains high concentrations of oxygen. This oxygen must be replenished from a separate supply to maintain the oxygen inventory.

To optimize the design of an oxygen-rich recycle system, the designer needs to know the influence that nitrogen, argon and carbon dioxide have on ozone production. The fragmentary data in the literature seem sufficient to support initial, first-cut estimates to be used until such time as reliable test data become available. The following approach has proven useful for preliminary estimates.

High concentrations of nitrogen in oxygen cause substantial reductions in the ozone output from a generator. This is proven by the fact that a generator operating on air yields about one-third to one-half as much ozone as one using pure oxygen. The addition of relatively small amounts of nitrogen to a pure oxygen stream, however, produces a rather unusual result. Cromwell and Manley [13] and Rosen [14] present data indicating that the ozone yield actually increases between 2 and 7% as the first 5 to 8 vol % of nitrogen is added to pure oxygen. The Cromwell and Manley data, represented by the "nitrogen" curve in Figure 12, were chosen for use because they suggest the least yield enhancement for low nitrogen concentrations and are considered more conservative. Following the initial increase in ozone yield with nitrogen addition, the yield decreases essentially linearly as more nitrogen is added. The yield extrapolates well to values expected for air feed (Y_{N_2} = 79 vol %). It should be evident that factor E in Figure 12 is the ratio of the ozone yield with the listed diluent divided by the yield with pure oxygen, all other generator operating parameters remaining the same. Therefore, E represents a performance degradation. Figure 12 also contains values derived from Cromwell and Manley for argon and carbon dioxide.

If one knows the performance of a generator with both oxygen and air feed, the Cromwell and Manley nitrogen data can be used to construct a graph, such as shown in Figure 13. These data are assumed for nitrogen

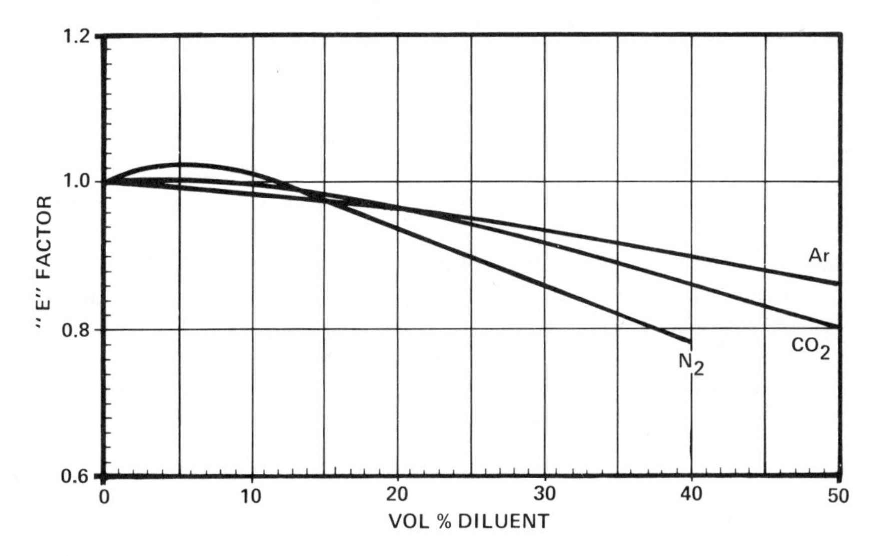

Figure 12. Relative effect of various diluents in oxygen on ozone generation efficiency [13].

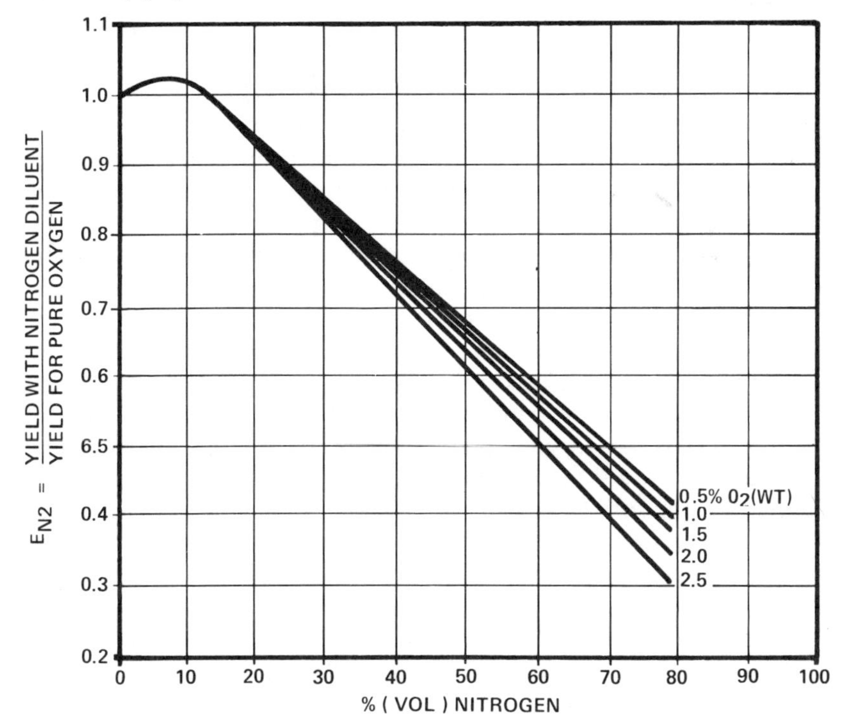

Figure 13. Effect of nitrogen on ozone generation.

concentrations up to 10 vol % for all ozone concentrations of practical interest. Further, for each ozone concentration, it is assumed that the ozone yield falls linearly from its value for 10 vol % added nitrogen to its known value for air feed, at 79 vol %.

For illustration purposes, assume that Figure 13 represents the generator in question. One can estimate the specific energy for an oxygen recycle stream containing 35 vol % nitrogen, 5 vol % argon and 4 vol % carbon dioxide for generating ozone at 1.5 wt %. Assume that 1.5 wt % ozone can be made from dry oxygen for 7.0 kWh/kg (corona power only). The nitrogen factor (0.77) would be read from Figure 13 at 35% N_2 at 1.5 wt % and the factors for argon (0.99) and carbon dioxide (1.00) come from Figure 12. The estimated specific energy (corona only) would then be:

$$\frac{7.0 \text{ kWh/kg}}{0.77 \times 0.99 \times 1.0} = 9.2 \text{ kWh/kg}$$

To this would be added the appropriate specific energy associated with cooling and feed gas preparation to yield the total specific energy for ozone generation.

ADDITIONAL SYSTEM DESIGN CONSIDERATIONS

There are a number of system design considerations that can affect the efficiency and reliability of an ozone generator. The generator system designer will want to address these potential problem areas.

The generator feed gas must be essentially free of hydrocarbons, corrosive vapors and any other substance that can react in the oxygen/ozone/corona environment to cause safety hazards or damage to the equipment. Of the three factors required for an explosion (fuel, oxidant and an ignition source), two are already present in the corona environment. Fuellike materials must be kept out of the feed gas stream. If hydrocarbons are potentially present, hydrocarbon analyzers should be installed to turn off the corona power if hydrocarbon concentrations approach 25% of the lower explosive limit (LEL). Fluorocarbons such as Teflon®* or refrigerants can be broken down in the corona to form fluorine, which can attack the glass dielectric material, potentially accelerating dielectric failure. Cooling fluids circulated around the outside of the corona cell may leak past seals and enter the corona space, resulting in the formation of a varnishlike coating on the dielectric surfaces. When

* Registered trademark of E. I. du Pont de Nemours and Company, Inc., Wilmington, Delaware.

this happens, the dielectrics must be cleaned periodically, as the coating reduces the efficiency of ozone production. The feed gas should be filtered to about 5 μ particle size to prevent small desiccant particles or other particles from entering the corona. Desiccant fines have been reported to cause a brown stain on dielectric surfaces [11].

The feed gas pressure should not be allowed to vary uncontrollably. As gas pressure influences the corona power draw and the voltage applied across the dielectrics, wide pressure variations can cause unreliable generator operation. Higher-than-expected corona power may cause fuses or breakers to open. Higher-than-expected applied voltages can cause premature dielectric failure.

The ozone system must be designed to prevent bulk water from entering the generator. Float valves for water-ring feed gas compressors or condensation traps on feed gas dryers have been known to stick, causing the ozone generator to become flooded with water. Bulk water in the corona cell leads to concentration of the corona, high current density and localized dielectric heating, causing premature dielectric failure. Even if a detection system interrupts the corona power before the water enters the corona cells, the debris in the water will be deposited on the cell surfaces. This debris must be removed prior to continued operation. Malfunctions or operating errors have been known to force effluent from the ozone contact chamber back into the generator. This can also cause at least corona cell contamination and perhaps dielectric failure. In addition, the system design and operating procedures must prohibit flammable, corrosive or even water vapors from migrating backward from the ozone contact chamber into the generator.

For water-cooled generators, the quality of the cooling water is important to minimize fouling of the heat transfer surfaces. Fouling will lead to reduced heat transfer efficiency and, therefore, reduced ozone generation and higher maintenance costs. Potable water is the technically preferred coolant. However, the consumption rates required for large commercial generators generally make potable water economically unattractive, except perhaps for systems in water treatment plants. On the other end of the water quality scale, treated sewage treatment plant effluent has generally been found to be an unsatisfactory coolant, as it often results in fouling. If high-quality water or other fluids is used in a closed primary cooling loop, effluent might be considered for final heat rejection if the final heat exchanger is specifically designed for minimum fouling and for ease of cleaning. Experience generally leads the system designer to use cooling-tower water or heat exchanger quality water (free of suspended solids, <5 mg/l chlorides) for the best balance between water costs and maintenance costs.

For air-cooled generators, the cooling air must be free of blown moisture, debris, aerosols of corrosive, oily or conducting materials, and visible dust. In general, filtered air is not required except perhaps in an extremely dusty industrial atmosphere.

COMMERCIAL OZONE GENERATORS—GENERIC REVIEW

Several types of commercial-scale ozone generators presently are available. The basic differences between these units are: corona cell geometry, power supply, heat rejection techniques, and operating conditions. Five basic corona cell types and three power supply types are briefly and generically discussed in this section. These types include:

1. corona cells: Otto plate (water-cooled), horizontal tube (water-cooled), vertical tube (water-cooled), vertical tube [double-fluid-cooled (oil and water)] and Lowther plate (air-cooled); and
2. power supplies: fixed low frequency (50–60 Hz), variable voltage; fixed medium frequency (400–600 Hz), variable voltage; and variable frequency (up to 3000 Hz) fixed voltage.

Table I gives a comparison of the various corona cell types of ozone generators and their associated power supplies. The two water-cooled tube type units have been combined in this table due to their many similarities. The comparative information presented is considered representative of the various generator types. One should consult generator suppliers to obtain more detailed information about specific units.

Otto Plate, Water-Cooled

The original design for this plate generator was developed by Otto in the early 1900s. The basic unit is still in use today (Figure 14) [15]. The cell consists of flat hollow blocks, separated by two glass plates and a gas space. The water-cooled blocks serve both as the electrodes and heat dissipators. A number of these generator elements are enclosed in one housing. In this unit, which is designed to operate at atmospheric or negative pressure, air is drawn into the housing, through the corona discharge area and out a central manifold passing through the glass plates and metal blocks. Ozone is produced as the air passes between the glass plates and is exposed to the corona discharge. Operation is designed to occur at slightly below atmospheric pressure, because the system is not designed to be leakproof. The dielectrics, for example, are made of ordinary window glass, which is not made to high tolerances. As a result of operating

Table I. Comparison of Commercial Ozone Generator Operating Parameters[a]

Parameter	Otto Plate	Tube (Water)	Tube (Double Liquid)	Lowther Plate
Feed Gas	Air	Air, oxygen	Air, oxygen	Air, oxygen
Feed Gas Dew Point (°C)	−60	−60	−40 to −51	−60
Cooling Fluid	Water	Water	Water and oil, typically	Air
Recommended Corona Cell Pressure (kPa, abs.)	100	100–200	170–200	170–240
Nominal Discharge Gap (mm)	≅3	2–3.5	1.3	1.0
Nominal Driving Voltage (kV, peak)	7.5–20	14–32	10	7–9
Nominal Frequency (Hz)	60/600	60/600	2500	2500
Dielectric Thickness (mm)	3–4.8	1–3	2.4	0.5
Corona Power Density (kW/m^2)	0.15–0.75	0.3–1.5	7.6	15.8
Power Supply Type				
Frequency (Hz)	Fixed 60/600	Fixed 60/600	Variable	Variable
Voltage	Variable	Variable	Fixed	Fixed
Typical Cooling Water Usage (liter/kg 0_3)	2500–4150	2500–4150	1200–2500	NA[b]
Cooling Air Usage (liter/kg 0_3)	NA	NA	NA	2.6×10^6
Ozone Production Rate (from air) for Commercially Available Generators (kg 0_3/day)	0.2–27	0.2–680	0.1–163	0.1–1500
Cooling Water Temperature (°C)	7–65	7–65	1–32	NA
Cooling Air Temperature (°C)	NA	NA	NA	<45

[a] Information in this table was supplied by generator manufacturers. Detailed information must be obtained from specific vendors.
[b] NA = not applicable.

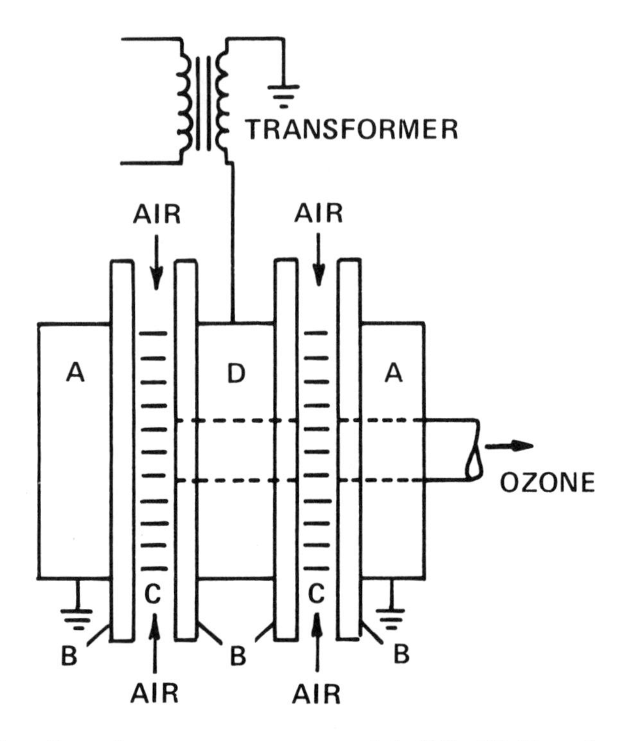

Figure 14. Otto plate ozone generator module [15]. (A) Ground potential, water-cooled block; (B) glass dielectric; (C) discharge gap; (D) high-tension, water-cooled block.

at subatmospheric pressures, ambient air can leak into the ozone generating cells.

Because of the low operating pressures of this cell, this type of ozone generating unit is limited to negative dissolution applications, such as vacuum injection or inductive turbine applications [14].

Horizontal Tube, Water-Cooled

This type of corona cell is currently the most commonly used for commercial ozone applications. The tube and a complete unit are shown in Figures 15 and 16. The corona cell consists of an outer, grounded stainless steel tube and a concentric glass dielectric tube having its inner surface coated with a metallic material. The two metal surfaces make up the two electrodes. Typically, a 3-mm gas space is maintained between the glass dielectric and the outer steel tube with electrode spacers. Power is supplied to the inner metallic coating through an axial bus bar containing

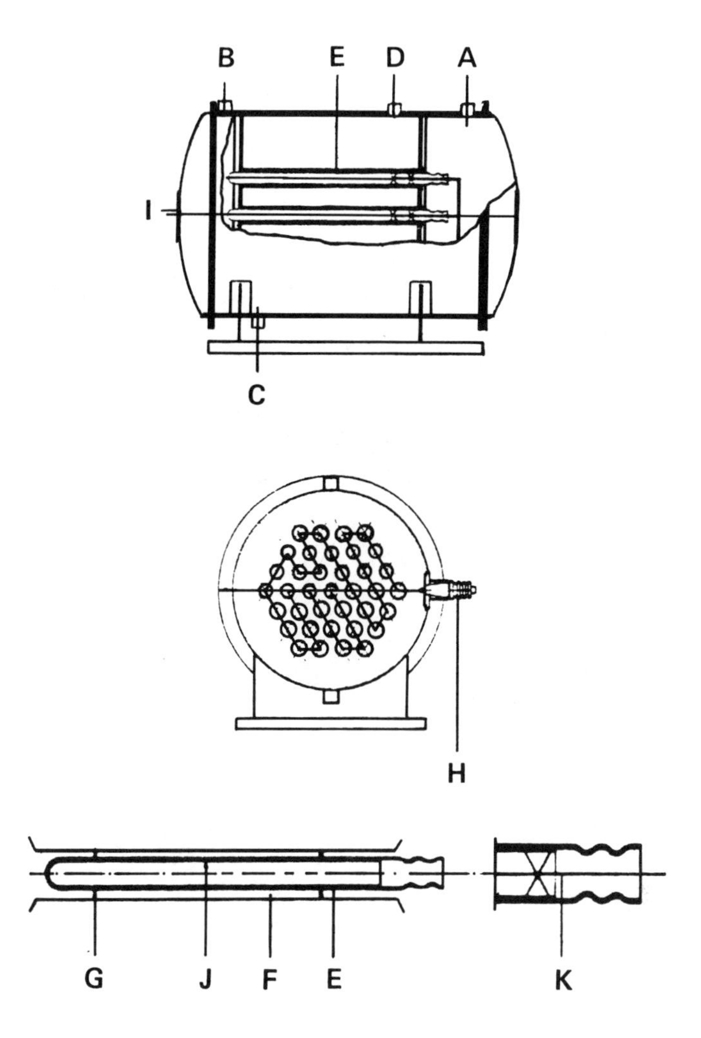

Figure 15. Corona cell details for a horizontal tube-type ozone generator [17]. A = air inlet; B = ozonated air outlet; C = coolant inlet; D = coolant outlet; E = dielectric tube; F = discharge zone; G = tube support; H = HV terminal; I = port; J = metallic coating; K = contact.

electrode brushes. Each corona tube can be individually fused to allow the generator to maintain continuous operation in the event of a single dielectric failure within the unit. Cooling of the corona tube is accomplished by passing water of potable or heat exchanger quality along the outside of the outer stainless steel tube. Flowrates of 2500–4000 liter H_2O/kg ozone (300–480 gal/lb) of 15–20°C cooling water are common.

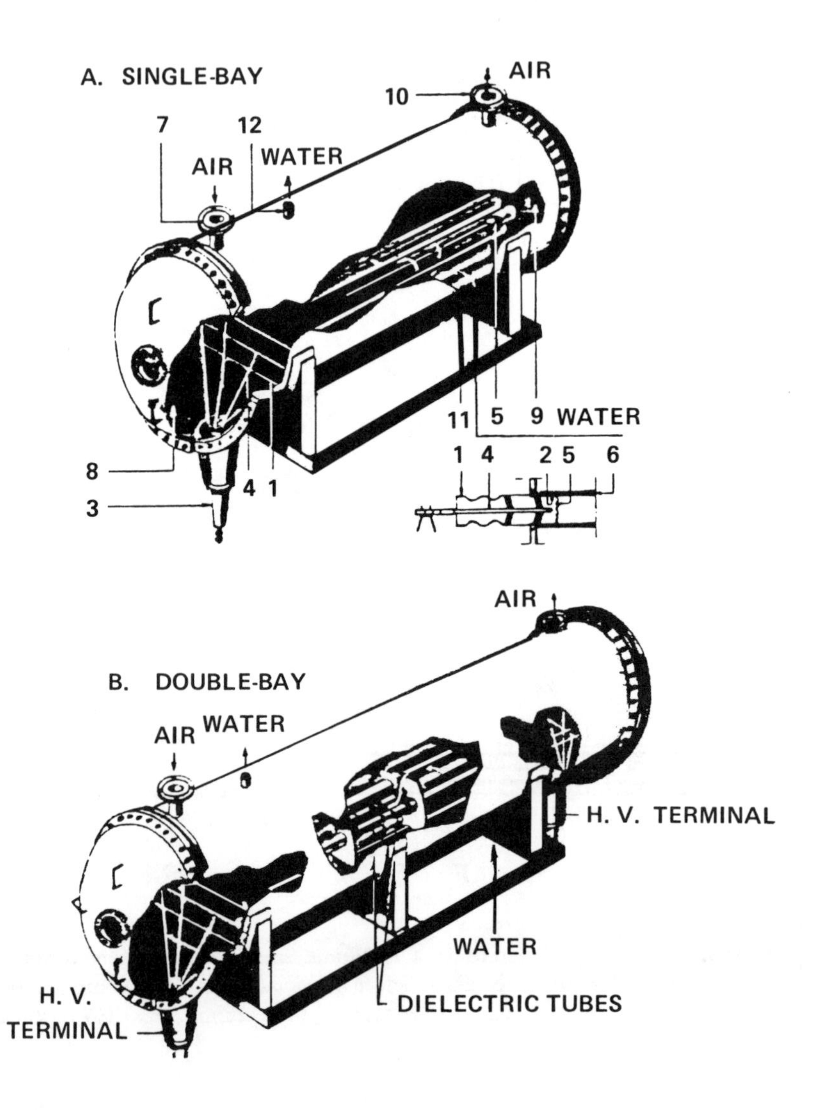

Figure 16. General arrangement of horizontal tube-type generators [16]. (1) dielectric tube; (2) metallic coating; (3) HV terminal; (4) contact; (5) centering piece; (6) ionization gap; (7) air inlet; (8) front chamber; (9) rear chamber; (10) air outlet; (11) water inlet; (12) water outlet.

Vertical Tube, Water-Cooled

This unit utilizes the cooling water as both the grounding electrode and the cooling (Figure 17). Three separate compartments are formed for the feed gas manifold, ozone-rich product gas manifold and cooling water. Dry air enters the top of the central metal tube, which also serves as the high-voltage electrode. The gas travels downward and emerges at the closed end of the glass dielectric tube and passes through the discharge gap. Each tube is almost entirely immersed in the cooling water of the lowest compartment.

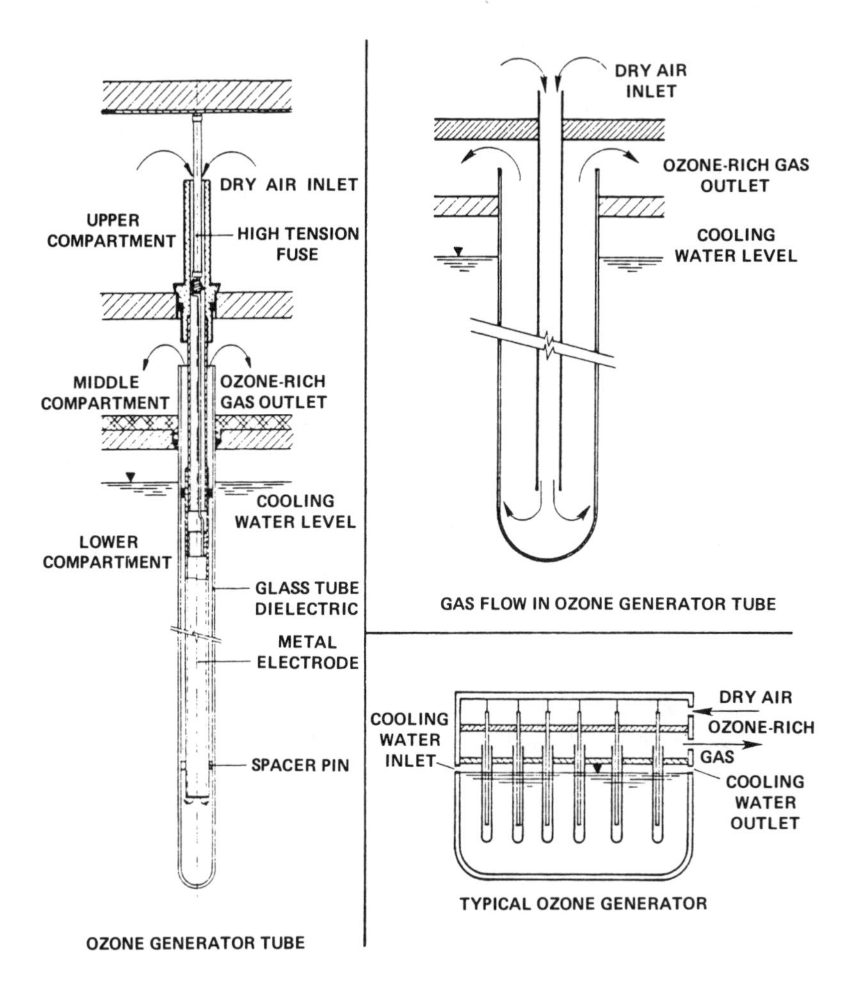

Figure 17. Details of vertical tube ozone generator [9].

Vertical Tube, Double-Liquid-Cooled

This unit utilizes both water and a nonconducting oil or similar fluid for cooling. This necessitates a more complex tube design (Figure 18). The corona cell consists of three annular tubes: the inner, ground electrode consists of a metal tube, and the outer, high-voltage electrode is a metal coating on the outer surface of the glass dielectric tube [18]. The ozone-producing corona discharge occurs between the inner electrode and the glass tube. Cooling water flows through the inner, metal electrode, and the cooling oil flows along the outer metal electrode. The oil, in turn, is cooled by an external oil-to-water heat exchanger for the final heat rejection. Typical water usage for this system is 2500–4000 liter H_2O/kg ozone produced (300–480 gal/lb).

A typical double-liquid-cooled generator is shown in Figure 19. The corona cells are located within the vertical cylinder on the left. The power supply components are located behind the front panel. The generator skid also contains an air compressor (at the left) and an air dryer (in the right rear corner) which comprise the air feed supply for the corona cells.

Lowther Plate, Air-Cooled

This generator, although a plate type, is significantly different from the Otto plate type. One major distinction is that the Lowther generator is designed to operate at slightly elevated pressure (10–15 psig), whereas the Otto generator operates at slightly below atmospheric pressure.

The basic cell (Figure 20) consists of air-cooled aluminum heat sinks, a steel electrode coated with a high-quality, thin ceramic dielectric material, a silicone rubber spacer to establish a uniform, thin discharge gap, a second ceramic-coated steel electrode with inlet and ozone-bearing outlet gas ports, and a second aluminum heat sink [20]. A module consists of 30–40 cells, stacked and manifolded together; a generator can contain 1–32 such modules. Heat rejection is accomplished by forced-air convective cooling through the finned extruded aluminum heat sinks, allowing for two-sided cooling of the cells.

A typical generator system is illustrated in Figure 21. Each generator has a front control panel and power supply components in the base. The corona cell modules are located in the midsection, behind the control cabinet. The cooling fan in the top section draws ambient air into the unit through the louvered panels to cool the power supply components and the corona cells. Cooling air is exhausted through the top of the unit. The feed gas and ozone manifolding is at the rear of each generator. For

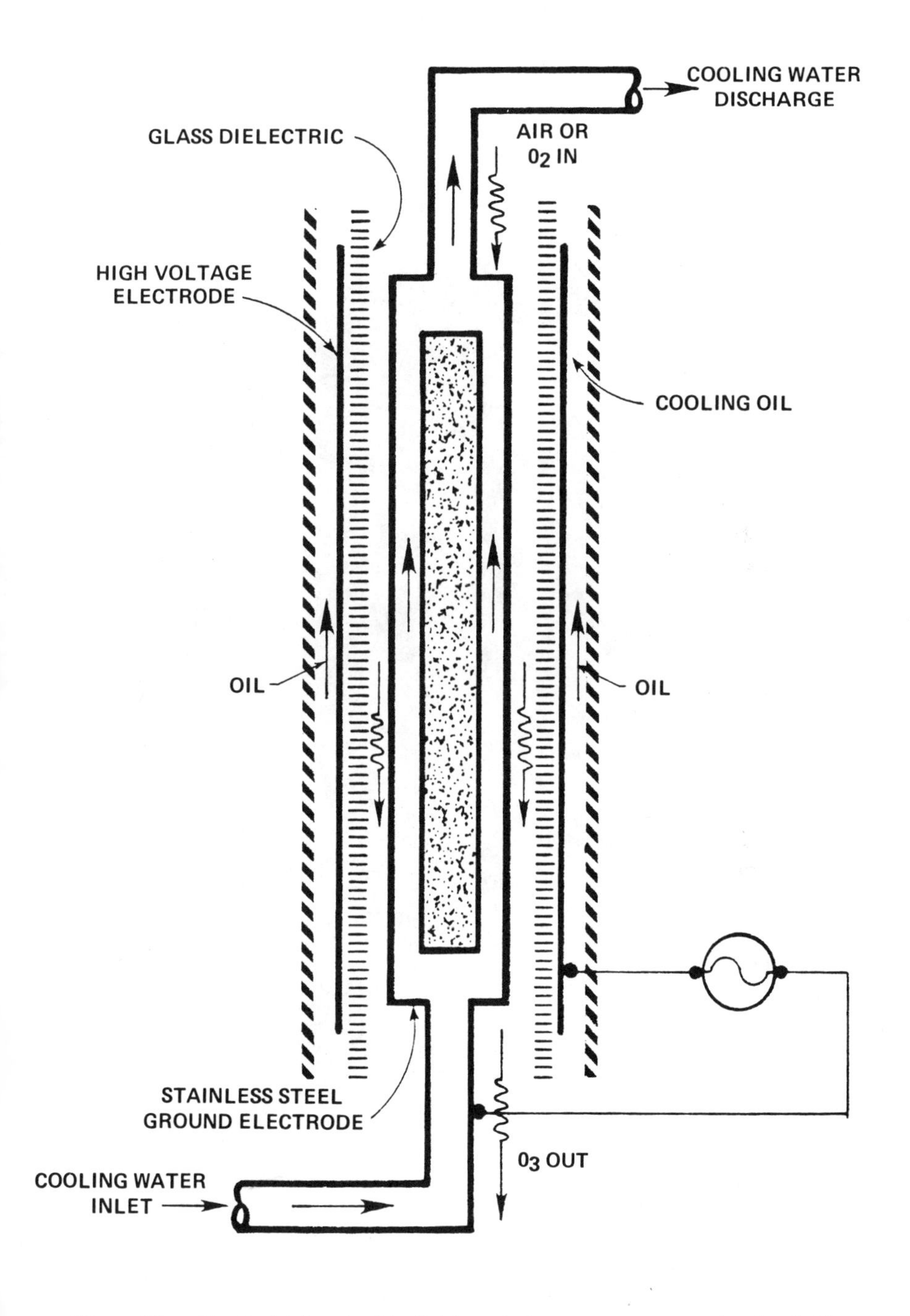

Figure 18. Vertical tube, double-fluid-cooled ozone generator tube detail [9].

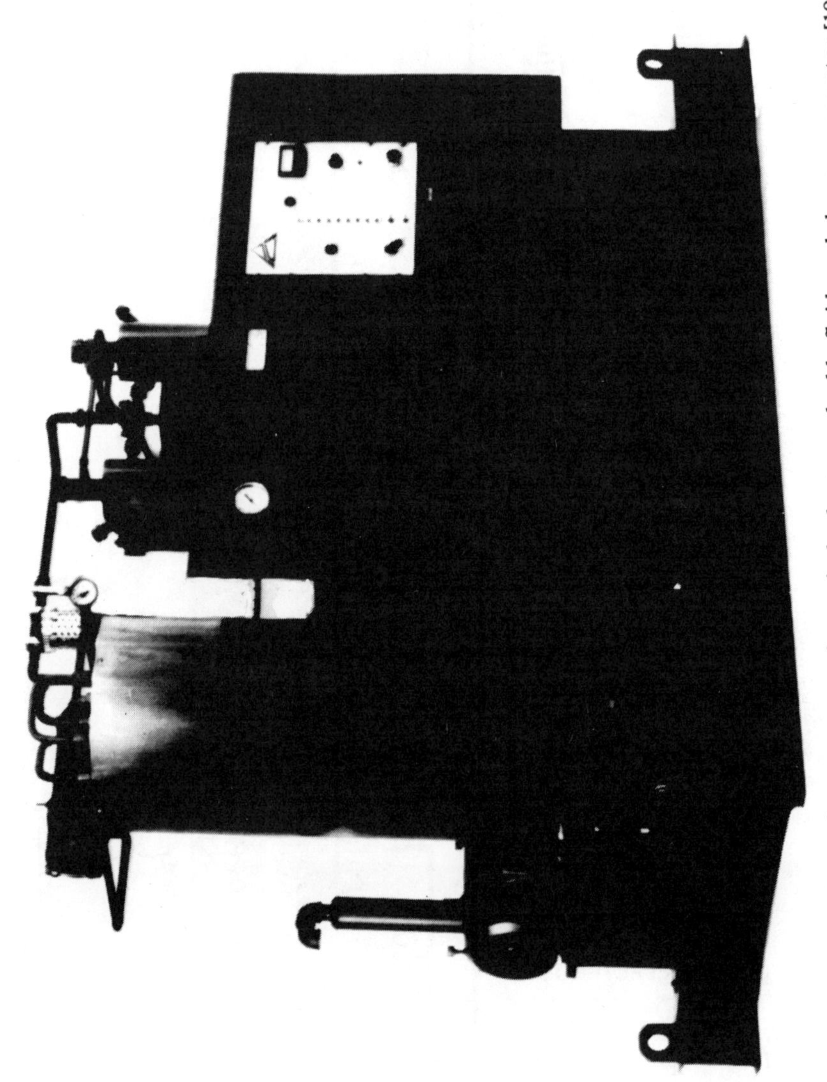

Figure 19. General arrangement of a vertical tube-type, double-fluid-cooled ozone generator [19].

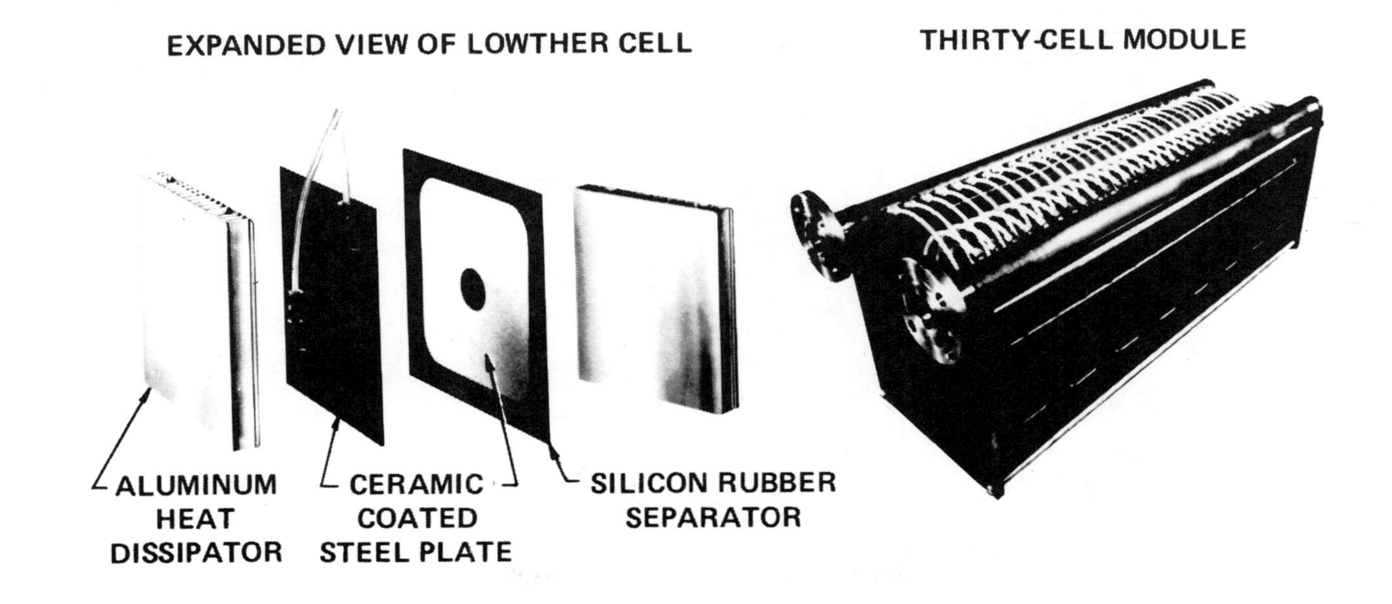

Figure 20. Corona cell and cell module arrangement for the Lowther plate–type, air-cooled ozone generator [20].

Figure 21. Typical installation of Lowther plate-type, air-cooled ozone generator [20].

the installation shown, the feed gas is oxygen produced by a cryogenic plant located outside the generator room.

Power Supply Types

Assuming that the gas pressure within the discharge gap and all corona cell dimensions are held constant during normal generator operation, Equations 5, 7 and 8 show that C_g, C_d and V_s must remain constant. Then, Equation 14 shows that power consumption is directly proportional to frequency f and peak voltage V_o. Therefore, two convenient methods of controlling ozone production are voltage and frequency variation. The three commonly used power supply control systems are shown diagrammatically in Figure 22. Figures 23 and 24 show typical, simplified electrical schematics for controlled voltage and frequency power supplies.

The fixed low-frequency (50–60 Hz), variable-voltage power supply is

A. FIXED LINE LOW FREQUENCY (60Hz), VARIABLE VOLTAGE

PS — POWER SOURCE
VT — POWERSTAT OR VARIABLE TRANSFORMER
HTT — HIGH-TENSION STEP-UP TRANSFORMER
OZONE GENERATOR

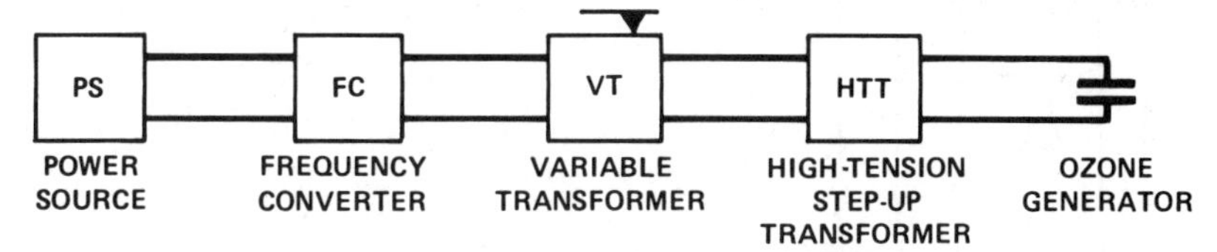

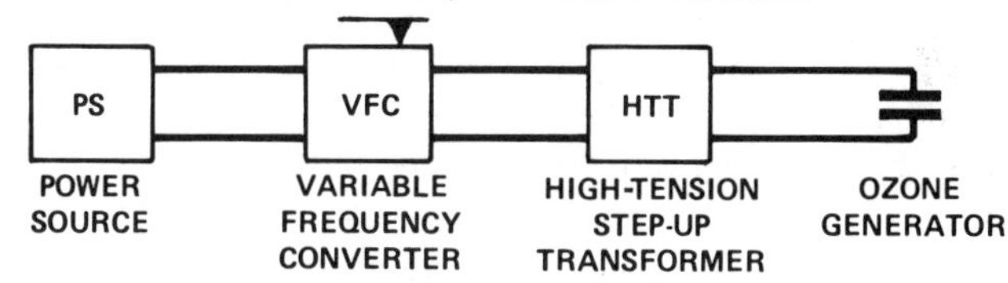

Figure 22. Representative types of power supplies for ozone generators [5,9].

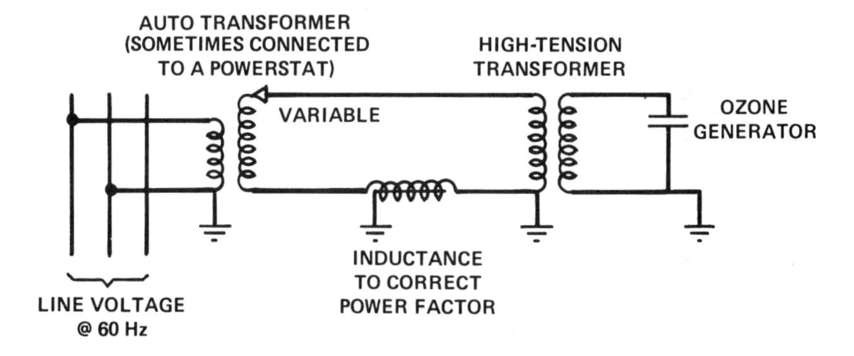

Figure 23. Typical controlled-voltage, 60-Hz power supply schematic [5,9].

controlled by a variable transformer preceding the high-voltage, step-up transformer (Figures 22A and 23). Electrically, the system has the advantage of simplicity. The system transforms single-phase, 60-Hz voltage and applies it to the corona cells. However, the low operating frequency requires high peak voltages to achieve the desired power draw.

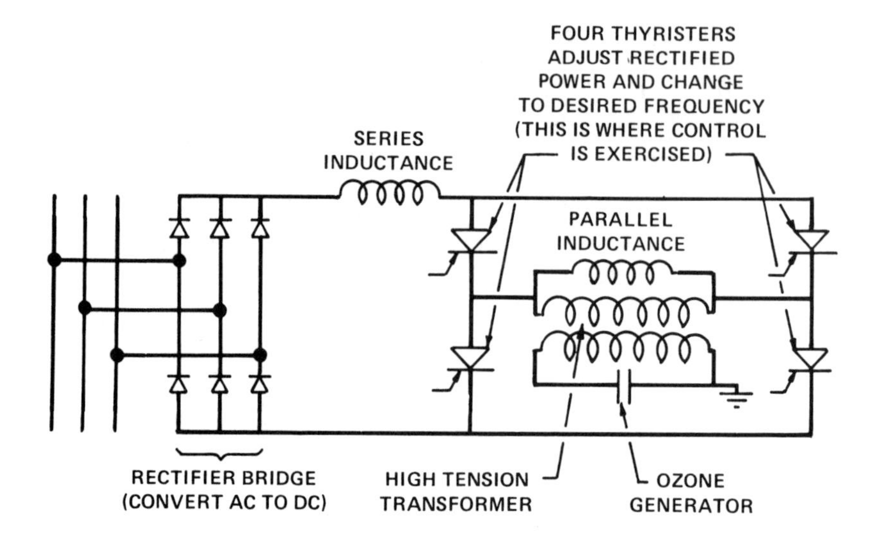

Figure 24. Typical variable-frequency power supply schematic [5,9].

The fixed medium-frequency (400–600 Hz), variable-voltage power supply adds a frequency converter to the previously described simple circuit. The frequency increase is added to boost the power density of the corona cells for a given peak voltage.

The variable-frequency, fixed-voltage power supply is controlled by a thyristor inverter bridge. In this manner, the direct current bus voltage is chopped to produce variable-frequency alternating current power, which then is applied to the load cells. The use of a high-frequency inverter allows a high power density to be obtained in the corona cell while maintaining a low peak voltage.

REFERENCES

1. Bensen, S. W. "Kinetic Considerations of Efficiency of Ozone Production in Gas Discharges," in *Ozone Chemistry and Technology,* Advances in Chemistry Series No. 21 (Washington, DC: American Chemical Society, 1959), p. 405.
2. Fraser, J. C. U.S. Patent 3,496,092.
3. Cobine, J. D. *Gaseous Conductors* (New York: Dover Publications, Inc.), p. 163.
4. Rosenthal, L. A., and D. A. Davis. "Electrical Characteristics of a Corona Discharge for Surface Treatment," *IEEE/IAS* 1A–11(3) (1975).
5. Larocque, R. L., Canozone, Montreal, Quebec. Personal communication (July 1979).

6. Reynolds, S. I. "Test Methods for Measuring Energy in A Gas Discharge," in *Symposium on Corona,* Special Technical Publication No. 198 (Philadelphia: American Society for Testing and Materials, November 1955).
7. Popovich, M. P., Y. N. Zhitnev and Y. V. Filippov. "Spectroscopic Investigation of Mixtures of Nitrogen with Oxygen and with Ozone in the Silent Electric Discharge," *Russian J. Phys. Chem.* 45(2) (1971).
8. Diaper, E. W. J. "Practical Aspects of Water and Waste Water Treatment by Ozone," in *Ozone in Water and Wastewater Treatment,* F. L. Evans III, Ed. (Ann Arbor, MI: Ann Arbor Science Publishers, Inc., 1972), pp. 145–179.
9. Miller, G. W., R. G. Rice, C. M. Robson, R. L. Scullin, W. Kühn and H. Wolf. "An Assessment of Ozone and Chlorine Dioxide Technologies for Treatment of Municipal Water Supplies," U.S. EPA Report EPA-600/2-78-147, Office of Research and Development, Cincinnati, OH (1978).
10. Liao, P. B., R. K. Allen and V. C. Oblas. "Ozone Disinfection for Olympia Wastewater Treatment Plant," paper presented at the International Ozone Institute Technical Symposium, Los Angeles, CA, May 1979.
11. Bean, E. L. "Ozone Production and Costs," in *Ozone Chemistry and Technology,* Advances in Chemistry Series No. 21 (Washington, DC: American Chemical Society, 1959), p. 430.
12. Bulletin 30.23, Cochrane Environmental Systems Division, Crane Company, King of Prussia, PA (1975).
13. Cromwell, W. E., and T. C. Manley. "Effect of Gaseous Diluents on Energy Yield of Ozone Generation from Oxygen," in *Ozone Chemistry and Technology,* Advances in Chemistry Series No. 21 (Washington, DC: American Chemical Society, 1959), p. 304.
14. Rosen, H. M. "Ozone Generation and Its Relationship to the Economical Application of Ozone in Wastewater Treatment," in *Ozone in Water and Wastewater Treatment,* F. L. Evans III, Ed. (Ann Arbor, MI: Ann Arbor Science Publishers, Inc., 1972), pp. 101–122.
15. Manley, T. C. and S. J. Niegowski. "Ozone," in *Encyclopedia of Chemical Technology, Vol. 14,* 2nd ed. (New York: John Wiley & Sons, Inc., 1967).
16. "Horizontal Tube Ozonizers," Technical Bulletin 1306C, Societé Degrémont, Reuil-Malmaison, France.
17. Johansen, R. P., and D. W. Terry. "Comparison of Air and Oxygen Recycle Ozonation Systems," paper presented at the Symposium on Advanced Ozone Technology, International Ozone Institute, Toronto, Ontario, November 17, 1977.
18. Bollyky, J. "Liquid Cooled Ozone Generator," U.S. Patent 3,766,051 (1973).
19. Fleischer, R. Personal communication (September 1979).
20. "LG-Model Ozone Generators," Brochure F-4266, Union Carbide Corp.

CHAPTER 3

OZONE GENERATION WITH ULTRAVIOLET RADIATION

Barry DuRon

Canrad-Hanovia Inc.
Newark, New Jersey

Absorption of light by oxygen has been observed from the far ultraviolet (UV) region into the near infrared (IR) region of the spectrum. Although the very weak absorption in the visible, near UV and near IR regions is of some interest in connection with photochemical reactions of oxygen with other molecules, for the purposes of this discussion, only the middle and far UV regions of the spectrum are of importance.

MECHANISM OF PHOTOCHEMICAL FORMATION OF OZONE

The formation of ozone from oxygen is endothermic [1,2]:

$$1.5\,O_2 \rightarrow O_3 \qquad \Delta H = 34 \text{ kcal} \tag{1}$$

From thermochemical considerations only, it would seem that light of any wavelength shorter than 842 nm could form ozone from oxygen. However, if we consider the probable mechanisms only, then the one which requires the least amount of energy:

$$O_2 + O_2 \rightarrow O_3 + O^{\cdot} \qquad \Delta H = 93 \text{ kcal} \tag{2}$$

corresponds to a wavelength of 307 nm [1,2].

The observed evidence points to a mechanism whereby an oxygen molecule in the ground state ($O_2\ ^3\Sigma_g^-$) dissociates on absorption of light into two oxygen atoms, the energy state of which depends on the wavelength of the absorbed light [3]. The oxygen atom then reacts with an oxygen molecule to produce ozone. The mechanism requiring the least amount of energy will, of course, produce oxygen atoms in their lower energy state [1,2]:

$$O_2(^3\Sigma_g^-) \rightarrow 2O(^3P), \qquad \Delta H = 118\,\text{kcal}. \tag{3}$$

The energy here corresponds to a wavelength of 242 nm.

The other observed reactions are [4–8]:

$$O_2(^3\Sigma_g^-) + h\nu \rightarrow O(^1D) + O(^3P) \qquad \lambda < 175\,\text{nm} \tag{4}$$

$$O_2(^3\Sigma_g^-) + h\nu \rightarrow O_2(^3\Sigma_u^-) \qquad 175\,\text{nm} < \lambda < 200\,\text{nm} \tag{5}$$

$$O_2(^3\Sigma_u^-) \rightarrow 2O(^3P) \tag{5a}$$

$$O_2(^3\Sigma_g^-) + h\nu \rightarrow 2O(^3P) \qquad 200\,\text{nm} < \lambda < 242\,\text{nm} \tag{6}$$

The energy state of the resulting oxygen atoms, whether triplet O (^3P) or singlet O (^1D) has no effect on the yield of ozone formation:

$$O_2 + O + M \rightarrow O_3 + M \tag{7}$$

where M is any inert body present, such as the reactor wall, a nitrogen molecule or a molecule of carbon dioxide.

The quantum yield of ozone production by irradiation of light at wavelengths shorter than 242 nm is 2.0, as seen from Equations 4 through 7, since every photon absorbed will produce two oxygen atoms, each of which will produce one ozone molecule.

However, the observed quantum yield is lower than 2.0, in part, due to the reaction:

$$O_3 + O \rightarrow 2O_2 \tag{8}$$

and, in part, to the photodissociation (photolysis) of ozone:

$$O_3 + h\nu \rightarrow O_2 + O(^1D) \qquad 200\,\text{nm} < \lambda < 308\,\text{nm} \tag{9}$$

The resulting singlet oxygen can further enter the "dark" reactions:

$$O(^1D) + O_3 \rightarrow O_2(^3\Sigma_g^-) + O_2^*(^3\Sigma_g^-) \quad (10)$$

where O_2^* is an O_2 molecule in a vibrationally excited ground state

$$O_2^* \rightarrow 2O(^3P) \quad (10a)$$

$$O(^3P) + O_3 \rightarrow 2O_2 \quad (10b)$$

The quantum yield of ozone photolysis is 4, as can be seen from the sequence of reactions given in Equations 9 through 10b [9,10].

PRACTICAL GENERATION OF OZONE BY PHOTOCHEMICAL MEANS

Since in practice it is very difficult, it not impossible, to produce light at a wavelength sufficiently short to produce ozone from oxygen without the simultaneous production of the longer wavelength that photolyzes ozone, the observed quantum yield of ozone production is a balance between the yield of production and the yield of photolysis. This balance depends, of course, on the ratio of "ozone-producing" to "ozone-consuming" wavelengths present. It also depends on the presence of other inert gases, which influence the yield of ozone formation via Equation 7. Different inert gases will have different effects. Observed relative yields of ozone formation by irradiation with the 185-nm wavelength radiation of a low pressure mercury lamp at 1 atm total pressure and 0.25 atm of oxygen pressure are [2]:

Added Gas	**CO_2**	**N_2**	**Ar**	**Ne**
Relative Ozone Yield	1.0	0.8	0.7	0.5

Experiments conducted at Canrad-Hanovia Inc. with low-pressure mercury lamps of 10 and 16 W show an observed quantum yield of 0.5 by irradiation of dry air, at 25°C, moving at a linear velocity of 150 ft/min in a direction parallel to the lamp axis. These lamps produce only two spectral lines in the region pertinent to ozone formation: one line at 187 nm, which is absorbed by oxygen and causes ozone production, and the other line at 254 nm, which is absorbed by ozone and causes its photolysis.

The relative intensities of these two spectral lines are inherent in the mercury spectrum, and although the ratio of 185 to 254 nm radiation can be somewhat improved by choice of the wall material and inert gases in the lamp, the change will not be so large as to produce quantum yields deviating significantly from 0.5 in air.

In practical terms, the information on quantum yields enables us to calculate the amount of ozone obtainable by irradiation with ultraviolet light. If the quantum yield is taken as 0.5—which is correct when low-pressure mercury lamps are used as the radiation source—this implies that two photons of light at 185 nm are necessary to produce a net one molecule of ozone.

The energy of a photon of light at a wavelength of 185 nm is:

$$E = h\nu = hc/\lambda$$

where h = Planck's constant, 6.63×10^{-27}
c = speed of light, 3×10^{10} cm/sec
λ = wavelength, 185 nm or 185×10^{-7} cm

$$E = (6.63 \times 10^{-27}) \times (3 \times 10^{10})/(185 \times 10^{-7}) = 1.1 \times 10^{11} \text{ erg}$$

The number of photons per second in one watt (10^7 erg/sec) of UV light at 185 nm is: $10^7/(1.1 \times 10^{-11})$, or 0.91×10^{18} photons per second. Since the quantum yield is 0.5, two photons are required to produce one O_3 molecule. The number of O_3 molecules produced per second from one watt of 185-nm radiation, therefore, is $(0.91 \times 10^{18})/2$, or 4.5×10^{17} O_3 molecules per second.

There are 6.02×10^{23} (Avogadro's number) molecules of gas in one mole of that gas. Therefore, 4.5×10^{17} molecules per second of O_3 correspond to $(4.5 \times 10^{17})/(6.02 \times 10^{23})$, or 7.5×10^{-7} g-mole of ozone per second, which, since the molecular weight of ozone is 48, corresponds to $(7.5 \times 10^{-7}) \times 48$, or 3.6×10^{-5} g of ozone produced per watt-second. This amount of ozone per watt-second corresponds to 0.13 g of ozone produced per hour per watt, or 130 g of ozone produced per kilowatt-hour.

It must be remembered that the figure of 130 g O_3/kWh refers to kilowatt-hours of actual UV light at 185 nm, and not to the power consumption of the lamp! Since lamp efficiencies in the 185-nm wavelength range lie between 0.6 and 1.5%, one will have to invest 67–167 kWh of electricity in the lamp in order to obtain that 1 kWh of UV light.

Actual ozone production per kilowatt hour, using low-pressure mercury lamps optimized for 185-nm radiation will be $130/67 = 1.94$ g/hr.

This calculation of the amount of ozone produced per kilowatt hour presupposes that all of the incident UV energy is absorbed by the oxygen. It is quite common, however, that due to physical limitations of the size of the reaction chamber, or the duct, a portion of the radiation will reach the reactor wall and be absorbed by it.

The minimum reactor size for efficient utilization of the UV radiation can be obtained by using the Beer-Lambert law for attenuation of light traveling through an absorbing medium:

$$\log I/I_o = -(a)(p)(l)$$

where
I_o = intensity of light entering the layer of absorbing medium
I = intensity of light leaving the layer of absorbing medium
a = absorption coefficient
p = partial pressure of absorbing species
l = absorption path

To find the absorption path within which 99% of the UV energy will be absorbed, the quantity $I/I_o = 1\% = 0.01$ is substituted into the Beer-Lambert equation:

$$\log I/I_o = \log 0.01 = -2$$

$$-2 = -(a)(p)(l)$$

therefore:

$$l = 2/(a)(p)$$

For oxygen and at a wavelength of 185 nm, "a" is approximately 0.1 $atm^{-1}cm^{-1}$ [3]. If air, in which the partial pressure of oxygen is 0.2 atm, is irradiated with UV light at a wavelength of 185 nm, the path for 99% absorption is:

$$l = 2/(0.1)(0.2) = 100\,cm$$

It is not always practical to produce ozone in a duct 200 cm in diameter. However, if 90% absorption is acceptable, $\log I/I_o$ becomes -1 and l becomes $1/(0.1)(0.2) = 50$ cm. If the partial pressure of oxygen is increased to 1 atm, that is, if pure oxygen is employed rather than air, l becomes 10 cm for 90% UV radiation utilization.

Passing air at 1 atm over a low-pressure mercury lamp which is placed conveniently in a 20-cm-i.d. duct, will result in a UV radiation utilization of:

$$\log I/I_o = -(0.1)(0.2)(10) = -0.2$$

$$I/I_o = 0.63$$

In other words, only 37% of the available ultraviolet radiation is absorbed by oxygen to produce ozone. The remaining 63% is absorbed in the reactor wall. Under such conditions, the ozone production rate will be $1.94 \times 0.37 = 0.72$ g/kWh. This production rate can be greatly improved if the reactor wall is made of, or is coated with, a reflective material, such as polished aluminum.

It is also important to keep in mind that the formation of ozone from oxygen by means of photochemical reactions is not affected by temperature. On the other hand, once formed, the decomposition of ozone is accelerated as the temperature is increased.

Figure 1 shows the absorption spectrum of gaseous oxygen in the region from 1200 to above 2400 Å (120 to above 240 nm). Figure 2 shows the absorption bands of ozone over the range 2000 to more than 2900 Å (200 to 290 nm). Figure 3 is a photograph of three types of ozone-producing mercury lamps.

CONCLUSIONS

It is apparent that UV light is not a very efficient means of producing large (lb/h) amounts of ozone, at least until lamps having higher short-wave UV output become commercially available. However, UV light is very suitable for producing ozone in small amounts—for laboratory pur-

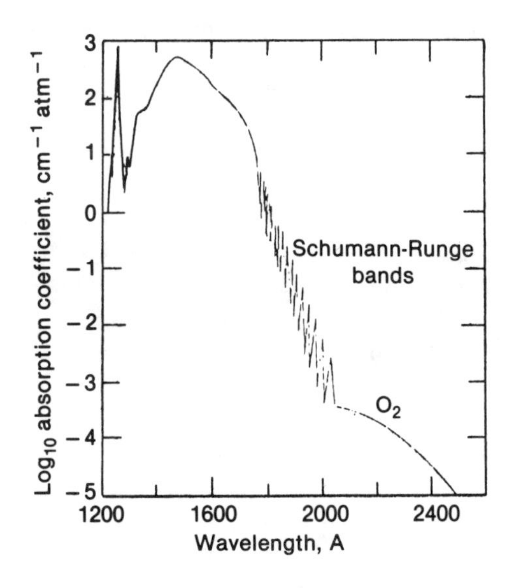

Figure 1. Absorption spectrum of oxygen (after Calvert and Pitts [3]).

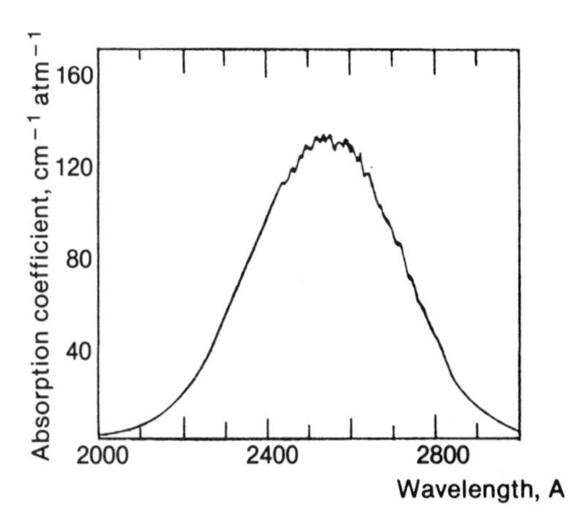

Figure 2. Absorption spectrum of ozone (after Calvert and Pitts [3]).

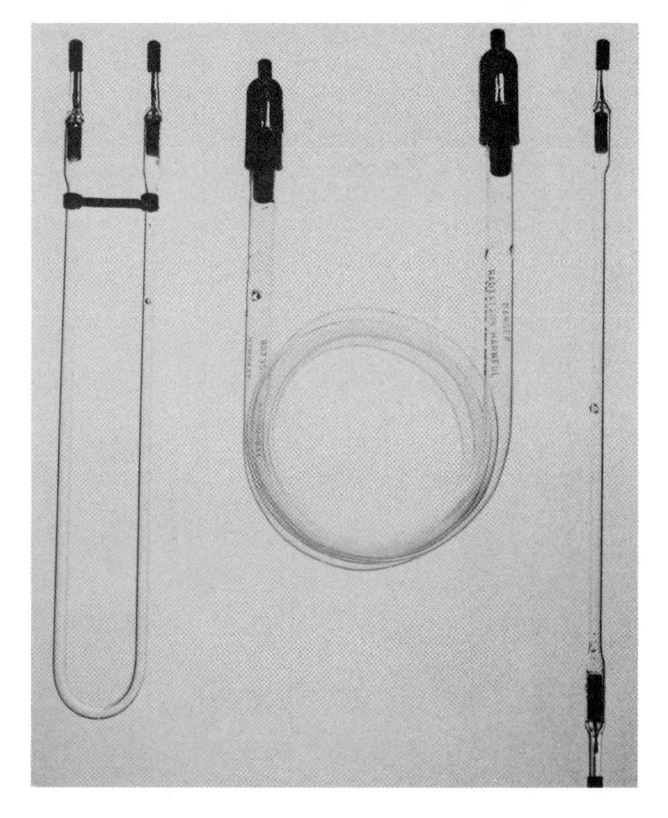

Figure 3. Three ozone-producing low-pressure mercury lamps (courtesy of Canrad-Hanovia Inc., Newark, NJ).

poses, germicidal activity in small samples, odor elimination, etc. The attractive features of ultraviolet are reproducibility, relative insensitivity to moisture and easy control of the rate of ozone production by linear control of lamp power.

REFERENCES

1. Volman, D. H. "Photochemical Gas Phase Reactions," in *Advances in Photochemistry,* Vol. I (New York: Interscience Publishers, 1963), p. 43.
2. McNesby, J. R., and H. Okabe. "Vacuum Ultraviolet Photochemistry," in *Advances in Photochemistry, Vol. III* (New York: Interscience Publishers, 1964), p. 157.
3. Calvert, J. G., and J. N. Pitts, Jr. *Photochemistry* (New York: John Wiley & Sons, Inc., 1966), pp. 205–209.
4. Warburg, E. *Sitzber. Preuss. Akad. Wiss., Physik-Math.* K1:216 (1912).
5. Warburg, E. *Z. Electrochem.* 27:133 (1921).
6. Groth, W. *Z. Phys. Chem.* (Leipzig) B37:307 (1937).
7. Vaughn, W. E., and W. A. Noyes. *J. Am. Chem. Soc.* 52:559 (1930).
8. Wulf, O. R., and E. H. Melvin. *Phys. Rev.* 38:330 (1931).
9. Castellano and Schumacher. "Kinetics and Mechanism of Ozone Decomposition," *Z. Phys. Chem.* (Frankfurt-am-Main) 83:1–4,54–63 (1973).
10. Popovitch, Popov and Tveritinova. "Problems of Quantum Yield of Ozone Photolysis," *Zh. Fiz. Khim.* 50(3):816 (1976).

CHAPTER 4

STATUS OF RESEARCH ON OZONE GENERATION BY ELECTROLYSIS

Peter C. Foller

The Continental Group, Inc.
Energy Systems Laboratory
Cupertino, California

DESCRIPTION OF THE PROCESS

Significant improvements to the electrochemical route for ozone generation have been demonstrated in recent research studies. Unlike conventional ozonators in which predried and compressed air (or oxygen) is passed through a high-frequency corona discharge, ozone is formed by electrochemical oxidation of water.

Certain aqueous fluoroanion electrolytes have been discovered, from which water may be oxidized to ozone at high current efficiency near room temperature [1-3]. Advances in anode material selection also have contributed to making near ambient electrolysis temperatures possible [4]. In previous work, attractive current efficiencies for ozone generation had only been observed at very low electrolyte (or anode surface) temperatures. Operation at such temperatures (−20 to −60°C), in addition to requiring costly refrigeration, disallowed the use of reduction of oxygen as the corresponding cathodic process during electrolysis. The kinetics of oxygen reduction from air become very poor as temperature is decreased. Hydrogen evolution had been considered as the only available cathodic process, even though theoretically an additional 1.23 V of cell potential is required. All economic projections for electrolytic production of ozone were most unfavorable.

The electrolytic process considered in this review (Figure 1) is composed of the following half-cell reactions. At the anode:

$$3H_2O \rightarrow O_3 + 6H^+ + 6e^-, \quad V^o = +1.51\ V$$

and parasitically

$$2H_2O \rightarrow O_2 + 4H^+ + 4e^-, \quad V^o = +1.23\ V$$

At present there is no direct evidence for the two-electron reaction:

$$O_2 + H_2O \rightarrow O_3 + 2H^+ + 2e^-, \quad V^o = +2.07\ V$$

At the cathode, not:

$$2H^+ + 2e^- \rightarrow H_2, \quad V^o = 0.0\ V$$

but

$$O_2 + 4H^+ + 4e^- \rightarrow 2H_2O, \quad V^o = +1.23\ V$$

The theoretical cell voltage for the production of ozone is 0.27 V. Nothing even close to this voltage is achieved in practice, because one must suppress oxygen evolution by employing anode materials that have very high oxygen overvoltages. Similarly, the oxygen reduction reaction, which has been studied extensively in the development of fuel-cells, is notoriously slow. Cell voltages on the order of 1.8–2.1 V are anticipated.

Overall, the process becomes: O_2 (air) $\rightarrow O_3$, and immediately certain inherent advantages may be pointed out. The air feed to the reactor need not be pretreated in any way. It need not be dried; in fact, slight humidification may be desirable to suppress water loss from the electrolyte. Compression also is unnecessary. Atmospheric CO_2 is rejected by the acidic electrolytes selected. On the anodic side, no NO_x is produced, only a mixture of ozone, oxygen and air serving as a carrier gas. Carrier gas (air fed to the electrolysis cells in excess of the stoichiometric requirements of the cathodes) is used to dilute the ozone formed as it evolves from the cells. Otherwise, ozone concentrations well over the explosion limit would be formed.

Here again an inherent advantage of electrolytic technology can be seen: the generated concentration of ozone is decoupled from power consumption, unlike in corona discharge technology. Ozone concentrations are determined first by current efficiency (which recent experiments [1,2] have shown may be obtained in the range of 30–50%) and second by the

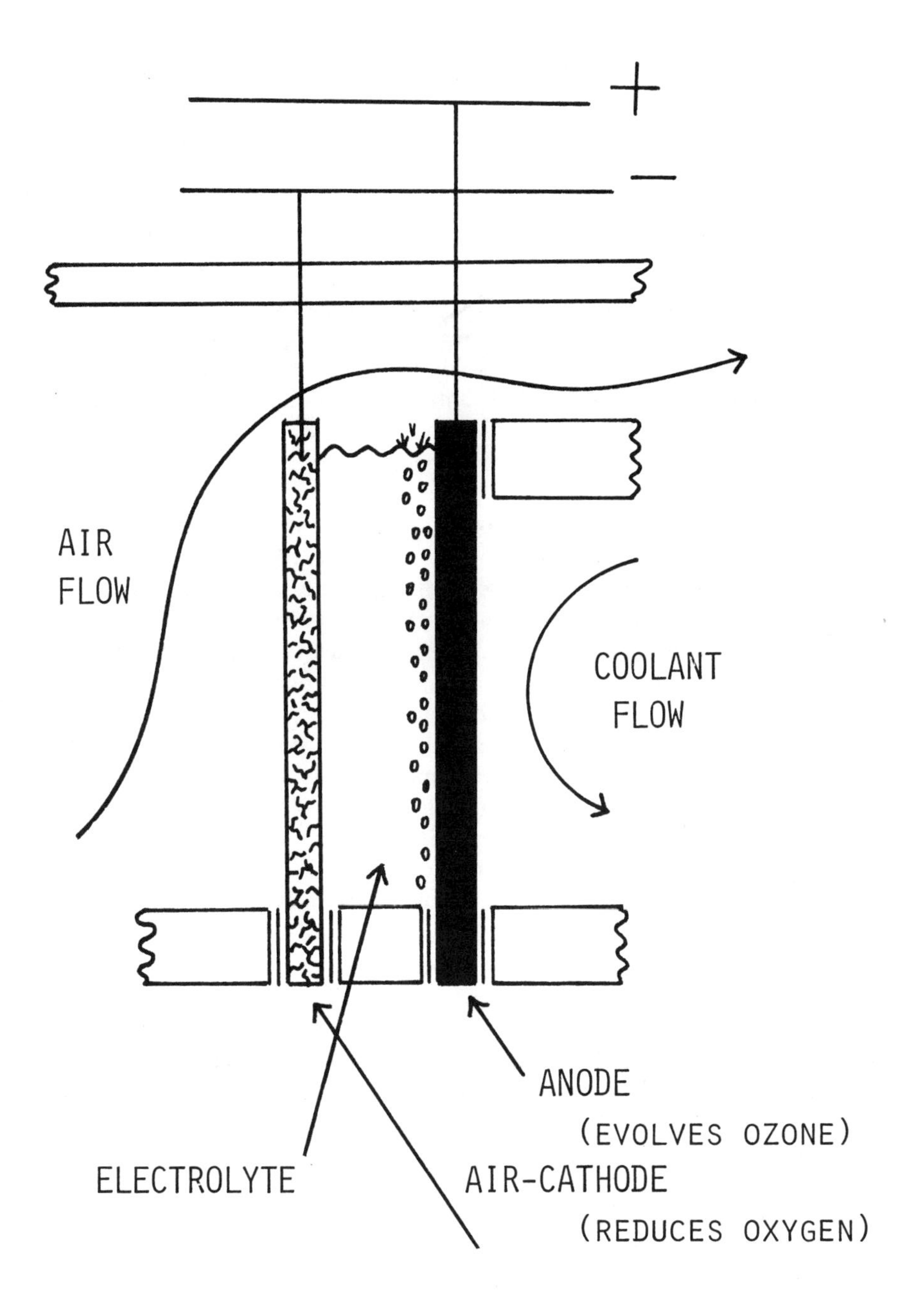

Figure 1. Schematic of ozonator electrolysis cell design.

flowrate of diluent gas. Ozone concentrations of at least 10% will be available using electrolytic technology. Air-fed corona discharge ozonators normally produce 2% ozone at best, many times at an energy efficiency lower than that found at concentrations approximating 1%.

HISTORICAL DEVELOPMENT AND RECENT RESULTS

Since ozone itself was first discovered by electrolysis of sulfuric acid in 1840 [5], approximately 25 publications have appeared dealing with its electrolytic generation. The field has developed slowly because until recently, the results have been uniformly discouraging and of academic interest only.

Work on electrolytic ozone generation may be characterized by electrolyte composition and by choice of anode material. The electrolyte must engage in no reactions other than oxygen and ozone evolution at the anode, and hydrogen evolution or oxygen reduction at the cathode. Chemical reactions with the ozone produced also must not occur. Such constraints led to the selection of acids of oxyanions and fluoroanions, as well as their alkali metal salts, as the most suitable electrolytes.

Very few anode materials are inert to ozone evolution conditions. Extremely high interfacial acid concentrations are produced during the anodic decomposition of water. High anodic potentials lead to dissolution or passivation in the case of most metals. Platinum has been used commonly, and proves to be sufficiently inert. Certain of its noble metal alloys have been used, although their oxygen overvoltages are reduced. Conductive oxides in their highest oxidation states have been used (e.g., the alpha and beta forms of PbO_2 and SnO_2) and show promise. Pyrolytic carbons also prove to be inert in certain electrolyte compositions.

The platinum/sulfuric acid anode and electrolyte composition have been the subjects of intense study in two electrolysis regimes. Early authors used narrow filaments of platinum to achieve current densities on the order of 50–100 A/cm^2 [6,7]. Current efficiencies (the fraction of ozone anodically evolved vs oxygen) of up to 27% were reported from 0°C electrolyte, however cell voltages of nearly 15 V were observed. A glow discharge mechanism seems likely due to the high electric field encountered and the gas-blanketing that must occur.

The second ozone generation regime explored in the platinum/sulfuric acid combination was the electrolysis of eutectic electrolyte compositions at the lowest temperatures possible [8,9]. Current efficiencies of up to 32% were reported; however, refrigeration costs (calculated as 1/3 to 1/2 of the energy consumed during the electrolysis itself) eliminated commercial consideration of the technology.

The platinum anode/perchloric acid combination was studied extensively in this same regime; however, maximum current efficiencies of 36% at −40°C still were inadequate for scale-up [10–12].

A major advance in electrolytic ozone generation came with the use of PbO_2 anodes by three different groups of workers. Semchenko et al. first electrolyzed phosphoric acid and found that yields of 13% current efficiency can be obtained at temperatures of 10–15°C [13].

Semchenko and co-workers next studied the use of perchloric acid, finding yields of 32% current efficiency at temperatures of −15°C [14,15]. In conjunction with the use of PbO_2 anodes, a small quantity of fluoride ion was added to the electrolyte with the apparent effect of raising anode potential (and therefore ozone current efficiency). As of 1975 these were the most encouraging results yet obtained. However, with PbO_2 anodes, some erosion is observed during ozone evolution, following a combined chemical/electrochemical mechanism advanced by Foller and Tobias [16].

Fritz et al. [17] continued the characterization of phosphoric acid–based electrolyte systems, notably a neutrally buffered system in which PbO_2 erosion is suppressed. Yields of 13% current efficiency were obtained at ambient temperatures.

Foller and Tobias [2] studied the use of fluoroanion electrolytes, and continued to find yields using PbO_2 anodes much greater than those obtained with platinum electrodes. Further, it was found that the electrolytes HBF_4 and HPF_6 were particularly well suited to ozone evolution.

Figure 2 illustrates the current efficiencies obtained during the electrolysis of various concentrations of HPF_6 with beta-PbO_2 anodes at 0°C. Although the circumstances of this electrolysis (rapid weight loss and high PF_5 vapor pressure) are not compatible with commercial development, these experiments illustrate that high current efficiencies for ozone generation may indeed be obtained. The research and development problem is to find alternative conditions in which to run the oxidation of water so effectively.

The platinum anode was found to give very high ozone yields in HPF_6 as well, which led Foller and Tobias to propose a rationale for electrolyte selection based on anion electronegativity. Electrolyte anion adsorption on anode materials also was found to correlate with ozone current efficiency [18,19].

Foller et al. [3] found that a certain form of carbon, known as glassy carbon, also was capable of producing relatively high ozone current efficiencies at temperatures above 0°C in fluoroanion electrolytes. Under conditions ordinarily corresponding to ozone evolution, pressed carbon blacks (high-surface-area carbons) rapidly degrade, exhibiting CO_2 evolution and structural disintegration. Graphite also undergoes disintegra-

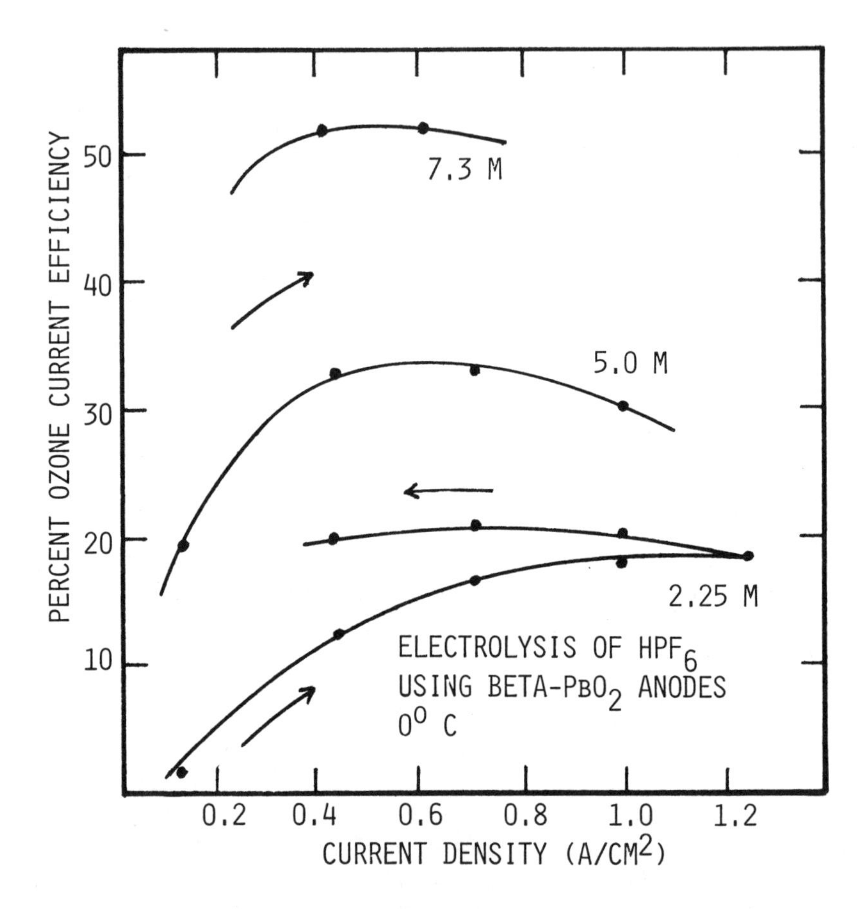

Figure 2. Current efficiency of the beta-PbO_2/HPF_6 anode/electrolyte combination as a function of current density and concentration at 0°C.

tion due to anion intercalation between its planes and consequent c-axis swelling.

Glassy carbon is much more resistant to oxidative processes and anion penetration due to its random, yet fully coordinated structure. This form of carbon is made by heat-treating certain resins under controlled inert atmosphere conditions. Attack is observed in oxyanion electrolytes and in low-concentration acids of the fluoroanions, however not at all in high-concentration electrolytes. The phenomenon is as curious as it is fortuitous, in that ozone yields reach their maximum at the highest concentrations of fluoroanion acid electrolytes.

Figure 3 presents ozone current efficiencies as a function of current density for the electrolysis of various concentrations of tetrafluoboric

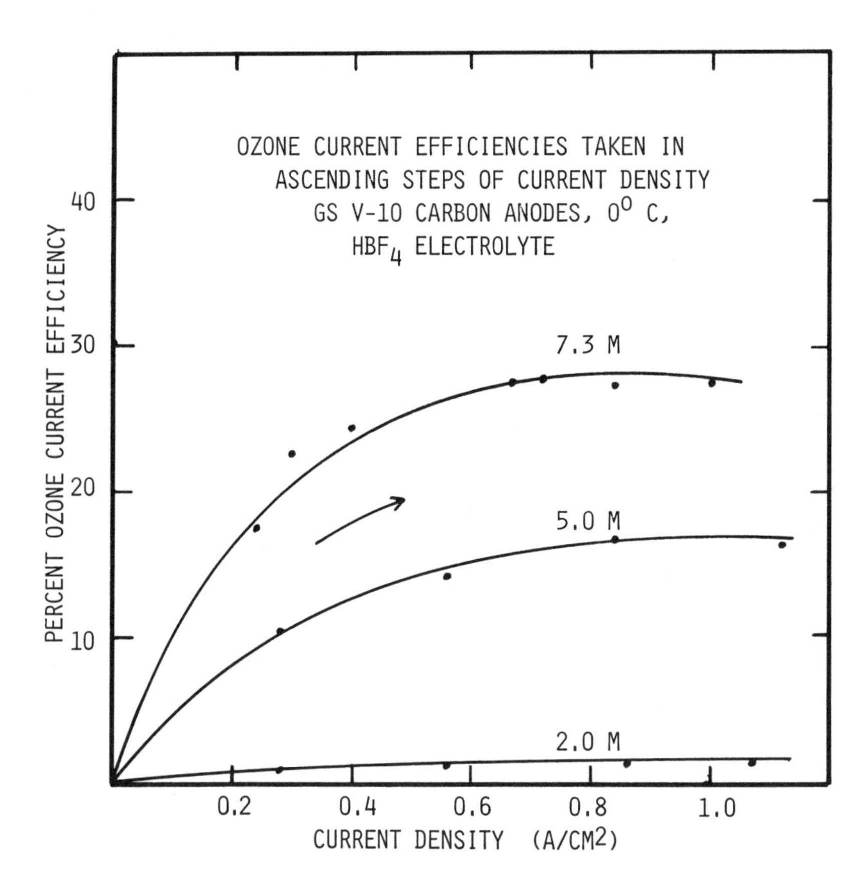

Figure 3. Current efficiency of the glassy carbon/HBF_4 anode/electrolyte combination as a function of current density and concentration at 0°C.

acid electrolyte with glassy carbon anodes at 0°C. The highest yields are found at the highest acid concentration commercially available (48 wt %). Figure 4 shows that these yields are stable over the periods of time investigated to date. No detectable weight loss is observed over 24 hours of accumulated running time in acid concentrations higher than 5 *M*.

There is a certain amount of confusion over ex situ versus in situ electrolytic ozone generation methods (when considering water treatment applications). What has been discussed to this point centers purely on gaseous ozone generation irrespective of contacting and end use. Methods have been advanced, however, that propose in situ ozone generation as an explanation of the efficacy of noble metal electrolysis as a treatment of potable water streams containing the chloride ion at levels on the

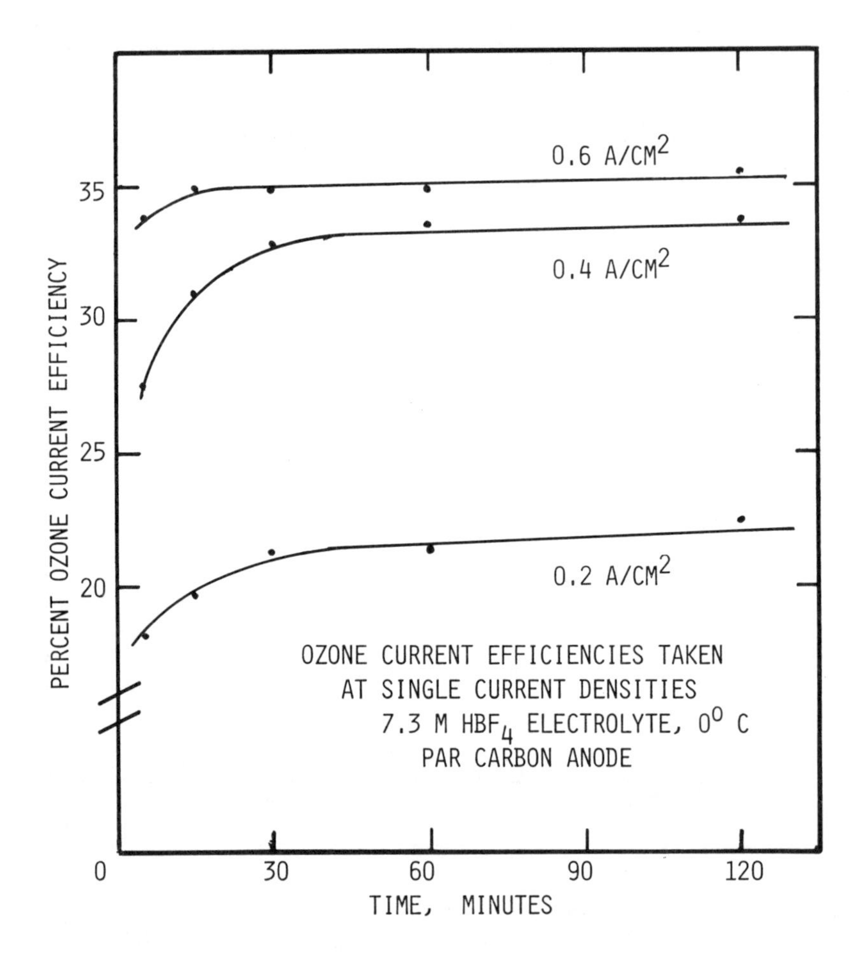

Figure 4. Current efficiencies of the glassy carbon/HBF_4 anode/electrolyte combination as a function of time at 0°C.

order of hundreds of parts-per-million [20]. Extraordinarily high voltages must be applied to pass minimal currents (due to poor solution conductivity). Actual anode potentials (independent of solution I-R) sufficient to oxidize chloride ion to chlorine (1.34 V) and hypochlorite are achieved. These then, in conjunction with the absorption and oxidation of organic substances on the electrodes themselves account for the levels of water sterilization observed.

From studies of ex situ electrolytic ozone generation, it is clear that levels of ozone production in dilute electrolytes are quite small, and indeed may be attributable to analytical difficulties in separating the effects of the other chlorooxidants, which most certainly are produced. In any event, electrolysis at such high voltages (no matter what the assumed reaction products or current efficiencies) cannot be economic in comparison to ex situ optimized ozone (or chlorine) generation processes.

PROJECTED COSTS

An accurate detailed cost estimate of electrolytic ozone generation technology is not yet possible. Projections, however, can be made, assuming that certain developmental milestones will be reached. Projections such as the following demonstrate why interest in electrolytic technology remains high.

Operating Cost

The operating cost of an electrolytic ozonator is almost entirely determined by the power consumption of the electrolysis cells. This power consumption may be derived from the current efficiency and cell voltage. Figure 5 plots the amount of ozone produced per direct current (dc) kilowatt-hour as a function of various current efficiency levels and cell voltages. Two regions of operation are indicated, which correspond to projected cell voltages for either oxygen reduction or hydrogen evolution as the cathodic process. The ranges of cell voltage chosen as representative of the two process configurations correspond to operation at 0.35–0.40 A/cm^2 (near the maximum of ozone current efficiency, but at the same time avoiding the higher levels of polarization at higher current density). An anode potential of 2.2–2.4 V vs a standard hydrogen electrode (SHE), and an air-cathode potential of 0.55–0.65 V vs SHE were chosen for the purposes of this comparison. Electrolyte conductivity and a projected interelectrode gap of 5 mm also were included in the calculations.

Several current efficiency levels are indicated in Figure 5, which then may be used to determine power consumption. Horizontal lines on the figure indicate the power consumptions of conventional corona discharge ozonators. A fairly broad range is defined when the power consumption of all auxiliaries such as air drying and compression are added in, considering the entire spectrum of capacities commercially offered.

The energy efficiency of ozone production at a cell voltage of 2.0 V (anode 2.4 V, cathode: 0.6 V, heat disippation (I-R loss): 0.2 V), and a

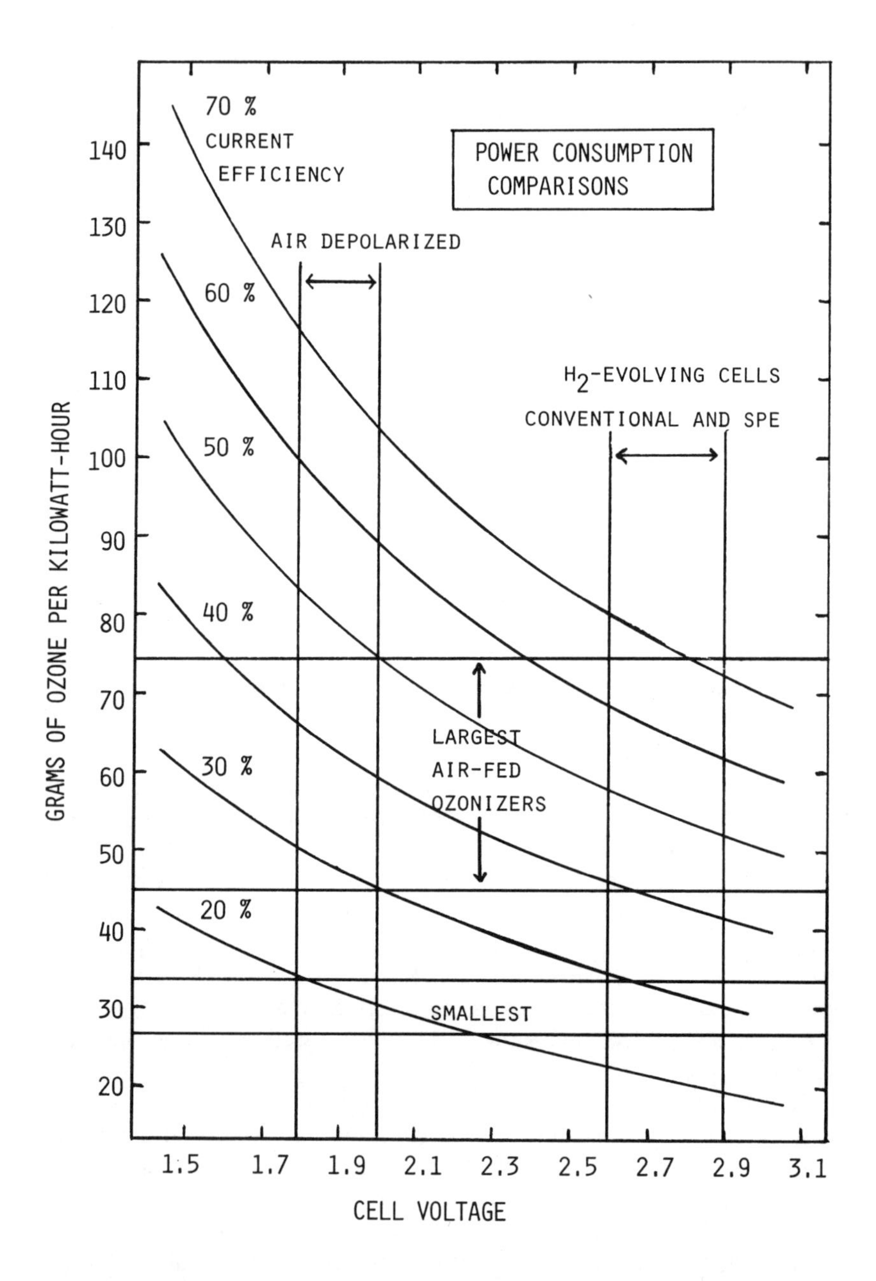

Figure 5. Analysis of the power consumption of electrolytic ozone generators.

current efficiency of 50% exceeds that of the best air-fed corona discharge ozone generators. Similarly, a current efficiency of only 17% at 2.0 V is required to undercut the energy consumption of some of the smaller air-fed units on the market today.

Projection of just where advanced electrolytic ozone generators will lie within this range of energy consumption when fully optimized is problematical. The 2.4-V anode potential and 50% current efficiency necessary to develop a 75-g/kWh ozone electrolyzer have been demonstrated with platinum anodes at temperatures compatible with the flow of cooling water. It is possible to achieve these performance levels under laboratory conditions. Stable current efficiencies of 35–40% also have been achieved with the much less expensive glassy carbon anodes in a less volatile elecrolyte (HBF_4), however, at somewhat higher anode potentials.

The optimization of energy efficiency in a commercially practical cell design will include the selection of a current density (the trade-off is that increasing current density increases current efficiency, but at the same time increases electrode potentials, I-R losses and heat generation), selection of an operating temperature (the trade-off is that increasing electrolyte temperature decreases air-cathode polarization, and increases electrolyte conductivity, but at the same time diminishes ozone current efficiency), and selection of anode, air-cathode, and electrolyte compositions. It is very likely that energy consumptions on the order of 45 to 50 g/AC kWh (95+% power supply efficiencies are common) can be achieved with nonnoble metal electrodes and cooling water compatible anode temperatures.

Operating costs also will include maintenance. Both anodes and cathodes probably will need replacement at certain intervals. Even platinum-clad anodes probably will be subject to slow erosion. Glassy carbon anodes so far have appeared extremely stable during 12- to 24-hour testing. The air-cathodes should exhibit lifetimes in excess of the 40,000 hours projected for high-temperature (190°C) municipal power generation fuel cells. In these fuel cells, catalyst area loss through sintering is a prime failure mode. At ambient temperature, longer lifetimes are expected, as this sintering is reduced. Periodic electrolyte rebalance through water or acid addition may also prove necessary.

Capital Costs

Electrolytic ozone generation should have initial cost advantages over conventional air-fed corona discharge technology for three reasons.

First, the cell stack can be assembled from injection-molded polypropylene framework, and nonnoble metal electrodes. The power supply required is very unsophisticated, a conventional dc source with minimal regulation. A 90-V, 3500-A unit for a 1000-lb/day ozonator can be purchased for $19,000 (1981). High-frequency and high-voltage power supplies for corona discharge ozonators are much more expensive. Contacting costs might be reduced because higher concentrations of ozone might reduce contactor sizes and/or increase the rate of water (to be treated) throughput. However, mass transfer studies at the higher ozone concentrations available by electrolysis must be conducted first, to prove this hypothesis.

The higher concentrations of ozone in air available by electrolysis imply that a given quantity of ozone can be applied using a much lower volume of air. This will provide savings, because of smaller gas-handling equipment.

The size and capital cost of electrolytic ozone generators may readily be estimated once some basic assumptions as to the progress of subsequent research are made. Assuming that 40% current efficiency can be achieved at cooling water temperature, and that a cell voltage of 2.0 V will be encountered at 350 mA/cm^2, a 1000-lb/day electrolytic ozonator may be sized.

A total current of 158,000 A is required, therefore, if a 90-V power supply is used, two parallel stacks of forty-five 1750-A cells may be envisioned. Each bi-cell would have an electrode area of 5000 cm^2 (50×100 cm) and a thickness of approximately 3–4 cm, counting air and coolant flow provisions. Thus cell stack dimensions of $1.5 \times 1.5 \times 2.0$ m appear likely.

Costs may be calculated on the basis of anode material ($50/$ft^2$), cathode material ($20/ft^2) and cell framing. A filter-press design seems most likely. Electrolyte, reservoirs, and auxiliaries such as monitoring equipment, air blowers and filters also must be added in along with assembly and overhead costs. Figure 6 compares the projected cost of electrolytic ozone generator with the costs of conventional air-fed ozone generators as determined in a 1979 study of the EPA Municipal Environmental Research Laboratory [21]. A dramatic reduction in initial cost is forecast due to the basic simplicity of electrolytic technology. Whether this will be, in the end, the 80% reduction exhaustively calculated in the preparation of Figure 6, or only a 50% reduction, it is clear that significant advantages in cost are promised.

The reduction of the capital cost of ozone generator is extremely important in that capital cost represents a very significant fraction of the total cost of ozonation. Amortization of equipment costs can outweigh operating cost (power consumption) for large installations. Figure 7,

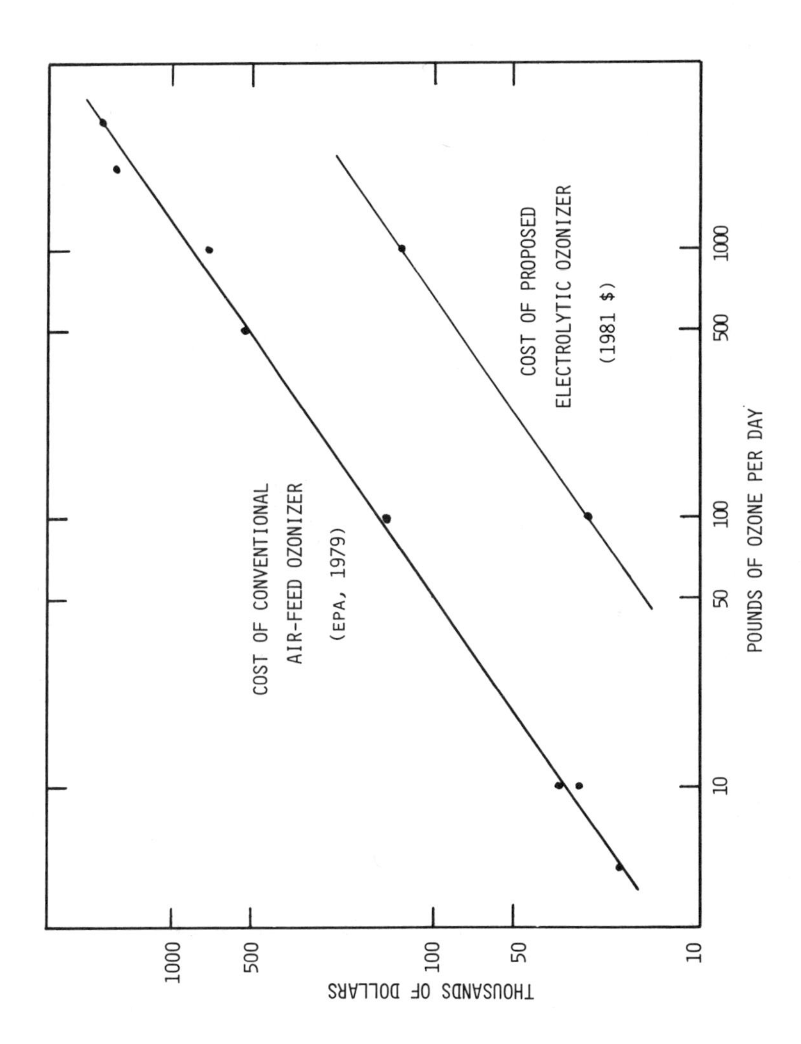

Figure 6. Comparison of capital costs [21].

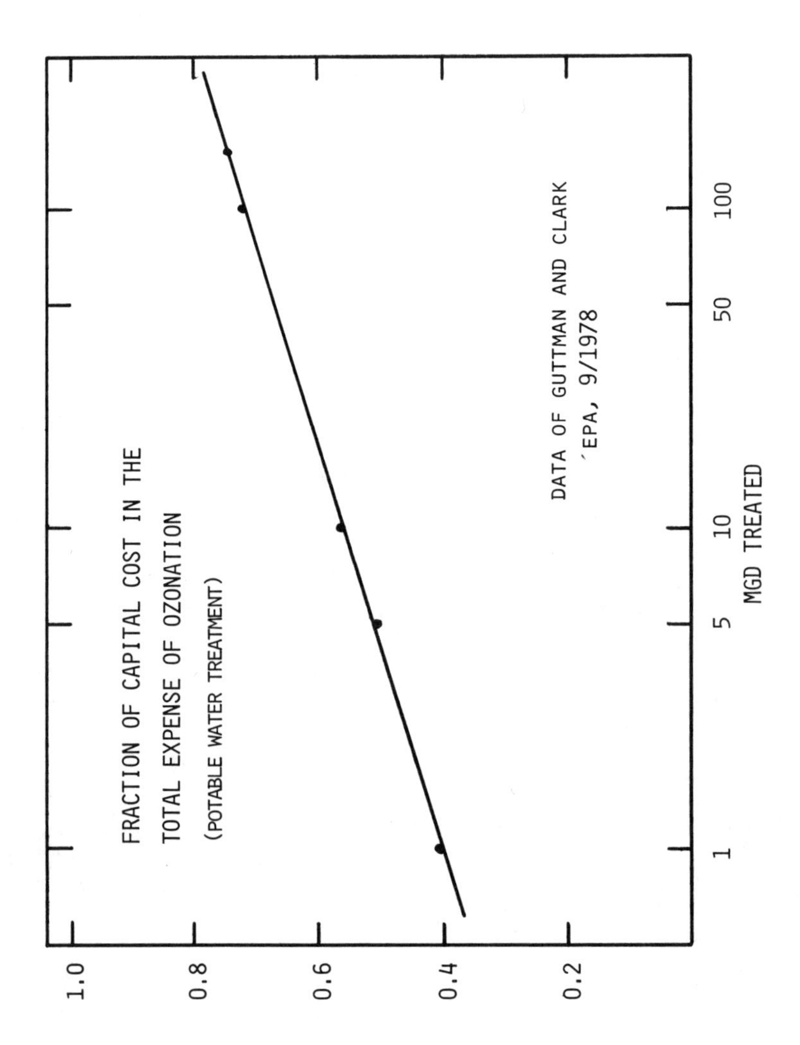

Figure 7. Fraction of capital cost in total expense of ozonation [22].

derived from the EPA-sponsored study of Gutmann and Clark [22], shows that even at 7% interest rates and 20-year amortizations, the fraction of capital-related costs in the total cost of ozonation (contacting costs included) rises rapidly.

Contactor costs may be reduced because mass transfer rates from the gas phase to solution phase are inversely proportional to one minus the mole fraction of ozone in the gas phase [23]. Therefore, at the higher ozone concentrations that are produced by electrolytic technology (at no energy penalty), contactor sizes might be reduced, and/or a greater volume of solution may be treated per unit time. Such potential advantages of electrolytic technology must be analyzed in greater detail with regard to specific applications of ozone.

DEVELOPMENT REQUIRED

To date, the electrolytic process has only been tested in small cells of 1–5 cm^2 electrode area. Further, testing has been performed only in half-cell configurations (i.e., hydrogen-evolving ozone cells and oxygen-evolving oxygen reduction cells). Only laboratory cell designs have been tested, but commercially practical configurations are on the drawing board.

Scale-up development involves co-opting the desirable features of modern fuel-cell design (fuel-cells employ oxygen reduction) and advanced water electrolyzer design (water electrolyzers produce H_2 and O_2). Both technologies are now highly developed.

Assuming the glassy carbon anode and tetrafluoboric acid electrolyte system will be selected for final scale-up, the following development steps must be undertaken.

First, an optimized air-cathode formulation must be developed for HBF_4 electrolyte and high current density operation. Air-cathode technology is quite advanced, and such formulations have been developed for acid electrolyte fuel-cells, as well as alkali chlorine/caustic cells and metal/air batteries. The air-cathodes will consist of a Teflon®-bonded Pt-catalyzed high surface area carbon. A catalyst loading of 1.0–1.5 mg Pt/cm^2 on a carbon such as Vulcan XC-72 bonded with approximately 20 wt % Teflon-30 seems likely. Techniques of cathode manufacture, however, are complex, and in many instances proprietary.

The glassy carbon anodes also will have to be optimized for ozone evolution. Sensitivity to production methods, such as heat treatment temperature and starting resin, has been noticed in ozone current efficiency data [4], and to some extent accounts for differences in yield seen between Figures 3 and 4.

Most importantly, integrated cell testing to co-optimize operational temperature and current efficiency for minimal power consumption must be performed in practical cell designs. At this point, long-term testing of the cells would be begun.

Mass transfer studies should be conducted, using ozone-air combinations that contain higher concentrations of ozone, so that optimally sized ozone contacting chambers can be designed.

In addition, higher concentrations of ozone in air should be tested for compatibility with materials of construction. Higher concentrations of ozone probably will result in shorter lifetime of certain components of ozone-handling equipment.

POSSIBLE APPLICATIONS

Electrolytic technology probably will find application in certain special-purpose fields well-suited to its particular characteristics in advance of its full optimization. These may be applications in which high concentrations of ozone are required (any concentration up to the limits of safety would be available), or in which relatively small quantities of ozone are needed (say 0.2–1.0 lb/day) at low initial cost. If 50% current efficiency at a 2.0-V cell voltage can be achieved in a commercial design at cooling water temperature, electrolytic technology will, of course, find the widest possible application.

Applications requiring very high concentrations of ozone are limited. Many current applications should benefit from increased concentration during contacting, which would, at the same time, require a fully optimized power consumption, especially on large scale.

Hazardous waste treatment is a likely application, in that power cost is not a central issue in disposal of certain highly toxic materials. The high concentrations of ozone (previously unavailable) that electrolytic technology can provide most certainly will improve the oxidation kinetics of organics. An advantage of ozone in this field is that it is nonspecific; it can decompose many unsaturated aliphatics and aromatic organics even when chlorinated. Known pesticide, phenol, cyanide, surfactant, nitrocompound, dye waste, higher alcohol, and organophosphate decomposition processes should be more rapid at higher ozone concentrations.

Applications requiring very low initial cost may also be amenable to unoptimized electrolytic technology. A low-maintenance, continuous-treatment process for swimming pools may be devised, for example. The electrode area required to treat a 20,000-gallon pool at 1-mg/l-day works out to less than 500 cm^2, and the power requirements lie in the range of 2–3 kWh/day (well below filtration pumping costs).

CONCLUSIONS

Recent developments indicate that the electrochemical synthesis of ozone may become an economically feasible alternative to corona discharge. Additional basic research is required, along with substantial engineering development. However, the needed development centers on the available technologies of fuel cell and water electrolyzer (products: H_2 and O_2) design.

The possible outcome of continued efforts in electrolytic ozone generator development is that high-concentration ozone generators of quite low cost may become available with power consumptions equal to those of the best air-fed corona discharge technology. This new breed of ozone generators also may enable contacting costs to be reduced. Further, the technology will scale-up and scale-down with equal ease. Research in this field undoubtedly will continue.

REFERENCES

1. Foller, P. C. PhD Dissertation, University of California, Berkeley (1979).
2. Foller, P. C., and C. W. Tobias. "The Anodic Evolution of Ozone," *J. Electrochem. Soc.* (in press).
3. Foller, P. C., and C. W. Tobias. U.S. Patent Application #154,854 (1981).
4. Foller, P. C., M. L. Goodwin, and C. W. Tobias. U.S. Patent Application #263,155.
5. Schönbein. *Pogg. Ann.,* 50:616 (1840).
6. McLeod. *Chem. Soc. J.* 49:591 (1886).
7. Fischer, F., and K. Massenez. *Z. Anorg. Chem.* 52:202–253 (1907).
8. Briner, E., R. Haefeli, and H. Paillard. *Helv. Chim. Acta* 20:1510–1523 (1937).
9. Seader, J. D., and C. W. Tobias. *Ind. Eng. Chem.* 44(9):2207–2211 (1952).
10. Putnam, G. L., R. W. Moulton, W. W. Fillmore, and L. Clark. *J. Electrochem. Soc.* 93(5):211–221 (1948).
11. Lash, E. I., R. D. Hornbeck, G. L. Putnam, and E. D. Boelter. *J. Electrochem. Soc.* 98(4):134–137 (1951).
12. Boelter, E. D. PhD Dissertation, University of Washington (1952).
13. Semchenko, D. P., E. T. Lyubushkina, and V. Lyubushkin. *Elektrokhimiya* 9(11):1744 (1973).
14. Semchenko, D. P., E. T. Lyubushkina, and V. Lyubushkin. *Otkryitiya, Izobret. Prom. Obraztsy. Tovarnye Znaki* 51(10):225 (1974).
15. Semchenko, D. P., E. T. Lyubushkina, and V. Lyubushkin. *Izv. Sev.-Kauk. Nauchn. Tsentra Vyssh. Shk. Ser. Tekn. Nauk* 3(1):98–100 (1975).
16. Foller, P. C., and C. W. Tobias. "The Mechanism of the Degradation of Lead Dioxide Anodes under Conditions of Ozone Evolution in Strong Acid Electrolytes," *J. Electrochem. Soc.* (in press).
17. Fritz, H. P., J. Thanos and D. W. Wabner. *Z. Naturforsch.* 34b:1617–1627 (1979).

18. Foller, P. C., and C. W. Tobias. "The Effect of Electrolyte Anion Adsorption on Current Efficiencies for the Evolution of Ozone," *J. Phys. Chem.* 85(22):3238 (1981).
19. Potapova, N., A. Rakov and V. Veselovskii. *Elektrokhimiya* 5(11):1418–1420 (1969).
20. Wilk, I. J. Paper presented at the 157th National Meeting, American Chemical Society, Minneapolis, MN, April 14–18, 1969.
21. Gumerman, R. C., R. L. Culp, and S. P. Hansen. *Estimating Water Treatment Costs, Vol. 2,* U.S. EPA, Municipal Environmental Research Laboratory, EPA-600/2-79-162b (1979).
22. Gutmann, D. L., and R. M. Clark. "Computer Cost Models for Potable Water Treatment Plants," U.S. EPA, Municipal Environmental Research Laboratory, EPA-600/2-78-181 (1978).
23. McCabe, W. L., and J. C. Smith. *Unit Operations of Chemical Engineering,* (New York: McGraw-Hill, 1976), p. 719.

SECTION 2

CONTACTING

CHAPTER 5

MASS TRANSFER OF OZONE INTO AQUEOUS SYSTEMS

Chiang-Hai Kuo

Chemical Engineering Department
Mississippi State University
Mississippi State, Mississippi

Floyd H. Yocum

Shell Development Company
Houston, Texas

Mass transport of ozone into aqueous solutions is important in both air and water pollution abatement. Absorption and reactions of ozone with organic and inorganic compounds in liquid droplets have been suspected to play a major role in the formation and accumulation of secondary aerosols in the atmosphere [1]. On the other hand, ozone has been used for the disinfection of drinking water and treatment of industrial wastewaters for many years [2] because of its high oxidation potential. All of these applications involve reactions of ozone in the gas phase with one or more constituents in the liquid phase. Before ozone can react with any substance in the liquid phase, whether the liquid is water or an organic solvent, it must pass through an interface between the two phases. This transfer of ozone from one phase to another is via diffusion and convective mass transport [3].

The overall mechanism of the gas-liquid reaction system can be visualized as consisting of several steps. These include diffusion of ozone through the gas phase into an interface between the gas and liquid phases, transport across the interface to the liquid phase boundary, and then transfer into the bulk liquid. The amount of dissolved ozone present may be depleted during each of these steps by decomposition or reactions with reactants diffusing from the liquid phase. Products formed in these reactions may penetrate across the phase boundary into the main liquid or may diffuse back into the gas phase if they are volatile.

This chapter examines the relative importance of diffusion in each step, and introduces fundamental concepts governing transport of a gas into the liquid phase. Effects of chemical reactions on the rate of mass transfer are predicted, and various theoretical and experimental investigations of ozone absorption into the aqueous solutions are reviewed. The chapter concludes with a brief discussion of the advantages and disadvantages of many gas-liquid contacting systems that have been used to conduct ozonation reactions.

MASS TRANSFER FUNDAMENTALS

The term "mass transfer" refers to the motion of molecules or fluid elements as a result of a potential or "driving force". The term encompasses both molecular diffusion and transport by convection. Mass transfer can be divided into four broad areas: (1) molecular diffusion in a stagnant medium, (2) molecular diffusion in fluids in laminar flow, (3) eddy diffusion in a free turbulent flow, and (4) mass transfer between two phases. Each area pertains to the environment in which the mass transfer is occurring. Fick's law of diffusion describes the molecular diffusion that occurs in a stagnant medium as a result of a concentration gradient. As this fluid (gas or liquid) begins to flow past a surface, mass transfer rates are increased from the movement or mixing (eddy diffusion).

In the laminar flow regime (flow characterized by a parabolic velocity profile), mass transfer is predominantly by molecular diffusion. Eddy diffusion is enhanced as the fluid velocity increases and flow becomes turbulent. In the turbulent flow regime, eddy diffusion can be considerably larger than molecular diffusion. Since the relative importances of molecular and eddy diffusion in an overall process are not known, the combined effects often are lumped together in the form of mass transfer coefficients. Relationships among the various mass transfer coefficients in the transport between phases will be developed. Several mathematical models [3–5] proposed to describe the behavior of mass transport in the liquid phase will be presented in this section.

Fick's Law of Diffusion

Fick's law of diffusion models molecular diffusion in a stagnant medium. Molecular diffusion is concerned with the movement of individual molecules through a substance by their thermal energy. Molecular diffusion may also result from temperature, pressure or electrical potential gradients applied to a system. In most cases, molecular diffusion is associated with concentration gradients where mass is transferred to equalize the concentration. When a homogeneous system, either gas, liquid or solid, contains more than one component with a concentration gradient, there is a tendency for the molecules to escape to the less populated region.

Considering one-dimensional diffusion in the z direction, the molar flux of component A relative to the average molar velocity of all constituents is given by Fick's first law of diffusion:

$$J_A = -D_A \left(\frac{\partial C_A}{\partial z} \right) \qquad (1)$$

where J_A = molar diffusion flux of A (mol/cm^2-sec)
D_A = molecular diffusivity of A (cm^2/sec)
C_A = concentration of A (mole/cm^3)
z = distance of diffusion (cm)

The negative sign indicates that diffusion occurs in the direction of least concentration. The diffusion coefficient (D_A) or diffusivity, is a proportionality factor and is a function of the system temperature, pressure and concentration.

If the mass transfer flux is expressed with respect to a stationary coordinate, the flux of A resulting from the bulk motion must be included. For diffusion in a binary mixture of the two species A and B, the molar flux of A is

$$N_A = J_A + (N_A + N_B)x_A \qquad (2)$$

where N_A = point rate of mass transfer of A per unit interfacial area (mol/cm^2-sec)
N_B = point rate of mass transfer of B per unit interfacial area (mol/cm^2-sec)
x_A = mol fraction of A

Equation 2 can be integrated for various conditions to describe the total mass transfer rate instead of diffusion flux at a single point. In the diffusion of A through the stagnant layer of B, N_B can be considered negligible. Under this circumstance, the steady-state rate of mass transfer of A is directly proportional to the concentration (or partial pressure) difference across the stagnant layer.

In general, the concentration varies with time and position. By considering a thin element of differential thickness, dz, normal to the direction of diffusion, the mass balance for species A yields:

$$\frac{\partial N_A}{\partial z} + \frac{\partial C_A}{\partial t} = 0 \tag{3}$$

where t = time of diffusion (sec)

In systems involving diffusion of ozone between the gas and liquid phases, the molar flux due to bulk motion can be ignored if the plane of transport is assumed to be stationary. The following form of Fick's second law then can be derived for cases of constant diffusivity by combining Equations 1, 2 and 3:

$$D_A \left(\frac{\partial^2 C_A}{\partial z^2} \right) = \frac{\partial C_A}{\partial t} \tag{4}$$

Equation 4 can be solved with associated initial and boundary conditions to yield the concentration of the solute as a function of time of diffusion and position. The instantaneous rate of mass transfer per unit surface area, then, can be derived from

$$N_A(t) = -D_A \left(\frac{\partial C_A}{\partial z} \right)_{z=0} \tag{5}$$

Knowledge of diffusivities is needed in calculating the mass transfer fluxes. Experimental values give the best results, but if these are not available, several estimation techniques can be used [6,7]. Expressions for estimating gas phase diffusivity are based on the kinetic theory of gases. The Wilke-Lee [8] modification of the Hirschfelder-Bird-Spotz method [9] is recommended. Liquid diffusivities may be estimated using the Wilke and Chang empirical correlation [10,11]. Liquid diffusivities vary considerably with concentration, and care should be taken in estimating values. A thorough review is given by Reid and Sherwood [12].

Mass Transfer Between Phases

The area of greatest importance to the engineer designing a gas-liquid system is mass transfer between the two phases. The transfer of a substance from one phase to another results from a concentration gradient as sketched in Figure 1. In the absorption of a gas by a liquid, it can be assumed that a fairly uniform composition is maintained in the bulk gas

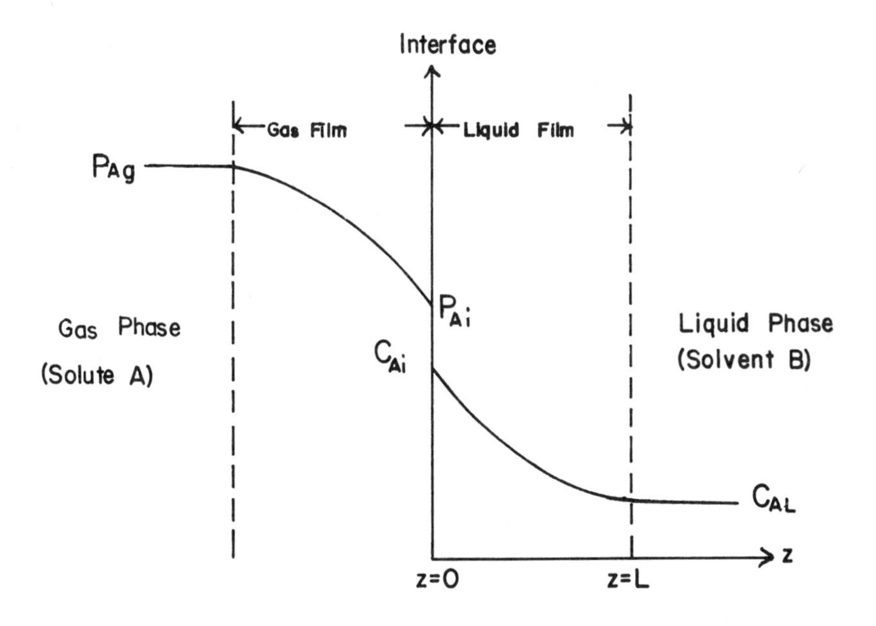

Figure 1. Mass transfer between phases.

because of turbulent motion. As suggested by the two-film concept of Lewis [13] and Whitman [14], the solute A must diffuse through a laminar layer between the gas bulk and the interface and another layer between the interface and the liquid bulk. Once the solute approaches the liquid phase boundary, it is carried away by eddy diffusion into the turbulent liquid, which is well mixed at a constant concentration. The steady-state rate of diffusion of the solute A from the gas to the interface is equal to the rate from the interface to the bulk liquid:

$$N_A = k_G'(p_{Ag} - p_{Ai}) = k_L'(C_{Ai} - C_{AL}) \tag{6}$$

where
k_G' = gas-side mass transfer coefficient for physical absorption (mol/cm²-sec-atm)
p_{Ag} = partial pressure of A in gas phase (atm)
p_{Ai} = partial pressure of A at the interface (atm)
k_L' = liquid-side mass transfer coefficient for physical absorption (cm/sec)

The mass transfer flux can also be expressed in terms of the overall mass transfer coefficients K_G' and K_L':

$$N_A = K_G'(P_{Ag} - P_{AL}) = K_L'(C_{Ag} - C_{AL}) \tag{7}$$

where K'_G = overall gas-phase mass transfer coefficient (mol/cm^2-sec-atm)
P_{AL} = partial pressure of A in equilibrium with the bulk liquid (atm)
K'_L = overall liquid-phase mass transfer coefficient (cm/sec)
C_{Ag} = concentration of A in the gas phase (mol/cm^3)
C_{AL} = concentration of A in the bulk liquid (mol/cm^3)

The steady-state equations describe a dynamic equilibrium that is established by the common substance between the two insoluble phases. For example, if an air-ozone gas mixture is brought into contact with distilled water, the ozone will be absorbed into the water until the water becomes saturated. This is shown in Figure 2 as the liquid-phase ozone

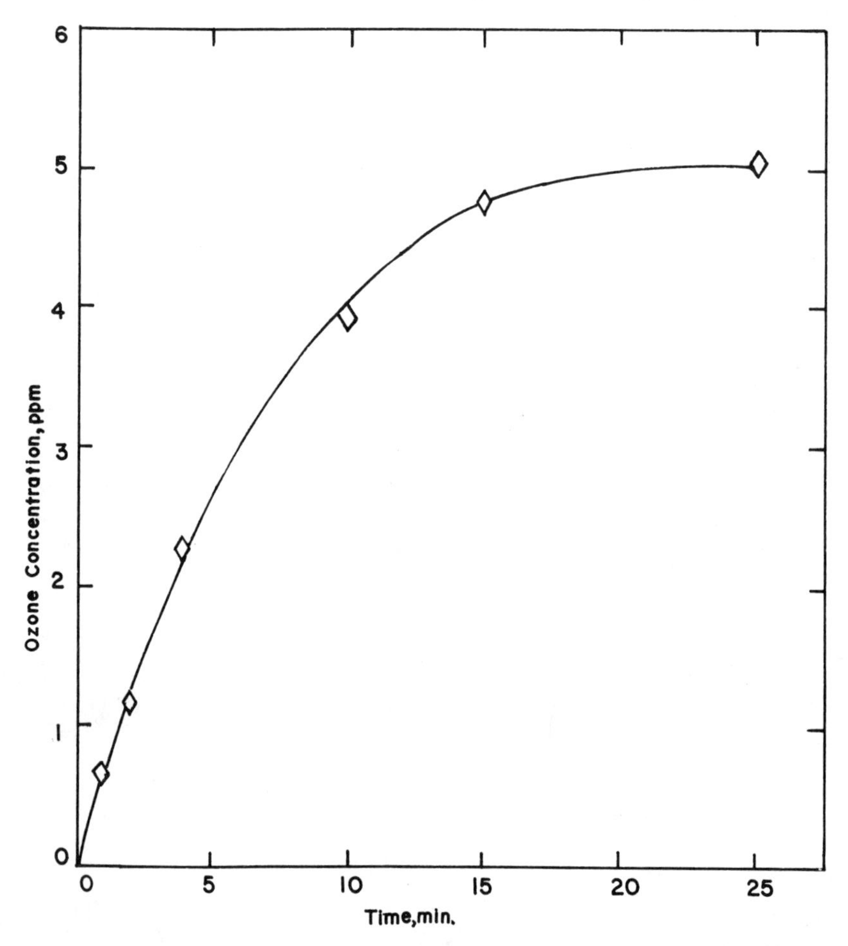

Figure 2. Concentration profile of ozone in absorption of ozone-air mixture (1.1 wt %) into 20 liters of distilled water. Gas flowrate = 0.41 ft^3/min; temperature = 16°C.

concentration increases asymptotically to the solubility limit. At this point ozone in the liquid phase is in equilibrium with the gas phase. If a mixture more concentrated in ozone is added to the system, a new equilibrium would be obtained. These equilibrium concentrations can be plotted to form an equilibrium distribution curve.

When ozone or another gas is relatively insoluble, a dilute solution is formed and the equilibrium (solubility) curve can be approximated as a straight line, in accordance with Henry's law. At a specific temperature, therefore, the mole fraction of A is related to the partial pressure of A by:

$$P_A = Hx_A \tag{8}$$

where H = Henry's law constant

Figure 3 shows ozone solubility and Henry's law constant as a function

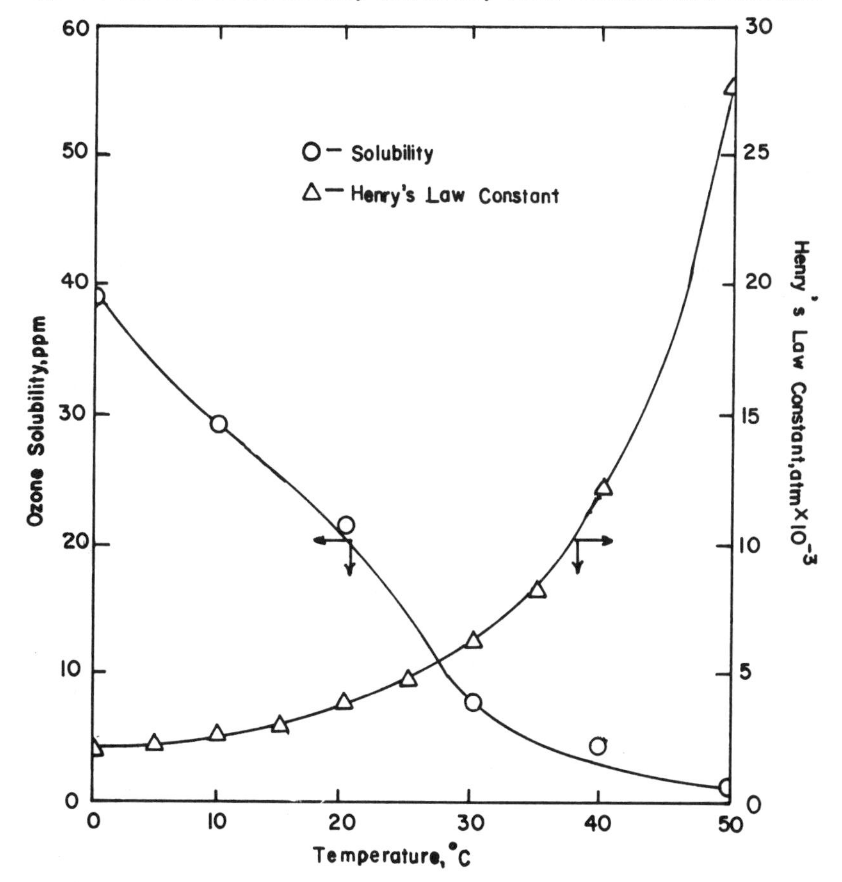

Figure 3. Solubility of ozone in water and Henry's law constant.

of temperature. Henry's law holds very well when the partial pressure of the solute gas does not exceed one atmosphere.

In the absorption of a gas with low solubility by a liquid, the rate of mass transfer is slow and the diffusional resistance at the actual interface seems to be insignificant. Therefore, the two phases can be assumed to be in equilibrium at the interface, and Henry's law is obeyed. By combining Equations 6, 7 and 8, the following relationships between the overall and individual mass transfer coefficients can be obtained:

$$\frac{1}{K_G'} = \frac{1}{k_G'} + \frac{H}{k_L'} \tag{9}$$

and

$$\frac{1}{K_L'} = \frac{1}{k_L'} + \frac{1}{Hk_G} \tag{10}$$

The above equations suggest that the overall resistance to mass transfer is the sum of the individual resistances represented by the terms on the right side. Since ozone is only slightly soluble in aqueous solutions, the Henry's law constant, H, is large, as indicated in Figure 3. Under this condition, the resistance in the gas film is negligible compared with that in the liquid film, and the process is controlled by the diffusion in the liquid film.

Mass Transfer Models

Many theories or models have been proposed to explain the phenomena of mass transport in the region adjacent to the liquid-phase boundary. The most noted ones are the film, penetration and surface renewal theories. These theories differ in two main aspects in postulating the theoretical mechanism for interphase mass transport. One aspect deals with diffusional behavior of the transporting species, and the other with hydrodynamic activities of the two phases near the interface.

As mentioned earlier, the film theory of Lewis [13] and Whitman [14] is based on the assumption that when two fluid phases are brought in contact with each other, there exists on each side of the phase boundary a thin layer of stagnant fluid. Accordingly, transport via convection is insignificant and mass transfer can be represented by molecular diffusion. This results in a simple model for a complicated transport process by ignoring the influence of hydrodynamic behavior on the mass transfer near the interface.

The film theory is formulated on the assumption that the film is very thin. Thus, the accumulation of the diffusing mass in the film is negligible and steady-state diffusion is attained. At the interface, the contacting phases are considered to maintain equilibrium, and the bulk of the liquid phase beyond the stagnant film is well mixed at a uniform concentration. In accordance with these propositions, the concentration distribution in the liquid film is a linear function of the distance, as can be seen from Equation 4, and the mass flux can be derived from the expression given in Equation 5. By defining the liquid-side mass transfer coefficient k_L' in terms of the concentration driving force $(C_{Ai} - C_{AL})$

$$\bar{N}_A = k_L'(C_{Ai} - C_{AL}) \tag{11}$$

where $\bar{N}_A$ = average rate of mass transfer of A per unit interfacial area (mol/cm^2-sec)
C_{Ai} = concentration of A at the interface (mol/cm^3)

the expression for k_L' can be derived for the film theory as

$$k_L' = D_A/L \tag{12}$$

where L = thickness of film (cm)

The penetration theory postulated by Higbie [15] assumes that turbulent eddies travel from the bulk of the phase to the interface, where they remain for a short time. If the exposure time is short, it is possible that a steady-state concentration gradient within a film may never have been achieved before the film is disrupted or replaced. Also, the transporting species may never approach the outer edge of the liquid film or element in the short contact time, and the thickness of the liquid film or element may be assumed to be infinite. The partial differential Equation 4 governing unsteady-state diffusion of the transporting species is applicable according to the penetration concept, and the instantaneous rate of mass transfer for an individual surface element can be derived from Equation 5. Higbie [15] assumed that every surface element is exposed to the solute for the same length of time, t_0, before being replaced. Therefore, the average mass flux can be derived to be

$$\bar{N}_A = \frac{1}{t_0}\int_0^{t_0} N_A(t)dt = 2(C_{Ai} - C_{AL})\sqrt{D_A/(\pi t_0)} \tag{13}$$

where t_0 = contact time at the interface (sec)

The expression for the liquid-side mass transfer coefficient is obtained from Equations 11 and 13 as

$$k_L' = 2\sqrt{D_A/(\pi t_0)} \tag{14}$$

Instead of a constant time of exposure, the surface renewal theory of Danckwerts [16] pictures the liquid phase as completely disturbed and with numerous infinitesimally small phase elements or eddies in the phase. These eddies are constantly changing the structure and position, resulting in turbulence in the interface boundary. Danckwerts proposed that the eddies may vary at locations but are continuously bringing microscopic masses of fresh phase from the bulk to the interphase surface. The chance of an element or eddy being replaced with fresh liquid is assumed to be independent of the length of time of exposure. If the fraction of the surface is replaced at a rate s, Danckwerts [16] showed that the distribution of the surface age can be represented by

$$\phi(t) = se^{-st} \tag{15}$$

where $\phi(t)$ = surface age distribution function
s = fractional rate of surface renewal

The Danckwerts surface renewal theory adopts this surface renewal concept to represent the hydrodynamic behavior and unsteady-state molecular diffusion for transport activities through elements or eddies. Thus, the concentration distribution and point mass transfer rate can be derived from Equations 4 and 5, respectively. The average rate of mass transfer per unit area is

$$\bar{N}_A = \int_0^\infty N_A(t)\, s e^{-st}\, dt = \sqrt{D_A s}\,(C_{Ai} - C_{AL}) \tag{16}$$

The expression for the liquid-side mass transfer coefficient is

$$k_L' = \sqrt{D_A s} \tag{17}$$

In addition to the above theories, other models also have been proposed in the literature [5] to describe the diffusional behavior or hydrodynamic conditions at the interface. For example, Dobbins [17] and Toor and Marchello [18] suggested in a film-penetration model that the equilibrium condition may be established in a surface element of finite thickness after a long residence time. They showed that the film and the penetration theories are not separate concepts, but merely limiting cases of the more general film-penetration model. Rate equations derived by these authors indicated the dependence of the mass transfer coefficient on diffusivity in the following form:

$$k_L' \propto D_A{}^{\nu} \tag{18}$$

where ν may vary from 0.5 to 1.0, as concluded by many experimental investigations [3]. According to the penetration and surface renewal concepts, ν is 0.5; on the other hand, the film theory yields $\nu = 1$. It should be noted that the rate equations derived from various theories contain at least one parameter that is not, in general, experimentally measurable. Therefore, quantitative verification of the various theories may not be permissible.

Factors Affecting Mass Transfer

Mass transfer rate is influenced by many factors, which can be summarized into hydrodynamic and physicochemical effects. Hydrodynamic behavior is concerned with the motion of the molecules. Turbulent flow results in increased contact between phases, and therefore high mass transfer rates. This is usually depicted in mass transfer equations as the interfacial area (a) between the two phases per unit volume of solution. For example, bubbles rising slowly to the surface in a column of liquid have small interfacial areas, and therefore, mass transfer rates are generally low. If intense mixing is added, the bubbles are sheared and mixed thoroughly, increasing both interfacial area and contact time. Mass transfer rates are considerably higher under these conditions.

Physicochemical effects must be taken into account for ozone systems due to the unstable nature of ozone. Ozone decomposes in the liquid phase, creating a continuous demand for ozone. Therefore, true equilibrium or solubility may not be attainable. Decomposition reactions of ozone will be discussed later.

Other physicochemical effects include temperature, pressure and chemical composition of a system. For example, an electrolyte must have both ions diffusing at the same rate to keep from producing an electric field. The intrinsic diffusivities of the two species are generally different, creating a charge which may, in turn, increase the diffusion rate of the slower diffusing ion. The effect of a charged solution may vary the diffusion of a third species, making it difficult to evaluate. Temperature and pressure affect both diffusivity and solubility. In general, the solubility of a gas in liquid increases as the pressure increases and as the temperature decreases.

MASS TRANSFER WITH CHEMICAL REACTIONS

Astarita [4] has divided chemical mass transfer processes into three major regimes controlled by gas- or liquid-phase transport or chemical reaction rate. Since ozone is only slightly soluble in the aqueous solution, resistance to transport of ozone from the gas phase into the interface is insignificant. Thus, the process is controlled mainly by molecular diffusion of the dissolved ozone and chemical reactions in the region between the interface and the liquid phase boundary. In this region, laminar flow behavior prevails and eddy diffusion is unimportant. Once the ozone approaches the phase boundary, it is carried away into the turbulent liquid by eddy diffusion. Since eddy diffusivity is much greater than molecular diffusivity, the liquid bulk provides little resistance to the overall mass transfer.

Decomposition of ozone in the liquid phase is an irreversible chemical reaction. Several complex mechanisms of decomposition have been proposed [19–23]. In spite of the complexity of the proposed mechanisms, the overall reaction can be expressed as

$$O_3 \rightarrow 1.5\, O_2$$

The kinetics of this decomposition reaction have been reported by various investigators [19–24] to be first-, 3/2- or second-order.

In a gas-liquid contacting system, constituents of the aqueous solution may diffuse from the liquid bulk into the interface and react with the dissolved ozone, as discussed earlier. Products of the reactions then may diffuse back into the liquid bulk, or may penetrate into the gas phase if they are volatile. In addition, ozonation reactions also may occur in the main liquid stream after ozone has diffused across the phase boundary.

Since ozone diffusion is accompanied by decomposition and ozonation reactions in the ozone absorption processes, the rate equations given in the previous section for physical mass transfer processes, in general, are not applicable. The equations for physical absorption may still be useful for certain ozone absorption systems in which the rates of chemical reactions are very slow compared with the rate of diffusion. Theories of simultaneous mass transfer and chemical reactions have been developed by many investigators based on both steady and transient models. The influence of chemical reactions on the absorption rate can be investigated in terms of the enhancement factor, E:

$$E = k_L / k_L' \tag{19}$$

where k_L = liquid-side mass transfer coefficient for chemical absorption (cm/sec)

In the above equation, the chemical mass transfer coefficient k_L is a measurable parameter for a given ozone absorption system. The physical mass transfer coefficient k_L' can be obtained from empirical correlations [3] or from physical absorption experiments conducted in the same gas-liquid contactor. Thus, the enhancement factor can be evaluated from the experimental data, and the result can be compared with values predicted by various theories. A few examples commonly encountered in the ozone absorption process are discussed in this section.

Diffusion with Irreversible Chemical Reactions

Absorption of ozone by water is a mass transfer process involving diffusion and an irreversible reaction for the decomposition of the absorbed ozone, as mentioned earlier. In general, diffusion is accompanied by complex reactions during the treatment of liquid wastes by ozone. Nevertheless, the rate of decomposition may be negligible compared with the rates of ozonation in some systems. For example, the reaction between ozone and formic acid in acidic solutions was much faster than the decomposition of dissolved ozone, as reported by Kuo and Wen [25]. Under this circumstance, the process may be treated as diffusion with an overall ozonation reaction by ignoring the effect of decomposition.

Processes Controlled by Diffusion

If the homogeneous chemical reaction is very rapid, the mass transfer rate of ozone may be limited by diffusion alone. In considering an instantaneous reaction of one mole of dissolved ozone (A) with b moles of a liquid reactant (B) to produce p moles of product (P):

$$A + bB \rightarrow pP$$

the dissolved ozone may be depleted completely before approaching the liquid phase boundary. The faster the reactant B is able to diffuse, the shorter the distance that ozone will have to diffuse into the liquid film to react with B. The rates of diffusion of A and B are governed by the partial differential equation:

$$D_A \left(\frac{\partial^2 C_A}{\partial z^2} \right) = \frac{\partial C_A}{\partial t}, \quad 0 < z < z'(t) \tag{20}$$

$$D_B \left(\frac{\partial^2 C_B}{\partial z^2} \right) = \frac{\partial C_B}{\partial t}, \quad z'(t) < z < L \tag{21}$$

where D_B = molecular diffusivity of B (cm^2/sec)
C_B = concentration of B (mol/cm^3)
$z'(t)$ = distance from the interface (cm)

The instantaneous reaction occurs only in the reaction zone at $z'(t)$, as illustrated in Figure 4. Analytical approaches to this system have been considered by Hatta [26], Danckwerts [5] and Sherwood et al. [3] from viewpoints of both steady-state and transient diffusion.

Hatta [26] applied the film theory concept of Lewis [13] and Whitman [14] by assuming that the reaction zone approaches its equilibrium position in a very short time and that steady-state behavior prevails for the absorption process. As sketched in Figure 4, the concentration distributions in the liquid film become linear in accordance with these propositions. The mass transfer rates of A and B can be expressed by:

$$\bar{N}_A = \left(\frac{D_A}{z'}\right)(C_{Ai} - 0) \tag{22}$$

$$\bar{N}_B = -bN_A = \left(\frac{D_B}{L - z'}\right)(0 - C_{BL}) \tag{23}$$

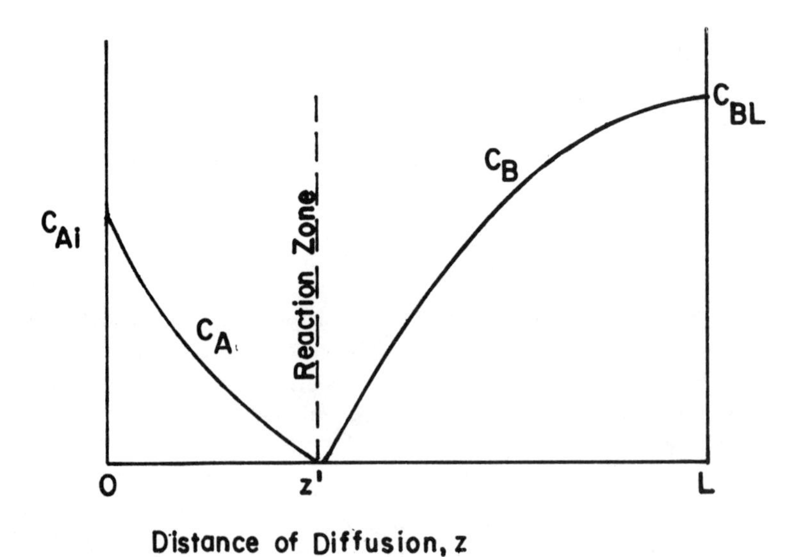

Figure 4. Concentration profiles for chemical absorption controlled by diffusion.

where z' = distance of reaction zone (cm)
$\bar{N}_B$ = average rate of mass transfer of B per unit interfacial area (mol/cm^2-sec)
b = stoichiometric ratio
C_{BL} = concentration of B in the bulk liquid (mol/cm^3)

By combining the above two equations, the absorption rate of A can be obtained:

$$\bar{N}_A = \left(\frac{D_A}{L}\right)\left(C_{Ai} - \frac{D_B\, C_{BL}}{b\, D_A}\right) \tag{24}$$

If the liquid-film coefficient is defined on the basis of the driving force, $(C_{Ai} - 0)$, we have

$$k_L = \left(\frac{D_A}{L}\right)(1 + q) \tag{25}$$

where

$$q = \frac{D_B\, C_{BL}}{b\, D_A\, C_{Ai}} \tag{26}$$

Thus, the enhancement factor is

$$E = \frac{k_L}{k_L'} = 1 + q \tag{27}$$

where the physical mass transfer coefficient is defined by Equation 12.

The effects of transient diffusion were considered by Danckwerts [5] and Sherwood et al. [3] by utilizing the penetration concept of Higbie [15]. They postulated that the reaction zone shown in Figure 4 may move away from the interface during the course of diffusion. Without attainment of a steady state, the diffusion equations may be solved by considering an infinite thickness for the liquid film, as suggested by Higbie [15]. Thus, the concentration profiles illustrated in Figure 4 may be expressed in the forms of error functions. The instantaneous rate of absorption of A can be derived as

$$N_A = -D_A\left(\frac{\partial C_A}{\partial z}\right)_{z=0} = \frac{C_{Ai}}{\mathrm{erf}(\alpha/D_A)^{1/2}}\left(\frac{D_A}{\pi t}\right)^{1/2} \tag{28}$$

where α (cm^2/sec) is defined by

$$\left(\frac{C_{BL}}{b\, C_{Ai}}\right)\left(\frac{D_B}{D_A}\right)^{1/2} \exp(\alpha/D_A)\,\mathrm{erf}\left(\frac{\alpha}{D_A}\right)^{1/2} = \exp(\alpha/D_B)\,\mathrm{erf}\left(\frac{\alpha}{D_B}\right)^{1/2} \tag{29}$$

The average rate of absorption, then, can be obtained as

$$\bar{N}_A = \frac{1}{t_0}\int_0^{t_0} N_A dt = \frac{2C_{Ai}}{\text{erf}(\alpha/D_A)^{1/2}}\left(\frac{D_A}{\pi t_0}\right)^{1/2} \tag{30}$$

On the basis of the driving force equal to $(C_{Ai} - 0)$, the liquid side mass transfer coefficient is found to be

$$k_L = \frac{2}{\text{erf}(\alpha/D_A)^{1/2}}\left(\frac{D_A}{\pi t_0}\right)^{1/2} \tag{31}$$

The enhancement factor is

$$E = \frac{k_L}{k_L'} = \frac{1}{\text{erf}(\alpha/D_A)^{1/2}} \tag{32}$$

where the expression for the physical mass transfer coefficient for the penetration theory is given in Equation 14.

Processes Controlled by Diffusion and Chemical Reactions

When the rate of chemical reaction between A and B is finite, the mass transfer process is controlled by both diffusion and chemical reaction. Under this condition, the chemical reaction will not be concentrated in a thin reaction zone (Figure 5) and the approximate treatments are not valid. The partial differential equations governing simultaneous diffusion and irreversible reactions in the liquid film are:

$$D_A\left(\frac{\partial^2 C_A}{\partial z^2}\right) - R = \frac{\partial C_A}{\partial t}, \quad z > 0 \tag{33}$$

$$D_B\left(\frac{\partial^2 C_B}{\partial z^2}\right) - bR = \frac{\partial C_B}{\partial t}, \quad z > 0 \tag{34}$$

The rate of chemical reaction can be expressed in the form:

$$R = k_i C_A{}^m C_B{}^n \tag{35}$$

where R = rate of homogeneous chemical reaction (mol/cm³-sec)
k_i = reaction rate constant for i^{th} order reaction $[(\text{mol/cm}^3)^{1-i}/\text{sec}]$
m, n = orders of reaction

k_i, m and n can be determined only by experimental investigation.

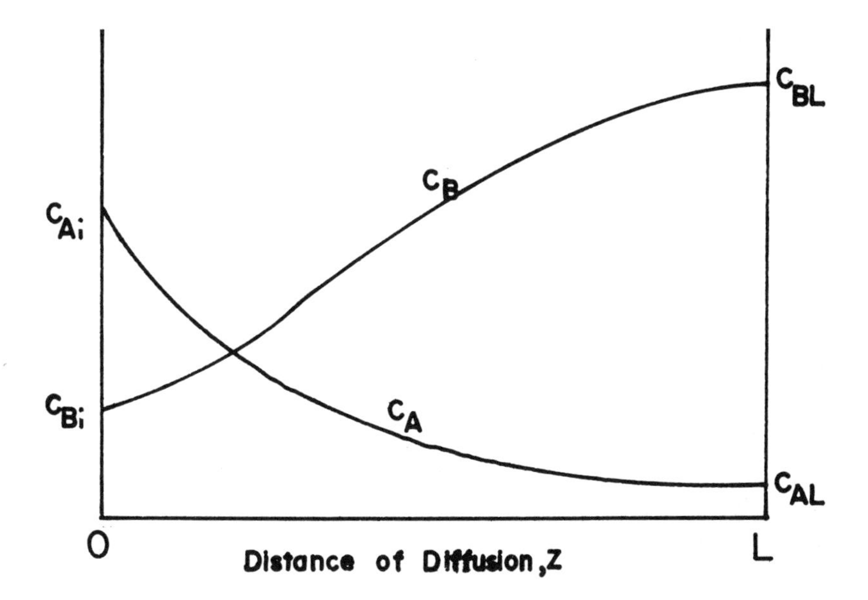

Figure 5. Concentration profiles for chemical absorption controlled by diffusion and reaction.

Since the mathematical system is nonlinear in general, exact solutions to the above differential equations may not be possible. Analytical solutions, however, can be derived for a special class of mass transfer processes in which diffusion is accompanied by a first-order chemical reaction. This corresponds to a first-order decomposition reaction of dissolved ozone in the absorption process, or a pseudo first-order ozonation reaction in which reactant B is present in large excess in the solution. For these cases of mass transfer with a first-order reaction, the concentration of ozone in the liquid film can be derived by solving the following partial differential equation:

$$D_A\left(\frac{\partial^2 C_A}{\partial z^2}\right) - k_1 C_A = \frac{\partial C_A}{\partial t}, \quad z > 0 \tag{36}$$

On the basis of the steady-state concept of the film theory, Hatta [27] obtained the rate of absorption of A as

$$\bar{N}_A = N_A = -D_A\left(\frac{dC_A}{dz}\right)_{z=0} = k_L' \sqrt{M}\,(C_{Ai} \coth\sqrt{M} - C_{AL} \operatorname{csch}\sqrt{M}) \tag{37}$$

where M = a dimensionless parameter, defined by

$$M = k_1 D_A / k_L'^2 \tag{38}$$

When the reaction rate is much faster than the rate of diffusion, the concentration of ozone approaches zero at the edge of the liquid film, and the chemical mass transfer coefficient and the enhancement factor can be obtained as

$$k_L = k_L' \sqrt{M} \coth \sqrt{M} \tag{39}$$

and

$$E = \sqrt{M} \coth \sqrt{M} \tag{40}$$

For the unsteady-state model of the penetration theory, Danckwerts [5,16] derived the following expression for the average rate of mass transfer:

$$\bar{N}_A = \frac{1}{t_0} \int_0^{t_0} N_A dt = k_L' \sqrt{M} C_{Ai} \left[\left(1 + \frac{\pi}{8M}\right) \operatorname{erf}\left(\frac{4M}{\pi}\right)^{1/2} + \frac{\exp(-4M/\pi)}{2M^{1/2}} \right] \tag{41}$$

The chemical mass transfer coefficient and the enhancement factor are:

$$k_L = k_L' \sqrt{M} \left[\left(1 + \frac{\pi}{8M}\right) \operatorname{erf}(4M/\pi)^{1/2} + \exp(-4M/\pi)/(2M^{1/2}) \right] \tag{42}$$

and

$$E = \sqrt{M} \left[\left(1 + \frac{\pi}{8M}\right) \operatorname{erf}(4M/\pi)^{1/2} + \exp(-4M/\pi)/(2M^{1/2}) \right] \tag{43}$$

Turning to the surface renewal model of Danckwerts, the average rate of absorption can be derived to be [16]:

$$\bar{N}_A = k_L' \sqrt{1 + M}\, C_{Ai} \tag{44}$$

Thus, the chemical mass transfer coefficient and the enhancement factor are

$$k_L = k_L' \sqrt{1 + M} \tag{45}$$

$$E = \sqrt{1 + M} \tag{46}$$

It should be noted that the partial differential equation governing the diffusion and first-order reaction also can be solved on the basis of the more general concept of the film-penetration theory as discussed by Huang and Kuo [28]. The equations derived on the basis of the film-penetration theory can be reduced to the equations above under limiting conditions postulated by the film, penetration and surface renewal concepts. Huang and Kuo [28] showed that the maximum deviation among the individual models is only 8.8% in the prediction of the enhancement factor that occurs at a moderate M value of 2.1. When the rate of reaction is much slower than that of diffusion, the rate of chemical absorption approaches that of physical mass transfer and the enhancement factor is $E=1$. On the other hand, the enhancement factors for all models approach $\sqrt{M}$ if the rate of reaction is much faster than that of the molecular diffusion ($M \gg 1$).

Concerning diffusion with bimolecular reactions of second or higher order, analytical approximate solutions have been developed, though exact solutions have not been possible. On the basis of the film theory, Van Krevelen and Hoftijzer [29] obtained an approximate expression for mass transfer accompanied by the second-order reaction:

$$R = k_2 C_A C_B \tag{47}$$

By assuming a step change in the concentration of B from C_{BL} at the bulk to C_{Bi} throughout the liquid film, simultaneous differential Equations 33 and 34 can be solved analytically to yield the following expression for the enhancement factor:

$$E = \{M[1 - q(E - 1)]\}^{1/2} \coth\{M[1 - q(E - 1)]\}^{1/2} \tag{48}$$

where M is defined by

$$M = k_2 C_{BL} D_A / k_L'^2 \tag{49}$$

Instead of assuming that C_B is constant throughout the film, Peaceman [30] derived another analytical solution based on linear variation of C_B with the distance of diffusion. The solution of Peaceman [30] deviated only slightly from that of Van Krevelen and Hoftijzer [29] with a maximum deviation of 8% at low values of q. Since the assumptions made in the two studies represent the upper and lower extremes of the concentration profile of B in the film, it can be concluded that the analytical approximate equations are correct to within 8% of the exact solution [30]. Analytical approximations based on transient absorption are difficult, but the partial differential equations have been solved numerically by several researchers, as reported in the literature [3–5]. However, the numerical results obtained by these investigators are not substantially different from those predicted by Equation 48.

For ozone absorption accompanied by a decomposition reaction of 3/2 order, Kuo et al. [24] obtained a numerical solution utilizing the film concept. The nonlinear differential equation

$$D_A\left(\frac{d^2C_A}{dz^2}\right) - k_0 C_A^{3/2} = 0 \tag{50}$$

was solved by finite difference methods to predict the mass flux and buildup of ozone in an aqueous solution. They found that the concentration of ozone in solution is influenced by two dimensionless parameters,

$$\beta = k_L' a D_A / V \tag{51}$$

$$M = k_0 C_{Ai}^{1/2} D_A / k_L'^2 \tag{52}$$

where a = specific interfacial area (cm^2/cm^3 solution)
V = liquid volume (cm^3)

At a given time, the ozone concentration increases as β increases and as M decreases. The concentration approaches an asymptotic value at large times, and the saturation or equilibrium concentration may not be attainable for large M and small β values. This implies that buildup to the equilibrium concentration may be impossible if the rate of decomposition is very fast or if the diffusion rate is very slow.

Diffusion with Complex Reactions

Complex kinetics, such as parallel, reversible and consecutive reactions, often are involved in ozone treatment of liquid wastes. In addition to the decomposition reaction, ozone may react with one or more constituents in an aqueous solution to form intermediate products, and the intermediates may react further with ozone or may be decomposed to yield final products. For example, the following reactions occur in the ozonation of methanol:

$$2O_3 \rightarrow 3O_2$$

$$O_3 + CH_3OH \rightarrow HCHO + H_2O + O_2$$

$$O_3 + HCHO \rightarrow HCOOH + O_2$$

$$O_3 + HCOOH \rightarrow CO_2 + H_2O + O_2$$

Since rates of reactions for these steps all are comparable in magnitude, as reported by Kuo and Wen [25], it may be necessary to account for each reaction step in the formulation of a mathematical model for an ozone injection process. This results in a set of highly nonlinear differential equations for which solutions are difficult to obtain. Therefore, it is desirable to examine carefully kinetics of ozonation reactions to simulate an absorption process by taking into account only the diffusion and controlling reactions.

If ozonation reactions can be considered to be instantaneous, the absorption is limited by diffusion. Under this circumstance, the approach similar to that of Hatta [26] may be utilized to derive the expressions for mass transfer coefficient and enhancement factor. Analytical solutions may be possible for mass transfer accompanied by complex reactions if the kinetics of reactions can be assumed to be pseudo first-order. For instance, on the basis of the film, surface renewal and film-penetration concepts, Kuo and Huang [31,32] developed theoretical equations for the enhancement factor for diffusion with the following complex reactions:

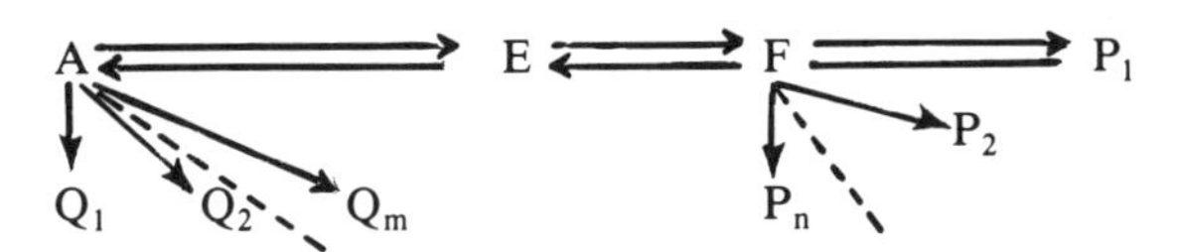

The problem of selectivity in consecutive reactions involving gas-liquid systems has been investigated [33–35]. Analytical approximations and numerical solutions to systems of molecular diffusion and high order complex reactions are discussed elsewhere [36–38].

In a recent study, Li and Kuo [39] investigated absorption of ozone by phenolic solutions. Although the kinetics of ozonation of phenol were very complex, the overall reaction was controlled by an irreversible reaction between ozone and phenol [40]. On the basis of this finding, the diffusion of ozone and the following decomposition and ozonation reactions were considered:

$$O_3 \xrightarrow{k_0} 3/2\, O_2$$

$$2O_3 + C_6H_5OH \xrightarrow{k_2} \text{products}$$

For steady-state transport, the absorption process is governed by the following differential equations:

$$D_A\left(\frac{d^2C_A}{dz^2}\right) - k_0C_A{}^{3/2} - 2k_2C_AC_B = 0 \tag{53}$$

$$D_B\left(\frac{d^2C_B}{dz^2}\right) - k_2C_AC_B = 0 \tag{54}$$

Where C_A and C_B are concentrations of ozone and phenol, respectively. The decomposition and ozonation reactions were 3/2- and second-orders, respectively, as determined from kinetic data [24,40]. The nonlinear differential equations were linearized by a method similar to that of Peaceman [30], and the concentration distributions and mass flux were obtained in terms of Airy integrals. An approximate equation then was derived by Li and Kuo [39] to predict depletion of phenol in the aqueous solutions. Their theoretical results indicated that the rate of depletion of phenol was influenced by the physical mass transfer coefficient as well as by the dimensionless parameter M, defined by Equation 49.

Ozone Absorption Experiments

Numerous studies have been conducted in gas-liquid contactors to investigate ozonation reactions. Many of these experiments were carried out to determine dosage requirements or design parameters for industrial applications. A few cases were undertaken to examine reaction kinetics by ignoring effects of diffusion on the gas absorption processes. Mass transfer phenomena were investigated by only a few researchers, and these will be discussed below.

The phenomena of diffusion in the gas phase and diffusion and chemical reactions at the gas-liquid interface were discussed in the previous sections. Once the unreacted ozone penetrates across the liquid phase boundary, it may react with constituents of an aqueous solution. Therefore, it is important to consider possible reactions in the liquid stream in studies of mass transfer coefficient or concentration behavior in the aqueous solution. To illustrate this, we consider the case of decomposition and ozonation reactions between the dissolved ozone A and the liquid reactant B in the absorption process:

$$A \xrightarrow{k_0} P$$

$$A + bB \xrightarrow{k_2} Q$$

The component mass balances at the phase boundary yield the following relationships if the main liquid is well mixed:

$$\frac{dC_{AL}}{d\theta} = a\left[D_A\left(\frac{\partial C_A}{\partial z}\right)_{z=L} - k_0 C_{AL}{}^{\ell} - k_2 C_{AL}{}^{m} C_{BL}{}^{n}\right] \tag{55}$$

$$\frac{dC_{BL}}{d\theta} = -a\left[D_B\left(\frac{\partial C_B}{\partial z}\right)_{z=L} + bk_2 C_{AL}{}^{m} C_{BL}{}^{n}\right] \tag{56}$$

By solving the differential equations discussed earlier for diffusion and reactions at the interface, the mass fluxes can be obtained, and the above equations can be integrated to predict the concentrations in the main liquid as a function of absorption time.

Absorption of Ozone in Water

Absorption of ozone gas in aqueous solutions of various pH values was studied by Kuo et al. [24] utilizing a bubble column. The experiments were conducted at 25°C in a semibatch mode by controlling flowrate and concentration of ozone in the gas stream. The concentrations of ozone in solution were measured at different absorption times. The experimental data were analyzed in light of theoretical predictions by numerical integration of Equation 55 for $\ell = 3/2$ and $k_2 = 0$ by utilizing the solutions of Equation 50 to compute the mass flux of ozone. It was found [24] that the theory agreed well with the experimental data in predicting the buildup behavior of dissolved ozone in solution. The asymptotic concentration achieved after long absorption time decreased as the solution pH value increased and as the gas flowrate decreased. Therefore, low pH values (slow decomposition rates) and high flowrates (fast diffusion rates) favor achieving a high concentration of ozone in the reactor.

Mass transfer of ozone into distilled water in a stirred tank reactor was investigated by Yocum [41,42]. The tests were performed at 12–25°C with various gas flowrates and speeds of agitation. Concentrations measured during the experiments were utilized to evaluate the volumetric mass transfer coefficient, which was defined in terms of the average driving force $(C_A - C_{AL})$. Equation 55 was integrated analytically for $\ell = 1$ and $k_2 = 0$ by replacing the mass flux with the following expression:

$$D_A \left(\frac{\partial C_A}{\partial z} \right)_{z=L} = k_L (C_{AE} - C_{AL}) \tag{57}$$

where C_{AE} = equilibrium concentration of A (mol/cm^3)

In the above equation, C_{AE} denotes the liquid concentration in equilibrium with a mean gas concentration between the entering and exit values. By expressing the equilibrium concentration and, in turn, mole fraction in terms of the inlet gas concentration and mass transfer rate, Yocum [41,42] showed that the volumetric mass transfer coefficient can be estimated from the slope of a semilogarithmic plot of the ratio of mole fraction against absorption time. The mass transfer coefficient was found to increase with turbine speed and superficial gas velocity [41,42]. Since these are factors contributing to the enhancement of molecular diffusion, the results obtained by Yocum [41,42] are in agreement with those obtained by Kuo et al. [24].

Absorption of Ozone in Alkaline Solutions

The rate of ozone absorption into aqueous solutions of potassium hydroxide was studied in a packed column by Rizzuti et al. [43]. By assuming negligible concentration of ozone in the liquid, a mass balance over a differential volume of the column can be taken to relate the volumetric mass transfer coefficient to inlet and outlet concentrations measured in an experiment. The mass transfer coefficients were reported by these investigators for experiments conducted at temperatures varying from 18 to 27°C in the pH range 8.5–13.5. These chemical mass transfer coefficients were much greater than the physical mass transfer coefficients, and a slope of 1/2 was obtained at high pH values in a semilogarithmic plot of $k_L a$ vs pH. On the basis of these results, these investigators suggested that the absorption of ozone in potassium hydroxide solutions was controlled by diffusion and a rapid second-order reaction with the chemical reaction rate governed by

$$R = k_2 [OH^-][O_3] \tag{58}$$

Since the mass transfer rate was limited by diffusion of ozone into the solutions, the enhancement factor (Equation 48) for diffusion and second-order reaction can be reduced to

$$E = k_L / k_L' \simeq \sqrt{M} = (k_2 C_{BL} D_A)^{1/2} / k_L' \tag{59}$$

or

$$k_L = (k_2 C_{BL} D_A)^{1/2} \tag{60}$$

where C_{BL} represents the average concentration of hydroxyl ion $[OH^-]$ in a solution. The reaction rate constant, k_2, was determined to be 1500 $M^{-1}sec^{-1}$ at 27°C by Rizzuti et al. [43] utilizing Equation 60.

It should be noted that the actual mechanism of reaction of ozone in the solutions might be quite complex, although the overall reaction was controlled by the second-order kinetics as considered by Rizzuti et al. [43]. For example, Hoigné and Bader [44] suggested that two reaction pathways, including direct reactions of ozone and free radical reactions involving hydroxyl free radical intermediates, may proceed simultaneously in aqueous media. In strong alkaline solutions, ozone decomposes rapidly resulting in formation of hydroxyl ions $[OH^-]$ at high concentrations. Under these circumstances a large fraction of the hydroxyl ions may be converted to hydroxyl free radicals, $OH^{\cdot}$ [19,23], and the reactions between these hydroxyl free radicals and constituents in the solutions may become predominant [44,45].

Absorption of Ozone in Phenolic Solutions

Absorption of ozone in aqueous solutions of phenol has been investigated by many researchers [39, 46–49]. A wetted-wall column was used by Rizzuti et al. [48] to measure absorption rates at various pH and temperature values of the aqueous solutions. Inlet and outlet gas concentrations were utilized to compute the mass transfer coefficients for different systems. These authors considered that the rapid ozonation reaction between ozone and phenol was zero-order with respect to ozone and first-order with respect to the concentration of phenol. A simplified equation (similar to Equation 59) was used by these investigators to evaluate the first-order rate constant at 25°C as 16 sec^{-1}. As discussed earlier, the kinetics of ozonation reactions between ozone and phenol in the liquid phase was studied by Li et al.[40] They reported the second-order rate constant at 25°C, for various pH values and at pH 5.2, to be 29,500 $M^{-1}sec^{-1}$. Since the concentration of ozone in the solutions was of the order of 1×10^{-4} M, the first-order rate constant of 16 sec^{-1} appeared to be of the same order of magnitude.

Rates of depletion of phenol in the bubble column were measured by Li and Kuo [39] in the absorption experiments conducted at 25°C with pH values of the aqueous solutions ranging from 2.2 to 11.4. The overall reactions in the absorption process were:

$$O_3 \rightarrow 1.5 O_2$$

$$2O_3 + C_6H_5OH \rightarrow \text{products}$$

No trace amount of ozone was detected in the aqueous solutions and, therefore, the ozonation reactions in the main liquid were assumed to be insignificant. The rate of mass transfer of phenol into the liquid film, then, can be considered to be proportional to the rate of mass transfer of ozone into the interface, less the rate of depletion by decomposition in the liquid film. On the basis of these assumptions, Li and Kuo [39] derived an approximate equation to show that the dimensionless concentration to the 2/3 power changes linearly with the ozone absorption time. This theoretical prediction was confirmed by experimental data measured by these authors. They also reported that the depletion rate of phenol increased with the gas flowrate (because of increasing $k_L'a$ value) and the pH (due to increase in ozonation rate and, in turn, the M value). At a fixed gas flowrate, however, Li and Kuo [39] found little additional depletion rate of phenol in the solutions with pH values greater than 7.

GAS-LIQUID CONTACTORS

This section presents a brief introduction to gas-liquid contactors, drawn primarily from their uses in chemical process industries. Not all types of contactors are suitable for ozonation reactions. Contactors may also be referred to as reactors, since we are considering the addition of an ozone-containing gas to a liquid for oxidation reactions. Major advantages and disadvantages of the various systems are cited. A detailed discussion of ozone contactors is presented by Masschelein [50].

Before a design engineer can specify a gas-liquid contactor (reactor), the overall process must be considered. Should the process be batch, semibatch or continuous? A batch process is one in which the reactants are charged to a reactor, and, after reaction, the products are discharged. The batch process is difficult to use for ozonation, as a continual supply of ozone is generally required. This leads to consideration of semibatch operation. The common semibatch ozonation procedure is to charge the liquid to the reactor, then continually charge ozone gas until the reaction is complete. Continuous processes have simultaneous introduction and withdrawal of reactants. An example of a continuous ozonation process is drinking water purification, where ozone gas is injected into the water as the water continuously flows through the reactor vessels.

The decision as to process type corresponds to the selection of an ozone reactor. The gas-liquid contactor (reactor) selected is governed to a large extent by the overall relationship between mass transfer and the chemical kinetics of the specific ozonation reaction(s). The controlling mechanism dictates, to some extent, the type of contactor that can be used. If ozone absorption is accompanied by fast reaction, a large interfacial area is desired to promote ozone mass transfer; therefore, the packed column may be preferred. If, on the other hand, the reaction rate is slow, so that a large liquid phase volume (holdup) is advantageous, a bubble column is more efficient. Table I presents a summary of commonly used gas-liquid contactors. (In Table I, the term "conversion" refers to that fraction of reactant transformed into intermediate or final products, not to the conversion of ozone from gas to liquid phase.)

The design engineer must take into account many considerations when designing a gas-liquid system. These include gas and liquid rates to satisfy production criteria, mass transfer and chemical reaction relationships and, ultimately, choosing a gas-liquid contactor and mode of operation that will function in the most economical manner.

In selecting a gas-liquid contactor, the following parameters need to be examined for their effects on mass transfer: specific interfacial area a, the mass transfer coefficient k_L, solubility of the dispersed phase, diffusivity of solute and the dispersed phase holdup. Other factors affecting mass transfer indirectly are dispersed phase superficial velocity, and bubble diameter and velocity. Many review articles and texts have been written on designing gas-liquid contactors [2,3,51–57]. Several researchers [2,58–60] have reviewed contactors specifically for ozone use. These refferences should be consulted for design equations. The contactors listed in Table I will be discussed, and general design considerations will be presented.

Packed Column

Packed columns are vertical pipes filled with packing to distribute the gas and liquid, and to promote mixing. Packed columns used for gas purification are commonly called absorbers, and normally operate with countercurrent flow of gas and liquid. An example is the absorption of carbon dioxide and hydrogen sulfide from solutions of ethanolamines.

Some packings also act as a catalyst to promote the reaction. Chen et al. [61] showed that a ferric oxide catalyst used in a packed column enhanced ozone utilization to oxidize phenol in aqueous solution, as compared to an inert packing. Countercurrent operation was used, but both counter- and cocurrent flows can be used with catalytic operations. In the latter case, both up- and downflow operations are encountered. Upflow provides better mixing of gas and liquid, but higher pressure

Table I. Contactors of Gas-Liquid Systems and Their Characteristics

Type	Mode of Operation	Mass Transfer	Advantages	Disadvantages	Reaction Regime
Packed Column	Liquid and vapors pass countercurrent to each other through the same passages created by packing material. Continuous operation.	Good mass transfer, varies with packing type and gas-liquid flowrates.	Wide operating range, Can withstand highly corrosive systems.	Expensive, difficult to maintain temperature profiles. Easily plugged.	Gas- or liquid-phase mass transfer controlling.
Plate Column	Liquid and vapors flow countercurrent to each other over the trays. Continuous operation.	Good mass transfer, proportional to interfacial area, which depends on gas flowrate.	Wide operating range and easy to clean.	Expensive and complex to design. Easily plugged.	Slow reactions where staging and large liquid volumes are required.
Bubble Column	Gas is dispersed into bubbles which rise through a column of liquid. Capable of continuous counter- or cocurrent, alternating countercurrent, or multiple counter- or cocurrent operation. Can be semibatch.	Low mass transfer, depends on interfacial area, which is a function of gas flowrate.	Low energy requirements.	Sparger may plug, giving channeling of gas bubbles. Poor mixing. Long contact time.	Reaction rate-controlled system requiring large liquid volume.
Spray Tower	Liquid is dispersed into a volume of gas containing ozone.	Moderate mass transfer with a large interfacial area.	Uniform gas phase.	High energy consumption. Solids can plug spray nozzle.	Fast reactions requiring little holdup time.
Agitated Vessel	Can be operated continuously, semibatch or batch. Tank with mechanical agitation applied.	Moderate to good mass transfer with interfacial area and gas holdup; dependent on gas flowrate and agitation.	High degree of flexibility and can handle solids. Good heat transfer characteristics.	Energy is required for agitation; stirred tank reactors require maximum theoretical volume to obtain desired conversion.	Mass transfer-controlled reactions.

Injectors and Turbines	Gas and liquid are forced or drawn cocurrently through a small opening.	Mass transfer and interfacial area are high.	Good mixing and very short contact times. Small contactor chambers.	Requires energy.	Mass transfer-limited reactions requiring short liquid holdup.
Tubular Contactor	Can be either cocurrent (usually vertical tube) or counter-current operation.	High mass transfer may be obtained at high gas velocities if water flow-rates are high.	Temperature control is easy to maintain; low cost; ease of operation.	Energy is required and static mixers may be needed to promote gas-liquid contact	Mass transfer-limited reactions requiring short liquid holdup.

drops and incidents of flow restriction are encountered. Cocurrent downflow over catalytic packing, with a continuous gas phase and dispersed liquid, is commonly referred to as a "trickle bed reactor."

Packed columns provide high interfacial areas; thus, they are good for mass transfer–limited reactions. They do not require a large pressure drop, but are somewhat limited in operating range. Since liquid and gas travel essentially in the same channels through the column packing, the ranges of liquid and gas loads are narrow for efficient operation. Isolated temperature excursions can occur in the column.

New plastic packings have reduced the cost of packed columns, and are capable of withstanding exposure to corrosive atmospheres. There are numerous packing designs from which to choose. Packing companies are helpful in supplying $K_G a$ and other design data. However, it must be realized that ozone reacts with many plasticizers. Proposed plastic materials should be tested in the presence of ozone before they are specified for use.

Plate Column

Plate columns are more expensive than packed columns, but provide wider operating ranges. Gas and liquid flow countercurrent, with the liquid being redistributed on each tray. Since the liquid flow is evenly distributed over the complete height of the column, large diameters can be used for high water throughputs. The trays can be designed to provide liquid holdups necessary for slow reactions while providing high interfacial area for mass transfer. Valve trays, which are variable-orifice perforated trays, are the ideal type of tray for absorber service when the wide operating flexibility of a plate tower is sought.

Bubble Column

Bubbles of gas are sparged into a liquid-filled column. This is the most commonly used ozone reactor for drinking water disinfection. The degree of mixing is dependent on the bubble size and superficial gas velocities. Bubble columns are simple to operate and inexpensive. They have been shown to be excellent for high-pressure ozonation [62]. Heat exchangers may be inserted into the column for temperature control.

The bubble column also is excellent for chemical reaction rate–controlled ozonation reactions. Gas contact time is governed primarily by the rising velocity of the bubbles and the height of the liquid column.

Mass transfer efficiency is not strongly influenced by pressure in a gas-liquid contacting system as it is in a gas-gas contacting system.

In some chemical process reactions, catalyst particles can be injected to form a slurry reactor. The particles are kept suspended by the bubble movement. On the other hand, catalyst particles can lead to sparger plugging problems, particularly if the gas flow becomes intermittent or if the rate of bubble rise is too slow to maintain dispersion of the catalyst.

Spray Tower

In spray towers the liquid is dispersed into a volume of gas containing ozone. This promotes very high interfacial area at the expense of high pumping costs. Spray towers, because of low contact time and high interfacial area, are good for instantaneous or rapid reactions. They are used commercially in several European treatment plants [63], and have been found capable of destroying cyanide in laboratory experiments [64].

Agitated Vessels

Agitated vessels (stirred-tank reactors) are useful for intermediate reaction regimes in which mass transfer and chemical reaction rates are of the same order of magnitude. Agitated vessels economically provide a system for high gas and liquid holdup. All three operating modes (batch, semibatch and continuous) can be employed with agitated vessels. Semibatch operation with ozone gas fed continuously to a constant volume of wastewater was used successfully to treat several refractory industrial wastewaters [65].

Agitated vessels operated in a continuous mode are commonly referred to as back-mixed reactors. Perfect mixing is assumed, which results in uniform composition throughout the reactor. Accordingly, the exit stream has the same composition as that within the reactor. For a chemical reaction rate–limited regime in which mass transfer effects are insignificant, design equations for back-mixed reactors show that they require the maximum theoretical volume to obtain a desired degree of chemical conversion.

The major effect of agitation rate on gas-liquid equilibria is to vary the interfacial area. Agitated vessels can approach the interfacial area of a packed column by using intense agitation, and can approximate the interfacial area of a bubble column in the absence of agitation. The technique of varying agitation rates can be used to identify the reaction regime. A curve similar to Figure 6 will be obtained for intermediate reaction regimes as the reaction mechanism transforms from one that is mass transfer–limited to one that is reaction rate–limited with increased agitation (interfacial area).

The power consumed by an agitator is a disadvantage due to its significant addition to the costs of operation. However, the power required does decrease as gas is dispersed into the liquid, reducing the mixture's density. It is best to determine the power requirements for mixing using the ungassed liquid.

Advantages of an agitated vessel are good mixing and heat transfer. Mechanical agitation can keep an added catalyst suspended, thus improving slurry reactor operation [66]. Excellent heat transfer rates are possible because of the agitation. Both jacket and inserted tube heat exchangers can be used, with the latter providing better heat transfer.

A standard configuration was developed for stirred tank reactors from the mass-transfer studies of Westerterp et al. [67] and Prengle and Barona [52]. In the standard configuration, ratios are used to relate the various reactor dimensions to the reactor diameter. These ratios allow for easy

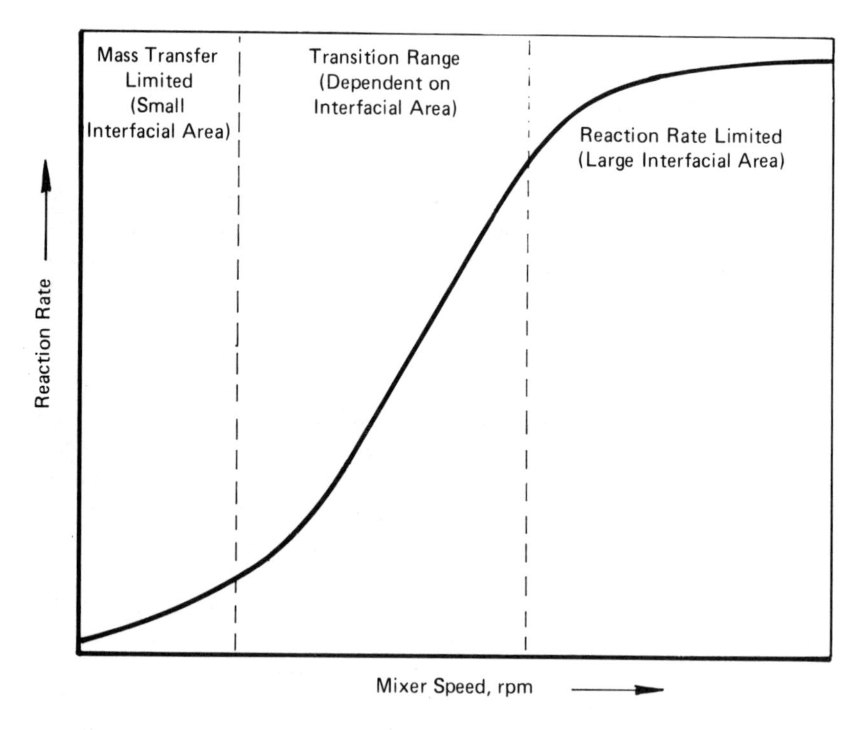

Figure 6. Effect of agitation on a gas-liquid reaction with slow to fast reaction.

scale-up of reactors from pilot-plant to commercial units. The standard stirred-tank reactor configuration [68] is given in Figure 7. Several researchers [3,67,69,70] studied the correlations between interfacial area and mass transfer in stirred tanks with many operating variables, including rate of agitation, gas velocity and reactor configuration.

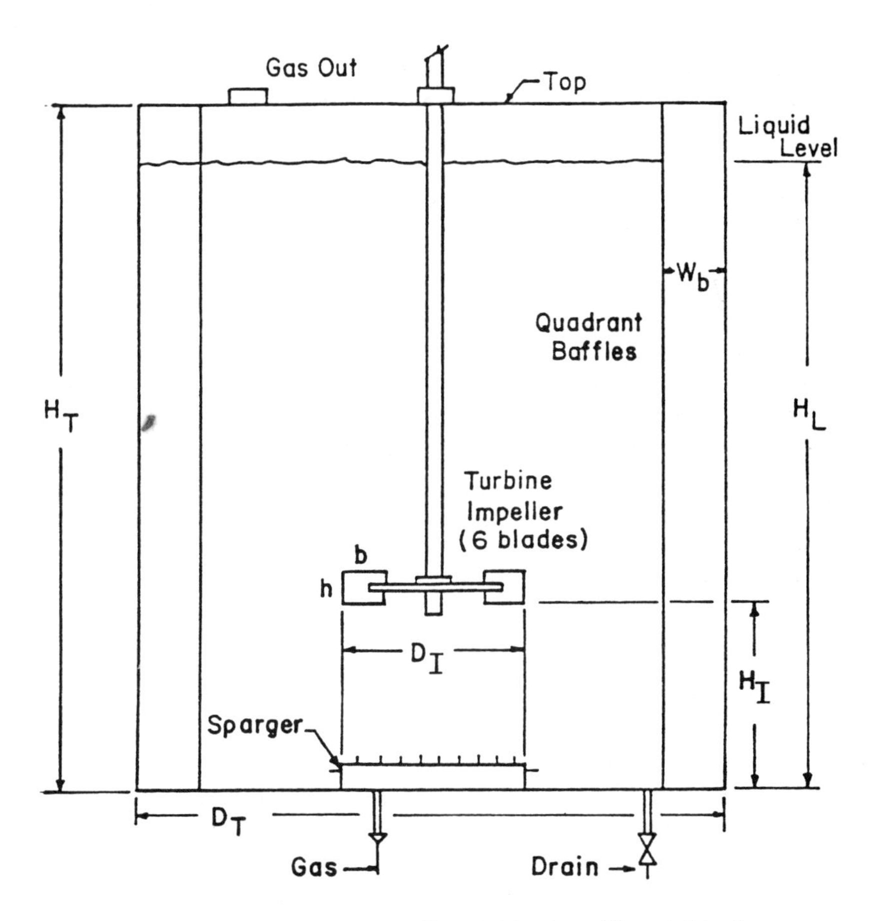

Dimensions: $D_T = 11.5$ inches; $D_I = 4.0$ inches; $W_b = 1.0$ inch
$H_T = 18.0$ inches; $H_L = 11.5$ inches; $H_I = 4.0$ inches

Sparger Holes: Three 1/16-inch holes per square inch of sparger cross-section.

Impellers: b = 1.0 inch; h = 13/16 inch

Volume: 20 liters

Figure 7. Dimensions of the stirred-tank reactor used by Prengle et al. [68].

Injectors, Turbines and Pumps

Injectors and turbines are used when a high degree of mixing is required. Moderate to high interfacial areas are obtained, and rapid reactions can be effected in small contact times.

Pumps provide a residence time of 1–10 sec. For a higher degree of mixing or sufficient shear to create interfacial area, in-line mixers may be used.

Venturi reactors are used to draw ozonated gases into liquids. The main advantages are ease of installation and low pressure drop. Plugging can create problems.

Tubular Reactors

Tubular reactors are constructed of either a single continuous tube (pipe) or several tubes operated in parallel. Most applications use cocurrent flow through the tubes to obtain plugflow. The profile of a plugflow reactor assumes complete mixing in the radial direction, but allows for no diffusion in the flow direction. This results in the velocity, temperature and concentration profiles being flat over any cross-sectional area taken perpendicular to the flow. Only the composition varies along the flow path. For a chemical reaction rate–limited regime in which mass transfer effects are insignificant, design equations for plugflow reactors show that they require the minimum theoretical value to obtain a desired degree of chemical conversion.

Tubular reactors are very difficult to design for gas-liquid reactions because the gas-liquid flow patterns are very complicated to define. Cichy et al. [53] and Rase [56] present discussions of the 25 possible flow configurations. Some success has been achieved using Baker and Govier charts to predict flow patterns [53].

Gas-liquid tubular reactors are used primarily because of their low cost, ease of expansion and excellent heat transfer characteristics. Turbulent flow usually is employed in horizontal tubular reactors to promote mixing and give plugflow conditions. In-line static mixers are becoming popular to redistribute the gas into the liquid in such an environment. These are stationary structures placed inside the tube, which force the cocurrently flowing gas and liquid to travel torturous paths, thus continuously increasing surface interactions between the two fluids. Since these inserts are stationary in the tube, the degree of mixing they provide and mass transfer enhancement thus are affected by the flowrates of both fluids through the tube. If the rate of liquid flow is very slow, the amount of mass transfer achieved approaches that obtainable in the tubular contactor operated with both phases flowing cocurrently. Richards et al. [71] studied the mass transfer of ozone into water using static mixers and reported improved transfer over standard tubular reactors.

The most common type of vertical tubular reactors used in the chemical process industries is the wetted-wall column, in which the liquid phase flows downward along the wall. Gas may flow co- or countercurrently in the center of the column. These reactors are excellent for very exothermic chemical reactions, but are limited to instantaneous reactions due to the limited interfacial area. Wetted-wall columns seldom are used for ozone applications, as very few ozone reactions are highly exothermic. This is because the reactions usually are present in small concentrations and are diluted in a solvent.

REFERENCES

1. National Academy of Sciences. *Ozone and Other Photochemical Oxidants* (Washington, DC: Printing and Publishing Office, National Academy of Sciences, 1977).
2. Rice, R. G. "Recent Advances in Ozone Technology," paper presented at the Institute of Water Pollution Control, Durban, South Africa (1975).
3. Sherwood, T. K., P. L. Pigford and C. R. Wilke. *Mass Transfer* (New York: McGraw-Hill Book Co., 1975).
4. Astarita, G. *Mass Transfer with Chemical Reactions* (New York: Elsevier Publishing Co., 1967).
5. Danckwerts, P. V. *Gas-Liquid Reactions* (New York: McGraw-Hill Book Co., 1970).
6. Treybal, R. E. *Mass-Transfer Operations,* 2nd ed. (New York: McGraw-Hill Book Co., 1968).
7. Ghai, R. K., H. Ertl and F. A. L. Dullien. *Am. Inst. Chem. Eng. J.* 19:881 (1973).
8. Wilke, C. R., and C. Y. Lee. *Ind. Eng. Chem.* 47:1253 (1955).
9. Hirschfelder, J. O., R. B. Bird and E. L. Spotz. *Trans. Am. Soc. Mech. Eng.* 71:921 (1949).
10. Wilke, C. R. *Chem. Eng. Prog.* 45:218 (1949).
11. Wilke, C. R., and P. Chang. *Am. Inst. Chem. Eng. J.* 1:264 (1955).
12. Reid, R. C., and T. K. Sherwood. *The Properties of Gases and Liquids,* 2nd. ed. (New York: McGraw-Hill Book Co., 1966).
13. Lewis, W. K. *Ind. Eng. Chem.* 8:825 (1916).
14. Whitman, W. G. *Chem. Met. Eng.* 29:147 (1923).
15. Higbie, R. *Trans. Am. Inst. Chem. Eng.* 31:365 (1935).
16. Danckwerts, P. V. *Ind. Eng. Chem.* 43:1460 (1951).
17. Dobbins, W. E. In: *Biological Treatment of Sewage and Industrial Wastes,* J. McCabe and W. W. Eckenfelder, Eds. (New York: Reinhold Publ., 1955).
18. Toor, H. L., and J. Marchello. *Am. Inst. Chem. Eng. J.* 4:97 (1958).
19. Alder, M. G., and G. R. Hill. *J. Am. Chem. Soc.* 72:1884 (1950).

20. Czapski, G., A. Samuni and R. Yelin. *Israel J. Chem.* 6:969 (1968).
21. Hewes, C. G., and R. R. Davison. *Am. Inst. Chem. Eng. J.* 17:141 (1971).
22. Kilpatrick, M. L., C. C. Herrick and M. Kilpatrick. *J. Am. Chem. Soc.* 78:1784 (1956).
23. Weiss, J. *Trans. Faraday Soc.* 31:668 (1935).
24. Kuo, C. H., K. Y. Li, C. P. Wen and J. L. Weeks. *Am. Inst. Chem. Eng. Symp. Ser.* 73(166):230 (1977).
25. Kuo, C. H., and C. P. Wen. *Am. Inst. Chem. Eng. Symp. Ser.* 73(166): 272 (1977).
26. Hatta, S., *Technol. Repts., Tohoku Imp. Univ.* 8:1 (1928–1929).
27. Hatta, S., *Technol. Repts., Tohoku Imp. Univ.* 10:119 (1932).
28. Huang, C. J., and C. H. Kuo. *Am. Inst. Chem. Eng. J.* 9:161 (1963).
29. Van Krevelen, D. W., and P. J. Hoftijzer. *Chem. Eng. Prog.* 44:529 (1948).
30. Peaceman, D. W. "Liquid-Side Resistance in Gas Absorption with and without Chemical Reaction," ScD Thesis, Massachusetts Institute of Technology (1951).
31. Kuo, C. H., and C. J. Huang. *Am. Inst. Chem. Eng. J.* 16:493 (1970).
32. Kuo, C. H., and C. J. Huang. *Chem. Eng. J.* 5:43 (1973).
33. Bridgwater, J. *Chem. Eng. Sci.* 22:185 (1967).
34. Van de Vusse, J. G. *Chem. Eng. Sci.* 21:645 (1966).
35. Kuo, C. H. *Am. Inst. Chem. Eng. J.* 18:644 (1972).
36. Brian, P. L. T., and M. C. Beaverstock. *Chem. Eng. Sci.* 20:47 (1965).
37. Onda, K., E. Sada, T. Kabayashi and M. Fujine. *Chem. Eng. Sci.* 27:247 (1972).
38. Li, K. Y., C. H. Kuo and J. L. Weeks. *Can. J. Chem. Eng.* 52:569 (1974).
39. Li, K. Y. and C. H. Kuo. In: *Water—1979, Am. Inst. Chem. Eng. Symp. Ser.* (1980).
40. Li, K. Y., C. H. Kuo and J. L. Weeks. *Am. Inst. Chem. Eng. J.* 25:583 (1979).
41. Yocum, F. H., "Oxidation of Styrene with Ozone in Aqueous Solution," DE Thesis, Tulane University (1977).
42. Yocum, F. H., In: *Water—1979, Am. Inst. Chem. Eng. Symp. Ser.* (1980).
43. Rizzuti, L., V. Augugliaro and G. Marrucci. *Chem. Eng. Sci.* 31:877 (1976).
44. Hoigné, J., and H. Bader. "Ozonation of Water: Selectivity and Rate of Oxidation of Solutes," *Ozone Sci. Eng.* 1(1):73–85 (1979).
45. Hoigné, J. "Mechanisms, Rates and Selectivities of Oxidations of Organic Compounds Initiated by Ozonation of Water," Chapter 12, this volume.
46. Eisenhauer, H. R. *J. Water Poll. Control Fed.* 43:200 (1971).
47. Gould, J. P., and W. J. Weber. *J. Water Poll. Control Fed.* 48:47 (1976).
48. Rizzuti, L., V. Augugliaro and G. Marrucci. *Chem. Eng. J.* 13:219 (1977).
49. Augugliaro, V., and L. Rizzuti. *Chem. Eng. Commun.* 2:219 (1978).
50. Masschelein, W. J. "Contacting of Ozone with Water and Contactor Off-gas Treatment," Chapter 6, this volume.
51. Barona, N., and H. W. Prengle. *Hydrocarbon Proc.* 52:63 (1973).
52. Prengle, H. W., and N. Barona. *Hydrocarbon Proc.* 49:65 (1970).
53. Cichy, P. T., J. S. Ultman and T. W. F. Russell. *Ind. Eng. Chem.* 61:7 (1969).
54. Levenspiel, O. *Chemical Reaction Engineering* (New York: John Wiley & Sons, Inc., 1967).

55. Smith, J. M. *Chemical Engineering Kinetics* 2nd ed. (New York: McGraw Hill Book Co., 1970).
56. Rase, H. F. *Chemical Reactor Design for Process Plants, Vol. 1* (New York: John Wiley & Sons, Inc., 1979).
57. Perry, R. H., and C. H. Chilton, Eds. *Chemical Engineer's Handbook* 5th ed. (New York: McGraw-Hill Book Co., 1973).
58. Stahl, D. E. "Ozone Contracting Systems," in *Proceedings of the First International Symposium on Ozone Technology* (Vienna, VA: International Ozone Association, 1975), p. 40.
59. Nebel, C., P. C. Unangst and R. D. Gottschling. "An Evaluation of Various Mixing Devices for Dispersing Ozone in Water," *Water Sew. Works* (1973).
60. Optaken, E. J. Paper presented at the American Institute of Chemical Engineering 86th National Meeting Houston, 1979.
61. Chen, J. W., C. Hui, T. Keller and G. Smith. *Am. Inst. Chem. Eng. Symp. Ser.* 73(166):206 (1977).
62. Hill, A. G., J. B. Howell and H. K. Huckabay. "Reaction of Ozone with Trace Organics in a Pressurized Bubble Column," *Am. Inst. Chem. Eng. Symp. Ser.* (1980).
63. Miller, G. W., R. G. Rice, C. M. Robson, R. L. Scullin, W. Kühn and H. Wolf. "An Assessment of Ozone and Chlorine Dioxide Technologies for Treatment of Municipal Water Supplies," U.S. EPA Report EPA-600/2-78-147 (Washington, DC: U.S. Government Printing Office, 1978).
64. Mathieu, G. I. "Application of Film Layer Purifying Chamber Process to Cyanide Destruction—A Progress Report," in *Proceedings of the First International Symposium on Ozone for Water and Wastewater Treatment,* R. G. Rice and M. E. Browning, Eds. (Vienna, VA: International Ozone Association, 1975), pp. 533–550.
65. Yocum, F. H., J. H. Mayes and W. A. Myers. *Am. Inst. Chem. Eng. Ser.* 74(178):217 (1978).
66. Chen, J. W., C. Hui and G. V. Smith. "Oxidations of Wastewater with Ozone/Catalyst and Air/Catalyst/Ultrasound Systems," in *Proceedings of the First International Symposium on Ozone for Water and Wastewater Treatment,* R. G. Rice and M. E. Browning, Eds. (Vienna, VA: International Ozone Association, 1975), pp. 120–131.
67. Westerterp, K. R., L. L. van Dierendonck and J. A. de Kraa. *Chem. Eng. Sci.* 18:157 (1963).
68. Prengle, H. W., Jr., C. G. Hewes III and C. E. Mauk. "Oxidation of Refractory Materials by Ozone with Ultraviolet Radiation," in *Proceedings of the Second International Symposium on Ozone Technology,* R. G. Rice, P. Pichet and M.-A. Vincent, Eds. (Vienna, VA: International Ozone Association, 1976), pp. 224–252.
69. Cooper, C. M., G. A. Fernstrom and S. A. Miller. *Ind. Eng. Chem.* 36:504 (1944).
70. Kawecki, W., T. Reith, J. W. van Heuven and W. J. Beck. *Chem. Eng. Sci.* 22:1519 (1967).
71. Richards, D. A., M. Fleischman and L. P. Ebersold. *Am. Inst. Chem. Eng. Symp. Ser.* 73(166):213 (1977).

CHAPTER 6

CONTACTING OF OZONE WITH WATER AND CONTACTOR OFFGAS TREATMENT

W. J. Masschelein

Brussels Intercommunal Water Board (CIBE)
Brussels, Belgium

Using ozone in water treatment involves a certain number of fundamental economic and engineering problems.

In practical use, ozone is not obtained as a pure gas. Therefore, the theoretical evaluation of its solubility depends on mixtures, in which the ozone concentration is often minor in comparison to that of the other constituents of the process gas.

Moreover, ozone is a labile, highly reactive, decomposing compound, which makes its use in experimental investigations outside of the laboratory difficult and even questionable, e.g., pilot-plant or full-scale uses. A most important point is the correct comparison of the flow at the input with that at the exit of the contacting system, in which appropriate corrections for the partial pressures of the gases must be made.

Dissolution of ozone is said to occur by gas transfer in an exchange process described by the double-film theory. Several fundamental aspects of this theory, having a practical incidence on the methods for transfer, are related here. A question for further development is the duality (or perhaps the complementary aspect of the action) of dissolved residual ozone and the concentrated ozonated bubble reaction mechanism. The former corresponds to the batch or plugflow kinetic model, while the latter is a "fully mixed reactor" concept. Further research on the subject is, however, necessary.

According to present evidence, the fast reaction of suspended matter and even bactericidal action could take place favorably in the mixed reactor, while the slower reaction with more resistant compounds requires dissolved ozone (batch-type reaction). The implications of this duality on the selection, operation and costs of the different contacting systems will be discussed on a comparative basis.

An important detail in establishing the mass balance is that of the correct measure of the gas at the inlet and outlet of the contacting system. During good practice the ozone concentration is accurately measured and the quantity of ozone passing within a given time is deduced as being the product of: ozone concentration × gas flow. Hence, this also requires accurate measurement of the gas flow.

The best method for comparison of the results is to use a standardized volumetric gas meter and to integrate during a measuring period. The most usual units are in normal cubic meters (Nm^3), that is, m^3 gas at 0°C and 1 atm pressure (1.02 bar).

In the case of the use of flowmeters, giving instant readings, appropriate corrections must be made as follows:

$$Q(\text{normalized}) = Q(\text{read}) \times F_P \times F_T \times F_D$$

where F_P = pressure correction factor (P absolute/1.02)$^{1/2}$
F_T = temperature correction (°K/273)$^{1/2}$
F_D = density correction (specific weight of the calibrating gas/ specific weight of the process gas)$^{1/2}$

To apply the correction of the pressure factor to the read value, the gas pressure must be measured at the flowmeter itself without the presence of any hindrance such as regulating valves, releases, automatic valves, restrictors, stoppers, etc. According to our observations, this precaution often may have been overlooked in practice. Special attention has been given to this question in comparison with experimental results presented here. Although at first sight it appears to be a technical question, it is thought that the correct flow measure in ozone contacting practice is of sufficient importance to lay stress on the point here in the introduction of this chapter.

The economic aspects of ozone contacting are of great importance, as ozone is produced through processes having high losses in energy. An appropriate cost-benefit appreciation of the contacting system requires thorough consideration of:

1. practical yield of dissolution;
2. efficiency of the contactor in terms of reaction with solutes and/or suspended matter;

3. energy necessary to operate the contacting or dispersing system;
4. maintenance inherent in each different contactor; and
5. costs of treatment of offgases.

Problems relating to the engineering aspects are deduced from former experience. In the choice of dimensions, concept and design, the objectives sought play an important role: batch-type reactivity or concentrated ozone bubble contact. In considering optimum dissolution through contacting, the packed tower or plate contactors may be preferred from a theoretical point of view. In this case, the theoretical yield of dissolved ozone is higher than that of bubbling contacting. However, in water treatment practice, attainment of the saturation yield of dissolved ozone (as a residual concentration value) is not only never reached, but is not really necessary. Therefore the practical dissolution equipment preferably must be selected on the basis of proven performance rather than on theoretical yields of dissolution of the pure gas.

The aim of this chapter is to cast a critical eye on the different parameters of ozone contacting systems so as to provide a basis for:

1. guidelines for further research on the subject;
2. criteria for selecting the most appropriate equipment in any particular water quality circumstance; and
3. stimulation to achieve still further progress in this field in which several "shadowy" points still remain regarding ozone reactivity and actual yield of dissolution attainable with existing contacting systems.

SOLUBILITY OF OZONE IN WATER

The driving force for the dissolution of a gas, e.g., ozone in water, is expressed by the difference between the actual concentration of the dissolved gas (C_L) and the saturation concentration (C_S) at the temperature and pressure of the experiment of application.

The dissolved concentration at the saturation of a gas in a liquid, C_S, has been expressed either in terms of the coefficient of solubility or solubility ratio S (in German, "die Löslichkeit"), or as an absorption coefficient β. The former is defined by the volume of gas dissolved per unit volume of liquid at the temperature and pressure under consideration and in the presence of an equilibrating quantity of the gas at 1 atm. The latter, often called the Bunsen absorption coefficient, is the volume of gas, expressed at normal conditions of temperature and pressure, dissolved at equilibrium by a unit volume of liquid at a given temperature when the partial pressure of the gas is the unit atmosphere. The latter condition is equal to the pressure of the gas, that is, the pressure of the gas itself, minus the vapor tension of the liquid.

The ideal expressions of the saturation concentration of dissolved gas are related as follows:

$$C_S\ (kg/m^3) = \beta M\ (kg\ gas/m^3\ gas)\, P\gamma$$

where M = mass volume of the gas
$P\gamma$ = its partial pressure in the given gas phase as a whole

For ozone, M is 2.14, and for oxygen it is 1.43.

At a pressure of 1.02 kg/cm^2 (760 mm Hg column) of pure gas at 0°C, the absorption coefficients for ozone and oxygen are 0.65 and 0.049, respectively. This corresponds to the statement that pure ozone would be about 13.3 times more soluble in water than is pure oxygen.

Expressed on this physicochemical scale, the relative absorption coefficients at 1 atm and 0°C are established as shown in Table I for different pure gases possibly involved during ozone contacting.

It is generally agreed in the literature that when ozone is dissolved in water, Henry's law is obeyed. This means that we should consider the C_S values as being proportional to the partial pressure $P\gamma$ of ozone when these values are at a given temperature (Figure 1). Hence, the more concentrated the ozone in the process gas contacted with the water, and/or the higher the total pressure of the gases with a given ozone content, the higher becomes the mass of dissolved gas at dissolution equilibrium.

Physical parameters other than the partial pressure also can have an impact on the solubility of ozone in water. The most important is the temperature of the liquid phase. The value of the Bunsen coefficient (m^3 gas/m^3 water) varies as a function of the temperature where ozone, oxygen and nitrogen are concerned, but also of that of the air mixture, as indicated in Figure 2 [1–5].

Even though the solubility of ozone in water is much higher than that of oxygen, concentrations above 1 g O_3/m^3 water are rarely necessary,

Table I. Solubilities of Gases Associated with Ozonation

Gas Solubility	Ozone	Oxygen	Nitrogen	Carbon Dioxide	Chlorine	Chlorine Dioxide
β (vol/vol)	0.65	0.049	0.0235	1.71	4.54	≃60
$\beta\ O_3/\beta$ gas	1	13.3	27.7	0.38	0.14	0.01
C_S (kg/m^3)	1.4	0.07	0.03	3.36	14.4	180

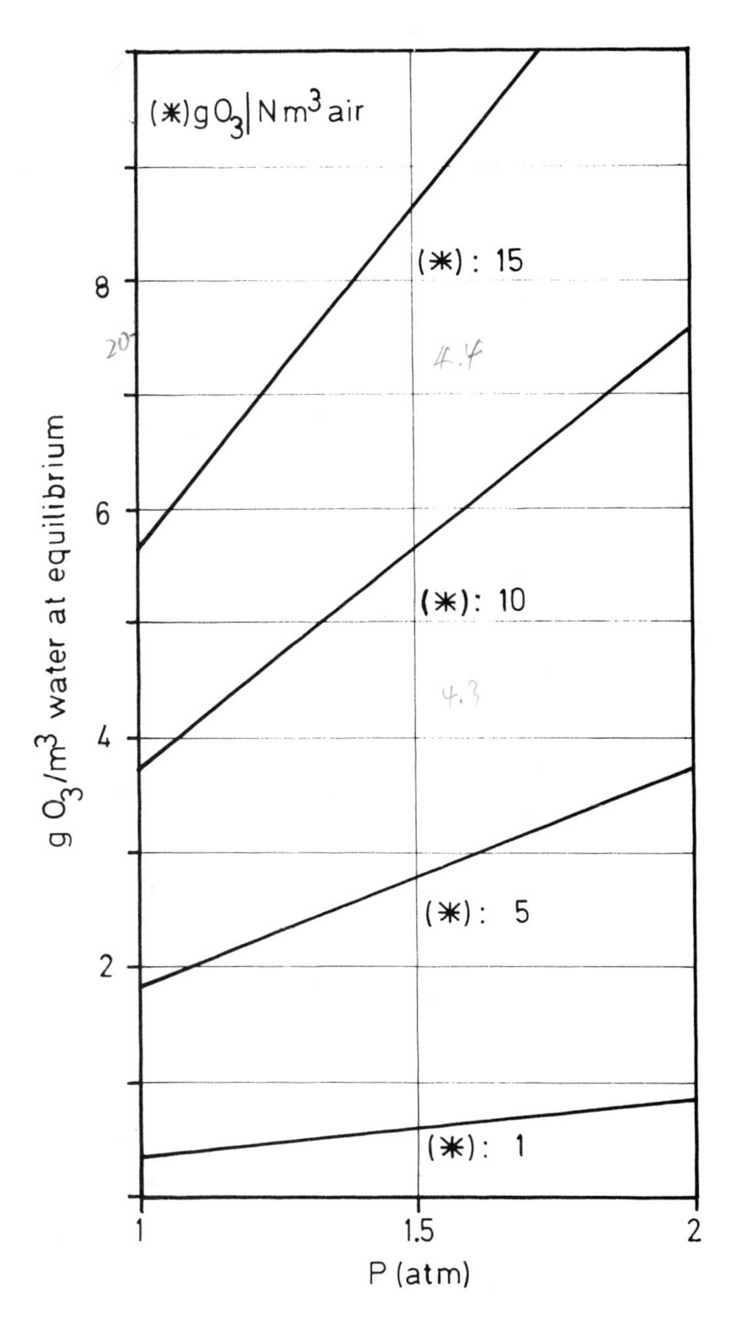

Figure 1. C_S of dissolved ozone at 10°C as a function of a partial concentration of ozone in the equilibrating gas phase.

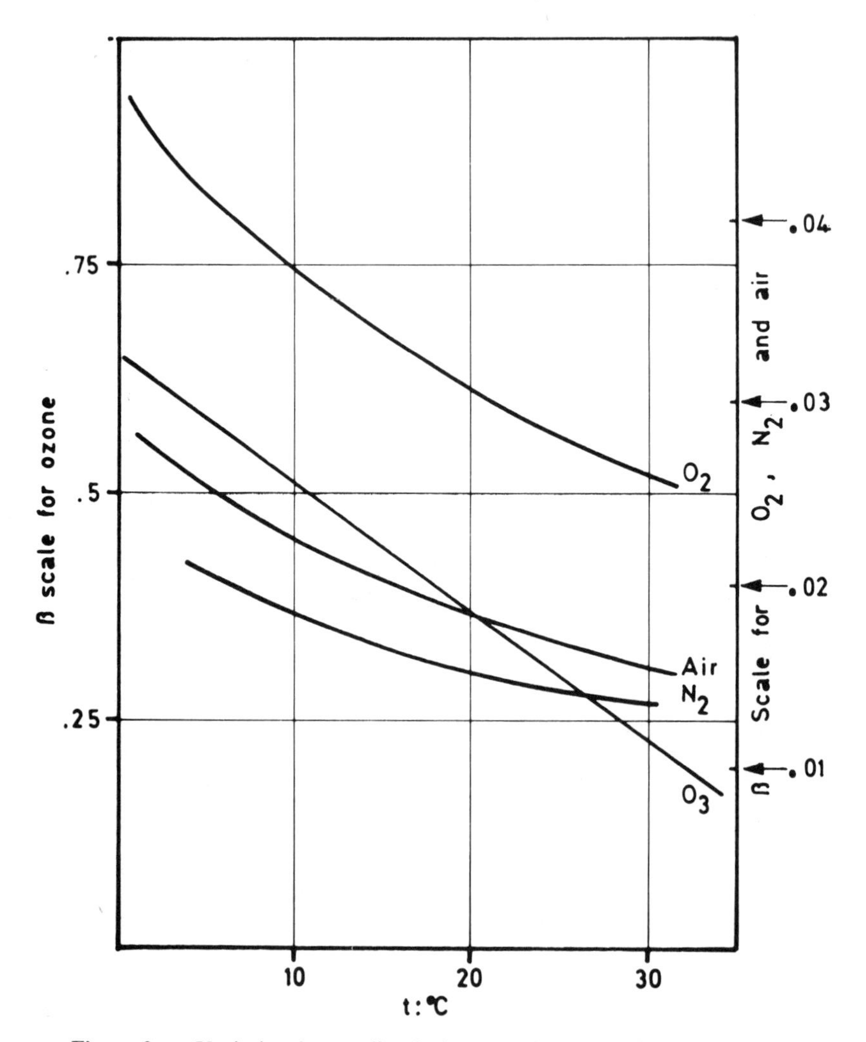

Figure 2. Variation in gas dissolution as a function of temperature.

and 2 g O_3/m^3 (dissolved residual) is rarely obtained. The reason is due to the fact that at present only dilute ozone concentrations are available for the process gas.

Rawson, one of many investigators of the solubility of ozone in water, states that the only general agreement we may come to is that Henry's law is obeyed, at least in the concentration range of 1–20 g O_3/Nm^3 gas and with partial pressures of up to 1% in air [5].

As the partial pressure of ozone in the process gas is low, the dissolved concentration at equilibrium with the gas phase remains low. Although controversy still may exist on the subject, the best available data for the equilibrium obtainable in practice are illustrated in Figure 3 [5]. In fact, the experimental data usually are reported in dynamic conditions, in which a process gas is dispersed or bubbled into a static or flowing liquid.

The correct measure of the incoming and outgoing flow is often questionable in reported experiments. Besides physical dissolution, several side effects may affect the practical equilibrium (e.g., the pH of the liquid phase affects both the solubility ratio and the decomposition rates of dissolved ozone; or the presence of catalysts or reacting compounds) hence the data of Figure 3, although generally accepted, still must be considered as tributary to particular conditions.

At any given ozone concentration (g/Nm^3) in the process gas, the pressure of the ozone equilibrated with the liquid phase increases with the increase of total pressure of the gas phase. For practical application, the approximation of an ideal gas-phase behavior is sufficient. Hence, if C is

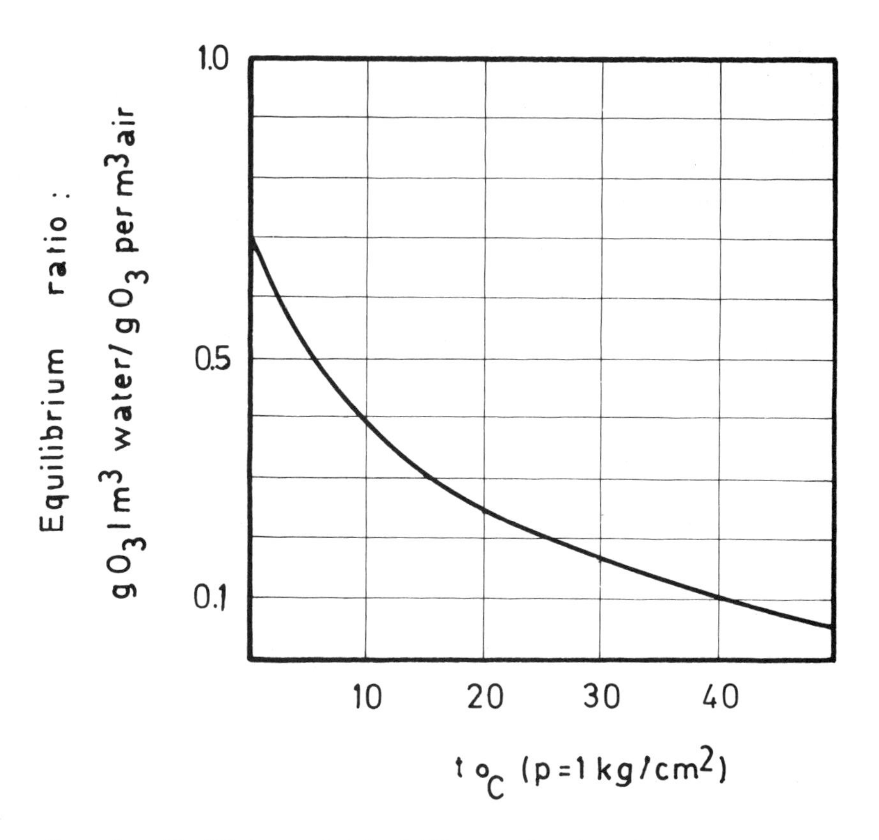

Figure 3. Equilibrium ratio of solubility as a function of water temperature.

the ozone concentration in the process gas at 1 bar absolute and P the total pressure of the gas phase, the working concentration of the ozone which can be equilibrated with the liquid phase is PC. In other words, an increase in pressure favors the dissolution capacity of a water.

Ozone in the gas phase is scarcely compressible at values higher than 2 bars under conditions compatible with the practical resistance of equipment used in large-scale applications. The technology most often encountered limits the total pressure of the ozone-containing process gas to 1.5 bars absolute. Indirect pressurizing, that is, change in pressure of the liquid phase into which the gas phase is dispersed, frequently is inherent to dispersing equipment.

When air is used as the process gas in ozonation, one obtains oxygen saturation of the water at equilibrium equivalent to gas mixtures containing 20.94 vol % O_2 at the given water temperature.

By dissolution of the ozone (and eventually of the oxygen and other minor accompanying gases such as oxides of nitrogen), the exhaust air is slightly enriched in nitrogen (0.5–1 vol %) as compared to the standard air composition. Therefore, the ozonated water probably is slightly oversaturated in dissolved nitrogen in comparison to the equilibrium solubility versus air. Progressive enrichment in nitrogen is one of the problems associated with the recycling of process gases.

This phenomenon is enhanced when oxygen or oxygen-enriched air is used as process gases. In these cases, the water is always significantly oversaturated with oxygen and the escaping gas is enriched with nitrogen. The equilibria for the dissolution of oxygen and nitrogen in moderately saline water, e.g., 1 kg dissolved solids per cubic meter, are well known in the literature [1–4].

Dissolved concentrations at equilibrium, at a gas pressure of 1 kg/m^2 are given in Figures 4 and 5 as a function of the water temperature. Here again Henry's law is followed and the correction for total pressure (if different from 1 kg/cm^2) applies linearly for values below 2 kg/cm^2. As established earlier [6,7], the practical yield of nitrogen stripping and oxygen dissolution must be considered systematically in any situation. For ozone contacting with a gas-water circulating turbine as used at the Tailfer plant [6], the yields are given in Figures 6 and 7.

When working at higher pressures, e.g., in packed columns, the solubility ratios of both oxygen and nitrogen increase linearly with the pressure up to 10 bar. By this technique, part of the purge of the exhaust gas for elimination of nitrogen can be avoided in the recycling gas. The condition necessary for this to be economically feasible is that the "extra dissolved oxygen" inherent to the pressure technique must be effective and useful in the treatment. This is the case in wastewater purification and is

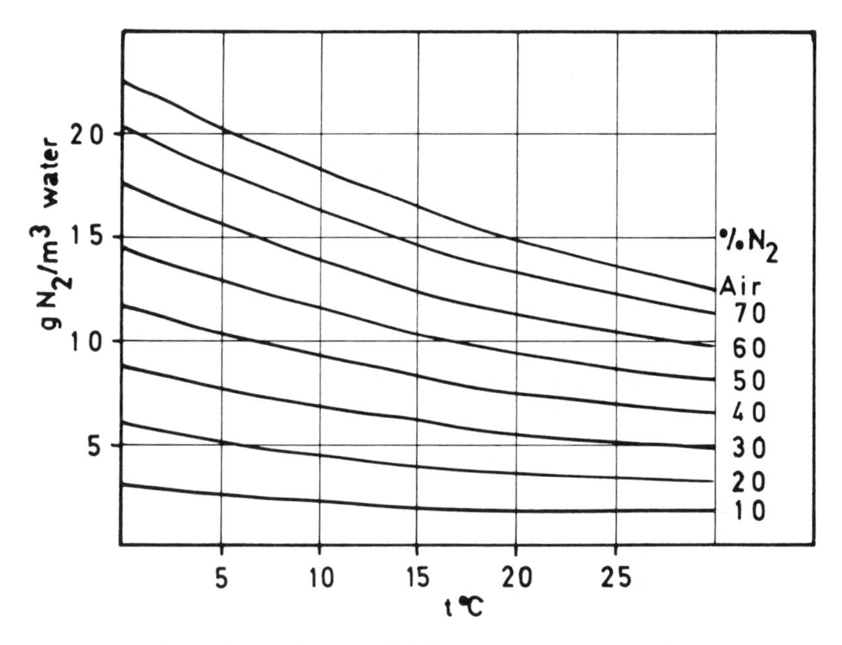

Figure 4. Nitrogen solubility in water in 1 kg/cm².

the basis of the Hoechst method (Federal Republic of Germany) for using oxygen-enriched process gas in the ozonation step. Besides the ozonation itself, the economics of the entire technique are carried out by selecting the relation between working pressure of the packed contacting columns, the N_2/O_2 ratio of the process gas, and the oxygen demand of the water. A similar approach is described in the application at Duisburg, Federal Republic of Germany [8].

Economic considerations resulting from the oxygen and nitrogen solubility rates, when using an oxygen-enriched process gas for the ozonation, often make it seem attractive to dissolve the ozone and accompanying gases in a substream of the water. This can be part of the water to be treated or an auxiliary flow of clear water. Except in the case of heavily charged wastewater, this practice is perhaps questionable, as will be discussed later on.

As a general rule, in the practice of ozonation, the volume of process gas (expressed in Nm^3) does not exceed 20% of the water volume contacted in one injection stage. In ozonation as a finishing step of drinking water treatment, the volume ratio Nm^3 gas/m^3 water is usually about 1/10. Therefore ozone dispersing and contacting techniques often have negligible effects on gas stripping, therefore requiring higher gas-to-

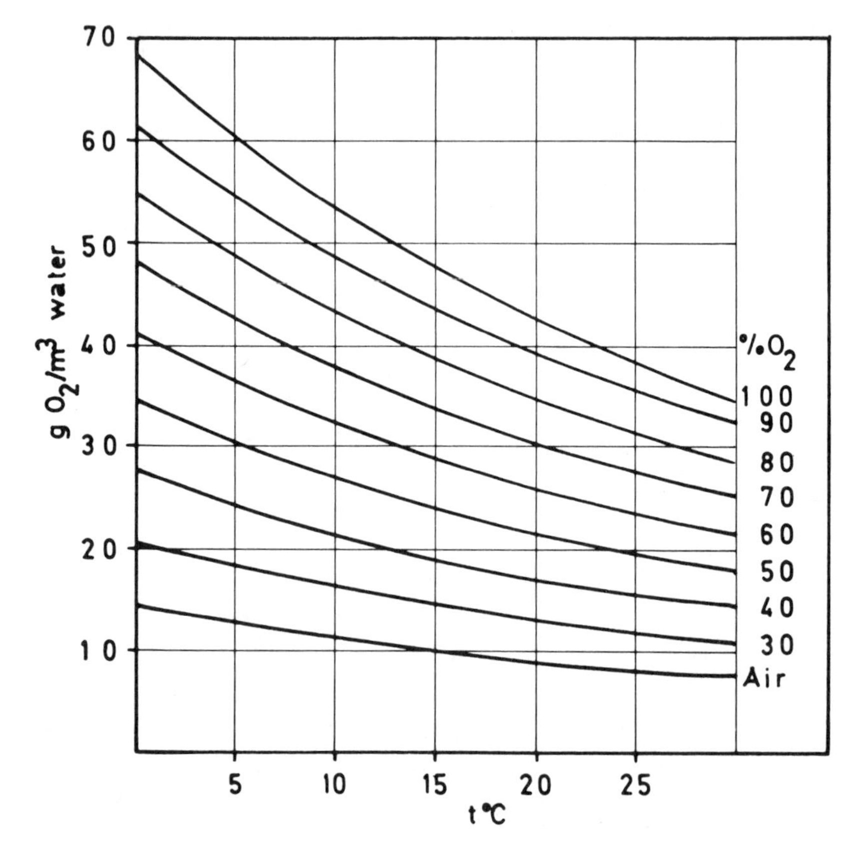

Figure 5. Oxygen solubility in water at 1 kg/cm^2.

liquid ratios. Hydrocarbons, if present, remain far below the low explosion level concentrations of the offgas. Chlorine, present as free chlorine at concentrations below 1 mg/l, can give rise to chlorine traces in the offgases, e.g., <0.05 ppm vol. At these levels the risk of toxicity is negligible; however, by accumulation, this gas is capable of poisoning palladium catalysts used for ozone destruction in offgases. At concentrations under 100 mg/m^3 of trihalomethanes (THM), the proportions of air to water used in ozonation scarcely remove 10 mg/m^3 by gas stripping. Moreover, the analysis is questionable owing to the possible effects of ozone during the reactions of formation and degradation of THM. In all contacting and dispersing systems, the offgas is saturated with water vapor according to the equilibrium conditions at the pressure and temperature of the operation.

From the preceding comments, it could appear that for ozone dissolu-

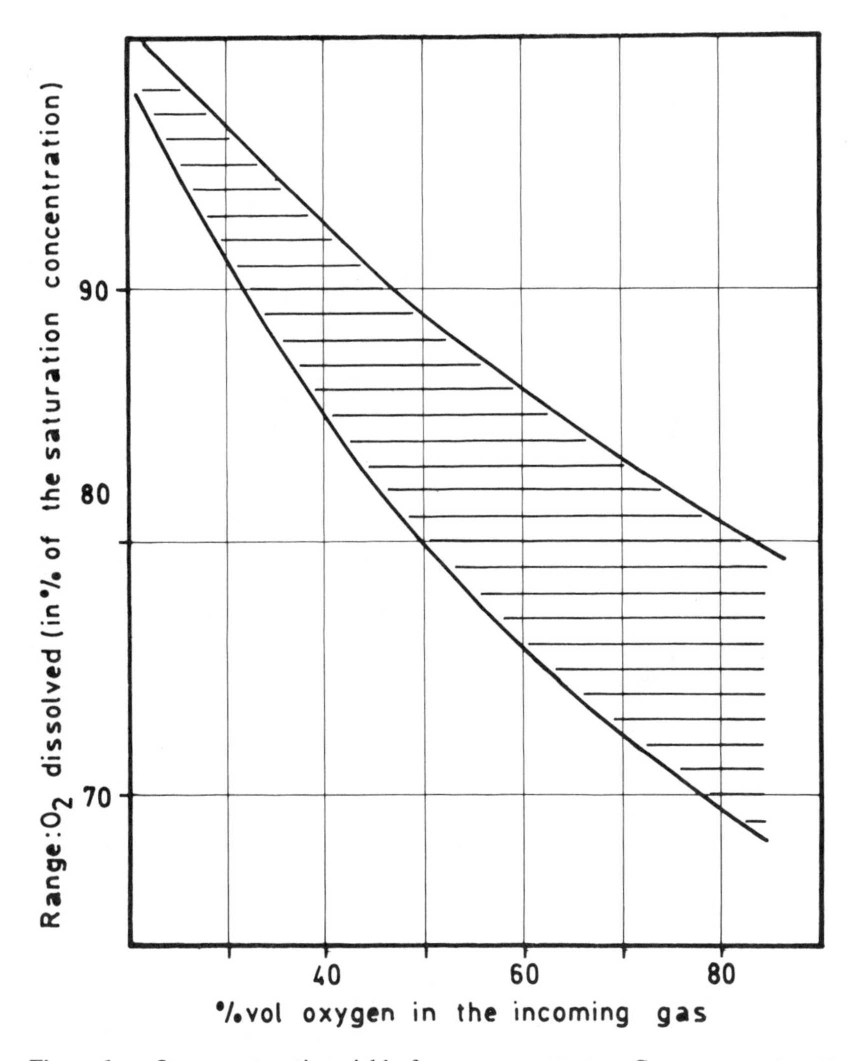

Figure 6. Oxygen saturation yield of an ozone contactor. Gas pressure at vent = atmospheric pressure.

tion at low pressure ($P \leq 0.5$ bar), even at low concentrations of ozone in the process gas, e.g., ≤ 15 g O_3Nm^3 air, the equilibrium C_S values could reach 10 g O_3/m^3 water. Actually, in practice these concentrations are neither attained nor sought when using existing contacting systems. This is due to several factors:

1. kinetics of ozone absorption and losses of ozone in the offgas;

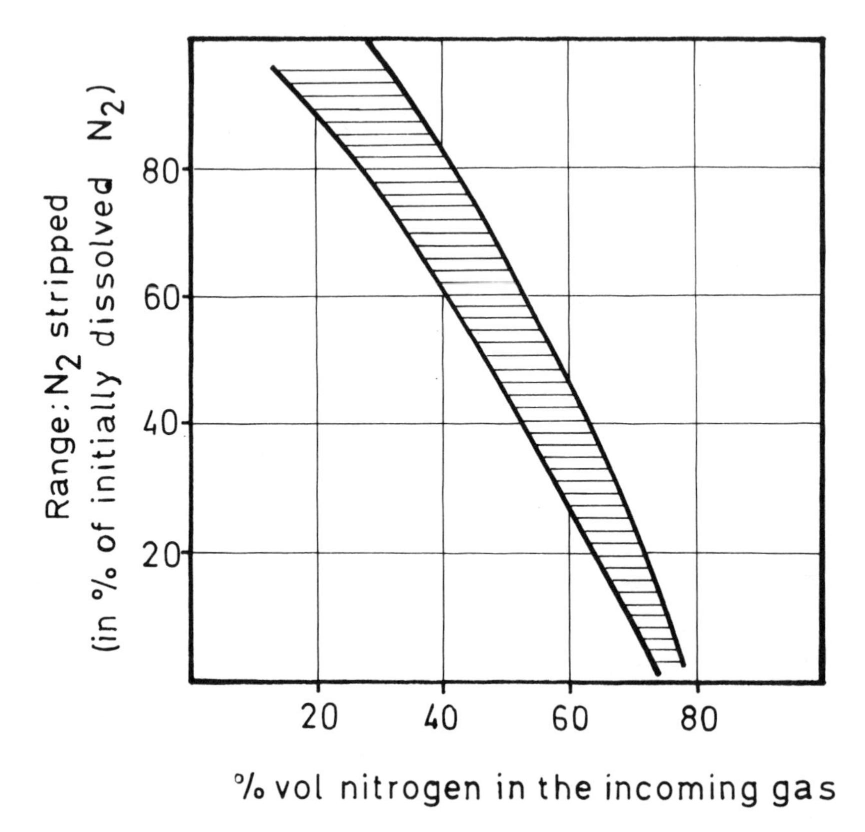

Figure 7. Nitrogen stripping yield of an ozone contactor. Gas pressure at vent = atmospheric pressure.

2. reaction of ozone during or immediately after the dissolving process;
3. decomposition of the dissolved ozone in water;
4. destruction of ozone by the action of the contacting device; and
5. reaction of ozone with substrates by the "concentrated ozone bubble reaction" during contacting.

In this contribution the "practical" yields of ozone contacting systems are always expressed as

$$\frac{100 \times \text{ozone concentration (g/Nm}^3\text{) in the offgas}}{\text{ozone concentration (g/Nm}^3\text{) in the input gas}}$$

In the existing plants with various contacting equipment and variable physical conditions of the process gas and water flow, this yield is the parameter used for ozone contacting control and survey of the security systems.

The measure of the residual concentration of the dissolved ozone after a sufficient reaction time, e.g., ≥ 6 min, is the usual parameter for monitoring the ozone dose.

However, the real chemical mass balance for ozone in the process should be:

$$\begin{aligned}\text{one dose of ozone} &= \text{ozone in the offgas} + \text{residual ozone in water}\\ &\quad + \text{reacted ozone} + \text{decomposed ozone.}\end{aligned}$$

The last two parameters are not often measurable in practical conditions of ozone contacting.

Decomposition of ozone into oxygen or more stable oxygen-containing molecules or radicals is produced by thermal decomposition and ionic or radical reactions.

Decomposition of ozone at residual concentrations in water depends on a series of parameters, the most significant of which is pH, besides temperature and possible presence of catalysts. At pH 7, the half-life of residual ozone can vary from 2 to 30 min in practice. Generally, decomposition is faster in less pure water. Also, the exact form of the kinetic equations may vary, but in general practice, a first-order relation describes the decomposition rate at the initial stage of reaction. Practical examples for the water of the Tailfer plant (Brussels) are given in Figure 8. Other examples are given in Hoigné [9].

The increase in decomposition velocity as a function of pH has been reported on several occasions [10]. Owing to present knowledge, waters encountered in practice having a pH value equal to or higher than 6 exhibit an ozone half-life that decreases linearly with increasing values of pH.

These general ideas are indicated here to support the appreciation of dispersing devices in practical systems. The physicochemical basis of the action of ozone at alkaline pH values may in fact be different from that at neutral pH values, both by the reacting species and the catalytic effects.

DYNAMICS OF OZONE TRANSFER TO WATER

Little experimental investigation has been carried out on the specific dynamics of ozone transfer from gas phase to water. Most of the concepts are based on the generalization of data known for oxygen transfer.

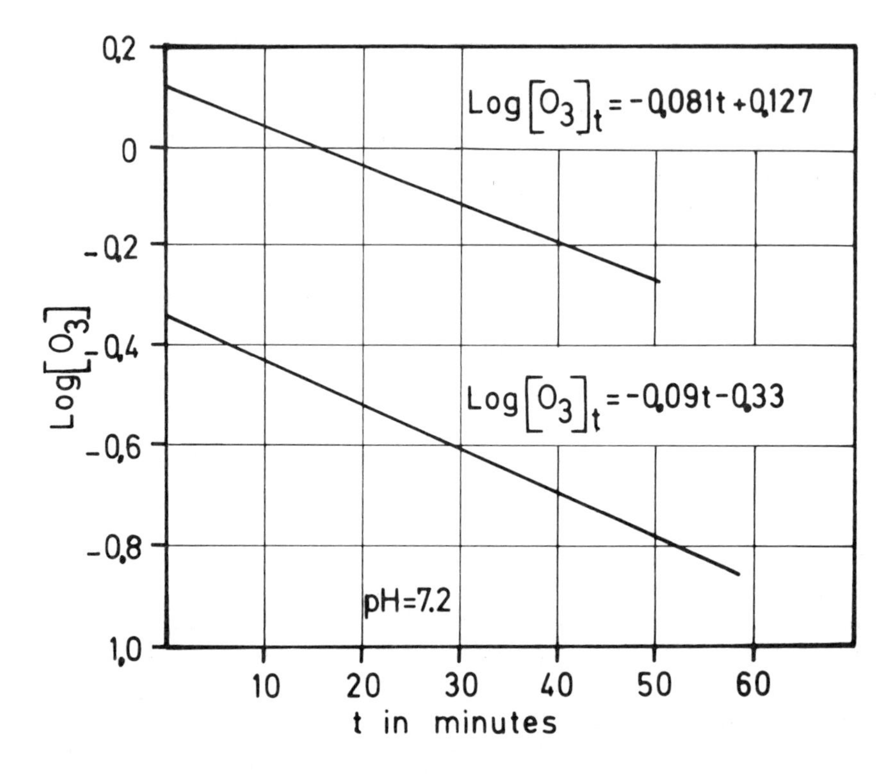

Figure 8. Residual ozone decomposition.

In the following comments, attention is drawn to simplified aspects of theoretical developments, with the intention of quantifying preliminary guidelines.

The diffusion of ozone in water obeys Fick's law for molecular diffusion, which is shown in Figure 9. Correction for the diffusion constant is given by the Nernst-Einstein relationship:

$$\frac{D\mu}{T} = \text{constant}$$

where μ = absolute viscosity
T = absolute temperature

Consequently, ozone transport in the gas phase is much faster than in the liquid phase, hence the latter process is overall transport rate-determining. Moreover, air-bubbling is more generally used than liquid spraying, as this technique is better adapted to transfers where the highest resistance is situated in the gas phase, e.g., ammonia absorption.

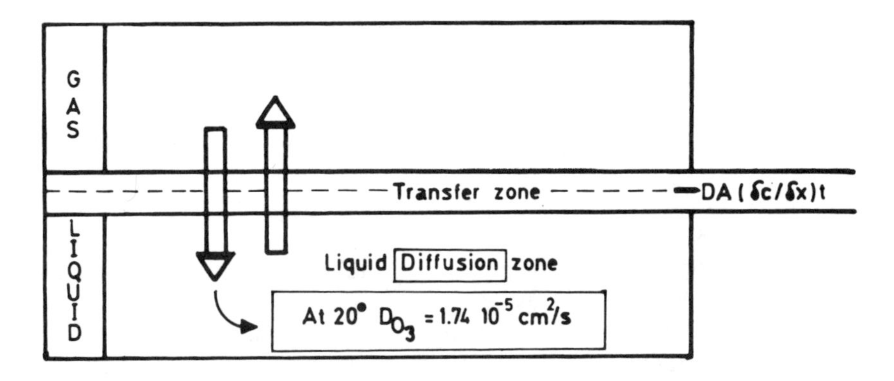

Figure 9. Ozone transfer.

The practical exchange rate between gas and liquid is proportional to the external surface of contact. Hence, to increase this surface, a bubble mechanism is adopted in one of two possibilities: (1) bubbling and/or dispersing the ozone-containing gas into the water, or spraying water into contactors, e.g., packed columns, containing the gas phase, or (2) contact of the liquid as a fine layer with the gaseous atmosphere. The choice of the system must be considered on an economic basis. Therefore, as long as the gas volume to be contacted with the liquid volume remains low (5–10 volumes of gas/100 volumes of liquid), the most widespread systems involve gas dispersing techniques.

From different empirical observations in gas transfer systems, the following rules can be accepted:

1. At any moment, the rate of gas transfer is proportional to the surface-to-volume ratio of the bubbles, except for very small, so-called rigid bubbles.
2. Transfer is favored by increasing the ratios of the water column above a diffuser and/or the relative downflow velocity of the liquid, to the upflow velocity of the gas in countercurrent contacting; the rate of transfer is in linear relation to $1/\sqrt{t}$, where t is the contact time.
3. The rate of transfer is highest during the (very short) period of bubble formation.
4. Mixing and turbulence promote ozone transfer within the limits of Re (Reynolds) ≤ 2000 in flow systems and $G \leq 150$ sec^{-1} for dynamic systems.
5. The driving force for transfer is the difference in concentration of dissolved ozone in the bulk of the liquid and the saturation concentration. Hence the water temperature, which influences the C_S value, is important.
6. Countercurrent dissolution first appears as more favorable than equi-

or cocurrent dissolution. However, practical parameters, such as pressure and relative velocity of the aqueous or gaseous streams, may invert the order. Therefore, existing evidence is conflicting and will be reviewed below [11].

Conditions for Bubble Formation with Gas Transfer

The limiting condition for pressureless bubble formation through gas emerging from the liquid using a capillary model (Figure 10) may be approached by the simple scheme relating the maximum possible radius of the bubble (assumed to be spherical) to the radius of the capillary, as long as $2\pi\,\tau < mg$ (τ = surface tension of the contact surface and mg = weight of liquid displaced by the gas bubble). At the free surface of the water without any counterpressure or vacuum, this condition is:

$$2\pi r(\max)\,72.8 = (\zeta_1 - \zeta_g)\,g\,\frac{4}{3}\,\pi\,r^3(\max)$$

$$r^2(\max) \simeq \frac{3}{2}\,\frac{72.8}{981}\ \mathrm{cm}^2$$

hence $r(\max) \simeq 3.4$ mm.

The largest dimensions of the gas bubbles at the injection point of the gas depend on the hydrostatic pressure in the system at that point (pi) compared to the pressure at the water surface (ps):

$$\frac{pi}{ps} = \frac{vs}{vi}$$

If ps = 1 and dispersing occurs at a depth of 5 m water column, then $1.5 \simeq (3.4)^3/[ri(\max)]^3$

$$ri(\max) = 3.4/(1.5)^{1/3} = \frac{3.4}{1.14} \simeq 3\ \mathrm{mm}$$

These bubble dimensions are limiting on the maximum side. If exceeded, the contacting system for dissolving ozone will not be operated efficiently because bumping and formation of large gas mats or torches will hinder the normal flow conditions and ozone exchange.

In practice, bubbles of much smaller dimensions are used to promote higher exchange rates by increasing the surface-to-volume ratio. A simple method for this purpose is to use gas diffusers with capillary

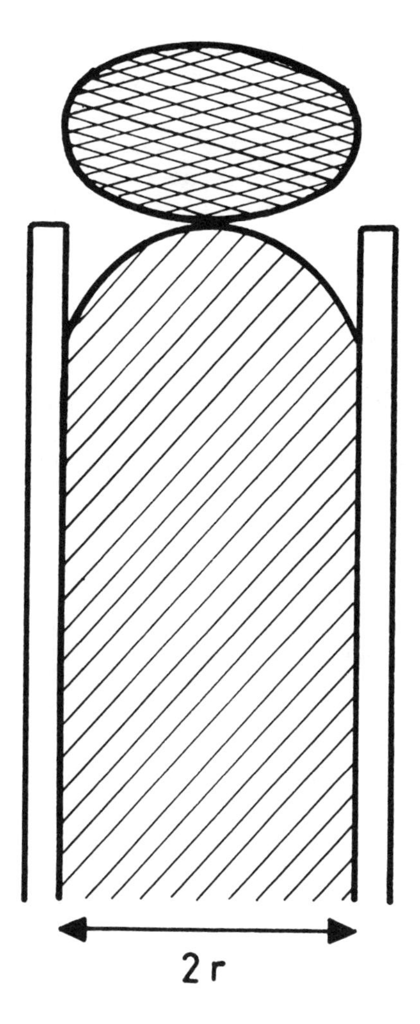

Figure 10. Schematic representation of capillary bubbling.

outlets. Typical pore sizes are smaller than 10^{-4} m, usually in the range of 50 to 100 μ. In fact, the efficiency condition of the pore size and relative pressure in the liquid and the gas must be such as to constitute a bubble volume within the limits of

$$\tfrac{4}{3}\pi\, 1^3 \quad \text{and} \quad \tfrac{4}{3}\pi\, 3^3\ \text{mm}^3$$

at the hydrostatic pressure at the injection point. The same holds for mechanical dispersing systems.

The energy necessary to fulfill these requirements depends on the hydraulic systems used to introduce the gas into the water, and only to a very small extent on the pressure conditions required for bubbling. The inside pressure of a bubble (P_I) is equal to the hydrostatic pressure above the liquid layer plus the weight of the liquid layer plus the surface tension at the gas-liquid interface

$$P_I \simeq P_{atm} + h\zeta g + \frac{2\tau}{r}$$

Consequently, the minimum "overpressure" required to form a bubble is $2\tau/r$ or only about 500–1500 dyn/cm^2, respectively, for bubbles having radii of 3 and 1 mm.

Rising gas bubbles are subjected to frictional forces due to the viscosity of the liquid. For spherical bubbles, these forces are approximated theoretically by Stokes' law:

$$f = -6\pi\eta r \vec{V}$$

The driving force for bubble rise is the Archimedian force:

$$F = (\zeta_l - \zeta_g)\frac{4}{3}\pi r^3 g$$

From these simple approximations one could conclude that when the value of r increases, the net driving force for bubble rise will increase. This conclusion is correct only in a limited range of bubble dimensions, that is, in the case of "rigid surface" bubbles. When the bubble sizes increase at a given relative water-to-bubble-velocity V, the frictional force increases, transforming the external surface of the bubble and the originating internal movement of the fluid inside the rising bubble. The critical radius at which this phenomenon starts is given by the empirical formula:

$$r_c = \frac{\tau}{(\zeta_l - \zeta)g} \simeq \frac{72.8}{981} \simeq 0.08 \text{ cm}$$

The approximate value of $r_c = 1$ mm is consistent with the preceding recommendations. Moreover, "rigid" or "dead" bubbles have a lower gas exchange rate than "moving-surface" or "soft" bubbles.

These frictional forces also counteract the bubble rise, so that during the contacting process, the bubble gradually expands by lowering of the hydrostatic forces, and becomes deformed into flat-shaped globules by the frictional forces and the internal circulation of the fluid. As a result, a macrorotational movement is obtained with secondary precession. These appropriate movements are necessary for renewal of efficient contact surface during gas-liquid exchange. Figure 11 gives a scheme of the movements during bubble rise. The experimental velocity of bubble rise in quiescent water is illustrated in Figure 12.

The bubble rise rate can be diminished in countercurrent contacting by increasing the downflow velocity of the liquid. This technique (called ballasting) is often used in practice, as it enables an increase in bubble-liquid contact time without having to increase investments in contact chambers. The new hydrodynamic conditions act in a very complex way on the different force relationships involved with the bubbles. Therefore, in practical systems, the average downflow velocity of the water should be kept below 20 cm/sec and even preferably below 15 cm/sec. On the other hand, to suck down the gas bubbles with the liquid flow, the velocity of the latter must be higher than 40 cm/sec, preferably above 45 cm/sec. In the intermediate range, instability occurs in the flow, with bubble agglomeration, hydrodynamic bumping phenomena and loss in exchange efficiency.

From these comments it is apparent that hydraulic flow conditions of the liquid to be treated in ozone contacting chambers are of primary importance. Under ideal operating conditions, no short-circuiting should exist and at no time should the systems be operated outside the critical range of velocities for bubble ballasting: 15–45 cm/sec. (The practical range is 20–40 cm/sec.)

Relative Performances of Counter- and Cocurrent Contacting

Since the driving force for ozone transfer is the difference between the saturation and actual dissolved ozone concentrations, the countercurrent process appears preferable as a technique for methodical exhausting of the gas: the less concentrated the gas in contact with the water, the higher the driving force for the dissolution of ozone. Moreover, ballasting (increase in contact time) of the gas bubbles occurs. In cocurrent or upflow injection, the gas is gradually exhausted of ozone, while the driving force diminishes concurrently. However, the initial dissolution is fast and, in subsequent contact, the residual concentration is completed by the slower gradual exchange of the already partially exhausted gas. This enables maintenance of a residual concentration of dissolved ozone which is sometimes even higher than that obtained by countercurrent contacting. Typical experimental data are given in Figure 13.

In fact, this example should only be considered as indicative, because

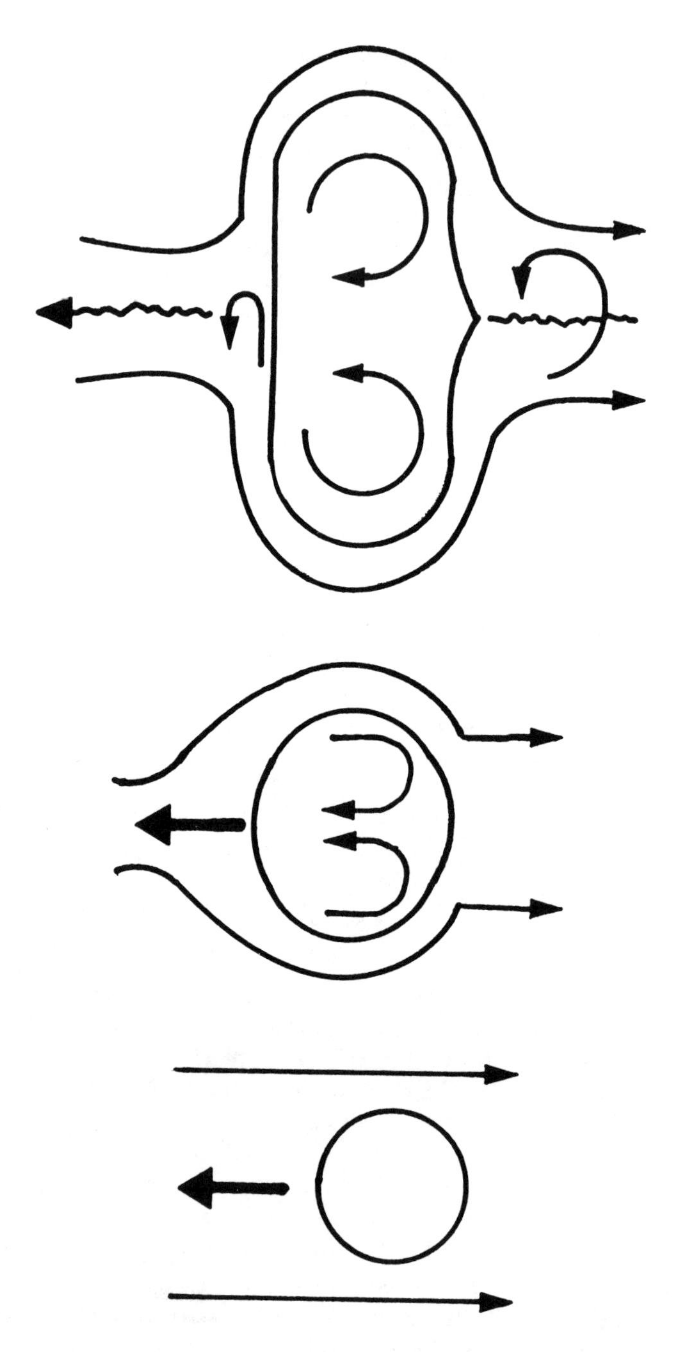

Figure 11. Scheme of movements during bubble rise.

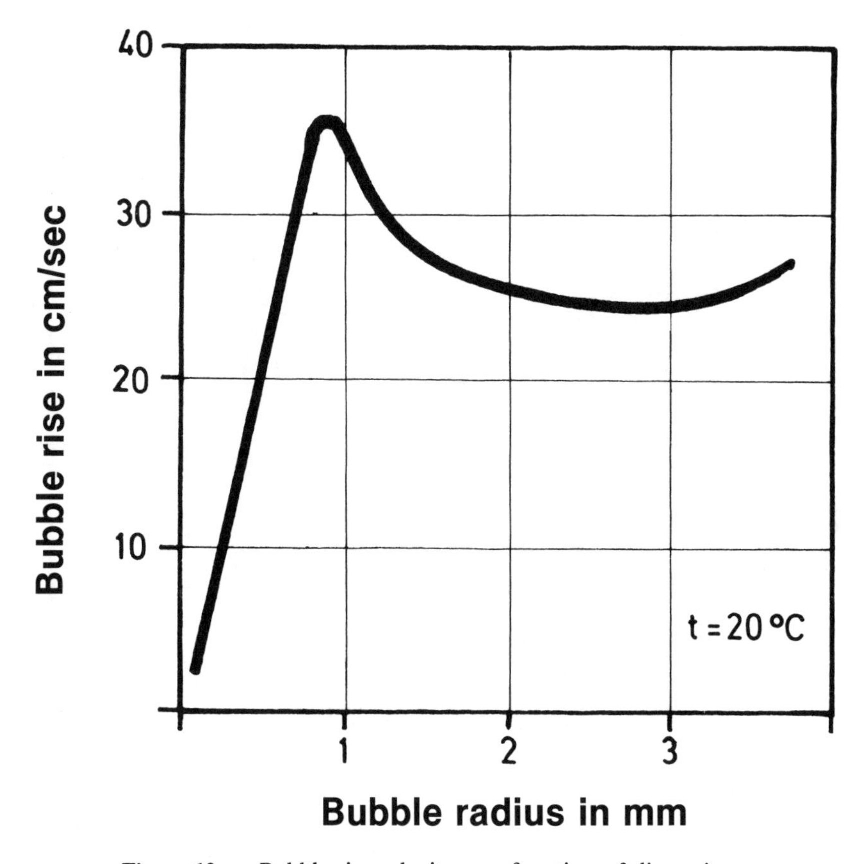

Figure 12. Bubble rise velocity as a function of dimensions.

if the height of the contact column, water velocity or proportion of ozone to water demand change, different results may be obtained. Therefore literature data are often conflicting where the comparison of the advantages of co- and countercurrent contacting are concerned [12,13].

The following conclusions, purely indicative, can be advanced:

1. To obtain the same residual ozone concentration after a prolonged contact time, countercurrent injection is more favorable than cocurrent injection.
2. The instantaneous dissolved ozone concentration reached at low values of water head (1–2 m) is higher in the cocurrent-upflow systems.
3. Both systems can be applied, but if maintenance of a residual dissolved ozone concentration is required, the countercurrent system is preferred, since the ozone-water contact time is increased by ballasting of the bubbles.

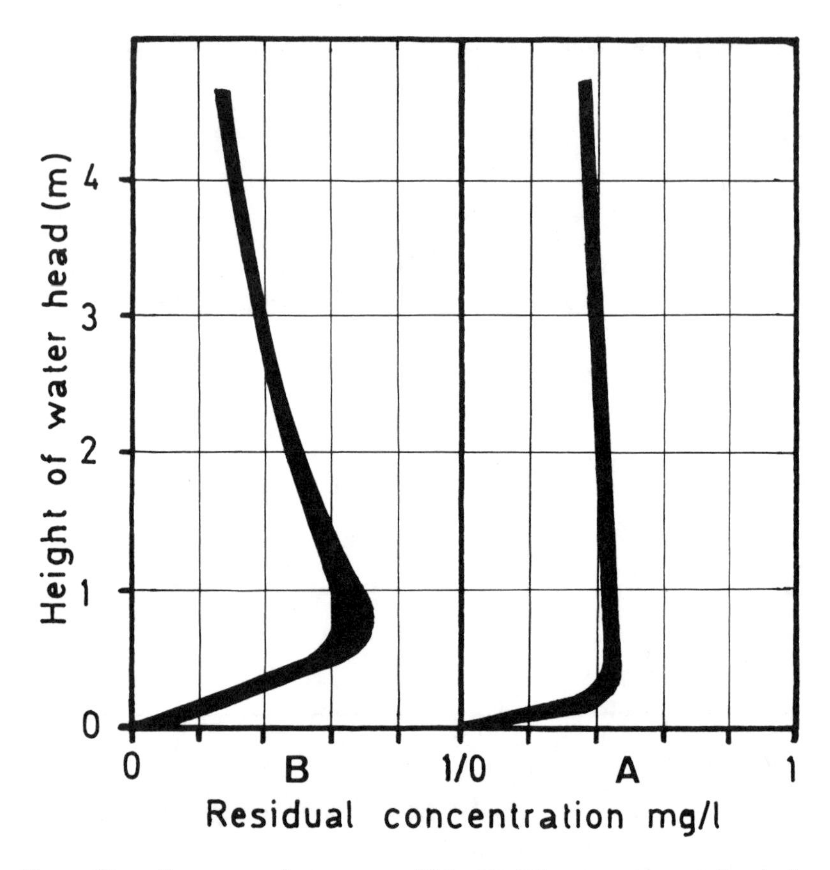

Figure 13. Counter- and cocurrent efficiency. (A) countercurrent dissolution (water flow downstream, upflow of air bubbles); (B) cocurrent dissolution (upstream flow of air and water). Conditions: water flowrate = 15 cm/sec; air/water flow ratio = 10%; ozone concentration in incoming air = 15 g/Nm³; ozone quantity contacted = 1.6 g/m³; water column head = up to 4 m; injection: porous ceramic diffuser (50–100 μm); inlet gas pressure = 0.5 bar; water temperature = 11–13°C.

As a guideline, the minimum height of an ozone-contacting column can be considered to be 2.5 times lower in the countercurrent contacting systems than in cocurrent columns.

Transfer of ozone without any reaction or decomposition occurs as a double-film process. By investigating the liquid phase and expressing the transfer in terms of ozone concentration in the liquid:

$$+\frac{dc}{dt} = +k_L S(C_s - C)$$

Considering $C=0$ at $t=0$ and integrating:

$$\log \frac{C_s - C}{C_s} = k_L St$$

where S is the specific exchange surface in the liquid film, which depends on practical conditions, e.g., agitation, pressure and the ratio of total volumes of gas and liquid.

In the gas phase, the practical concentration of ozone is expressed as a partial pressure ratio:

$$Y = \frac{p(O_3)}{p(T)}$$

Part of Y is transferred to the liquid to form C and another part remains in the gas phase:

$$Y^* = \frac{p^*(O_3)}{p(T)}$$

Hence, $(Y - Y^*)$ is equal to the driving force in the gas phase for the transfer of ozone to the liquid.

The practical exchange rate is proportional to the external surface area of the gas phase (S) in contact with the liquid. In that case, the number of molecules transferred as a function of the time is:

$$-\frac{dM}{dt} = k_{(g)} S (Y - Y^*)$$

The equilibrium condition is: what the gas phase releases, the liquid phase receives. Thus, in the liquid:

$$+\frac{dM}{dt} = +k_L S(C_s - C)$$

The whole process is schematized as in Figure 14.

Considering the integrated form of the transfer equation and plotting $\log[(C_s - C)/C]$ as a function of the time, an experimental value of $k_L S$ is obtained as the slope. Under comparable conditions of S, that is the specific exchange surface, relative values for k_L may be obtained for different gases, e.g., k_L is about 2.5 times higher for oxygen than for ozone (0.05 cm/sec for oxygen and 0.02 cm/sec for ozone) [14,15]. These discrete values result from the assumption that the nominal transfer ratio

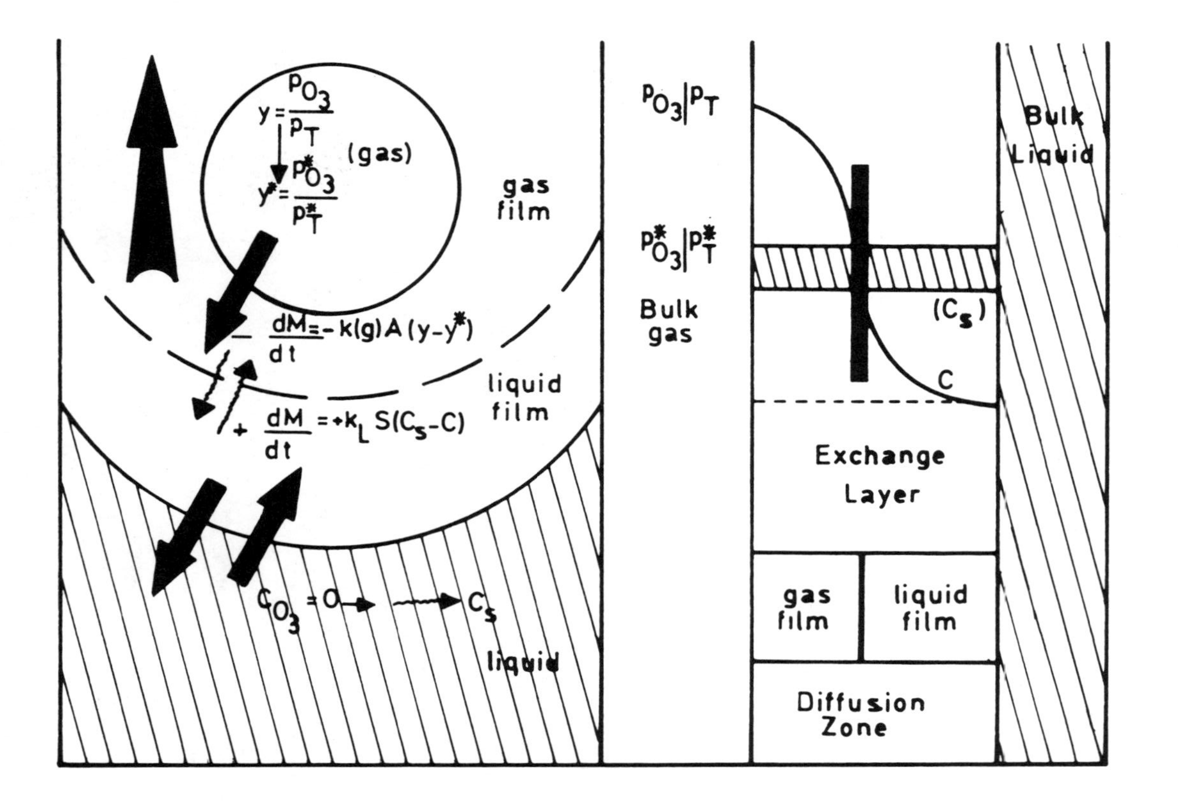

Figure 14. Scheme for double-film process.

can be expressed as $(k_L S C_s V)/W$ where V is the volume of the contactor and W the power dissipated in this volume. In practice,

$$k_L S = F\sqrt{W/V} \cdot \sqrt{\frac{\text{gas flow per minute}}{\text{unit liquid volume}}}$$

where F is a proportionality factor, depending on the concentration of the ozone in the process gas (often $\simeq 1.2$).

For evaluation of the specific exchange surface, several equations have been proposed. Besides the preceding approach, in a static water layer through which an ozonated gas is bubbled:

$$k_L = 1.13\sqrt{\frac{D\,V_R}{d_B}} \simeq 0.3\text{–}0.4 \text{ mm/sec}$$

where D = diffusion constant for ozone in the liquid (e.g., 1.74×10^{-5} cm^2/sec)
V_R = relative upflow velocity of the bubbles in the liquid;
d_B = average diameter of the bubbles.

The most reliable value for k_L for ozone is 0.2 mm/sec while the $k_L S$ value appears to be about four times higher in agitated vessels than in simple bubble rising systems [15].

CHEMICAL REACTIONS DURING OZONE TRANSFER

It has also been established [16] that if the rate of oxidation in a direct ozonation process is limited by the chemical reaction rate of the dissolved solutes (second-order reactions), the oxidation proceeds less efficiently in batch plugflow reactors than in stirred contactors. In practice, the oxidation of slowly reacting compounds kinetically takes place for the most part as a function of the dissolved ozone concentration. More reactive species react immediately during the contact process itself, and in this case the reaction is favored by mixing.

As a consequence, if the ozone transferred to the liquid is consumed by a chemical reaction, the specific transfer coefficient, k_L, no longer is influenced by the diffusivity only, since a significant part of the ozone dissolved in the liquid phase is exhausted continuously. Hence:

$$k_L(R) > k_L \quad \text{and} \quad \frac{k_L(R)}{k_L} = B$$

where $k_L(R)$ is the transfer coefficient in the presence of the chemical reaction. An approximation for B is [16]:

$$B = \sqrt{1 + \frac{D k_1}{k_L^2}}$$

where k_1 is the first-order reaction constant of the substrate oxidized by the dissolving ozone (Figure 15).

The reaction is of the "slow" type if $D k_1/k_L^2 < 1$. If there is no residual ozone present ($C = 0$) the transfer equation is:

$$dM/dT = k_L S C_s V$$

where V = volume of the liquid. If the reaction is very fast, as with the oxidation of iodide, $k_1 \gg k_L$. Then the total quantity of ozone can react in the exchange layer without diffusion into the medium and, in that case, $C_S \rightarrow 0$ and

$$B \simeq \sqrt{\frac{D k_1}{k_L^2}}$$

Thus, $dM/dt = CS\sqrt{D k_1}$ and $k_L S = S\sqrt{D k_1}$

In this case, the rate of ozone transfer is influenced only by the value of k_1. Also, the bulk volume of the liquid phase may be neglected as long as k_1 is sufficiently high. Mixing improves the reactions during the diffusion.

OZONE DESTRUCTION BY THE CONTACTING SYSTEM

On several occasions it has been advanced that part of the ozone could be decomposed in the bubble-forming process itself without efficient reaction. Experiments with "superozonated" water, without any further ozone demand, have given the following ozone losses during dispersion in the order:

packed columns $\simeq$ bubble columns $<$ gas disperser ($<$ vortex impeller)

No comparative data are known for ejectors, static mixers or gas-liquid circulating systems.

At alkaline pH (>8), the same order is maintained, but the packed column has lower decomposition losses than the countercurrent bubble column.

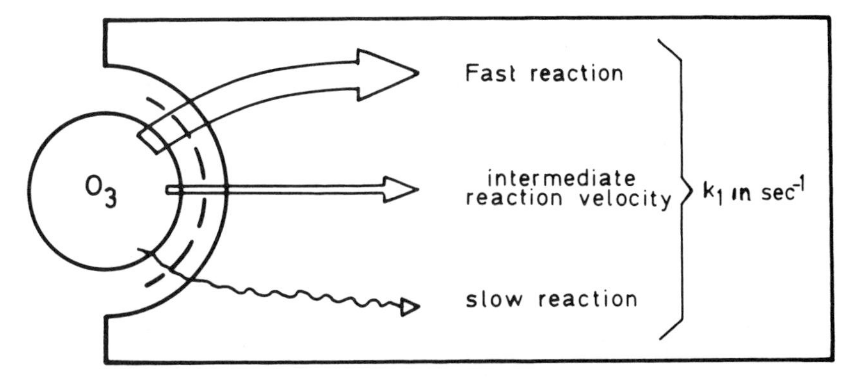

Figure 15. Schematic representation of effect of reactions on ozone transfer.

Obviously, this situation only occurs in the absence of substances reacting with ozone and for waters free of ozone demand. In such cases, the losses can reach 5–10% for the static systems; and up to a maximum of 25% has been advanced for certain dispersers (vortex impellers exhibit variable losses of ozone with time due to the instability of the method). This destruction of ozone may result partly from compression of the ozonated gas or a mixture of gas-water containing the ozone.

Part of the specification for mechanical ozone contacting systems must be: the concentrated ozonated gas must not be compressed at a pressure exceeding 3 bars (absolute) and preferably less than 2.5 bars. Therefore, although appearing similar at first sight, a difference exists in this respect between the gas dispersers and the liquid circulating turbines. The latter are in fact high-performance multiejectors (see below).

In these experiments, the process gas flow often is not measured at the exit of the contacting chambers and, therefore, the data are often arguable (see above). The pressure applied for flow correction at the exit of full-scale contacting chambers often is questionable.

Similar problems may be associated with the sampling method used in the analytical determination of ozone concentration in the offgas. To obtain comparative results for different contacting systems, the counterpressure at which the gas flow is measured must be constant or corrected for normalization.

In the analytical determination of ozone, another method is advanced for ozone decomposition: bubbling through sintered glass (Figure 16), e.g., in a washing flask. Consequently, bubbling through a tube or into a Muencke flask may be preferred. To investigate the question of ozone decomposition by the bubbling systems, Masschelein et al. have conducted some investigations in an experimental column [17].

In these preliminary experiments [17], the air containing ozone is cir-

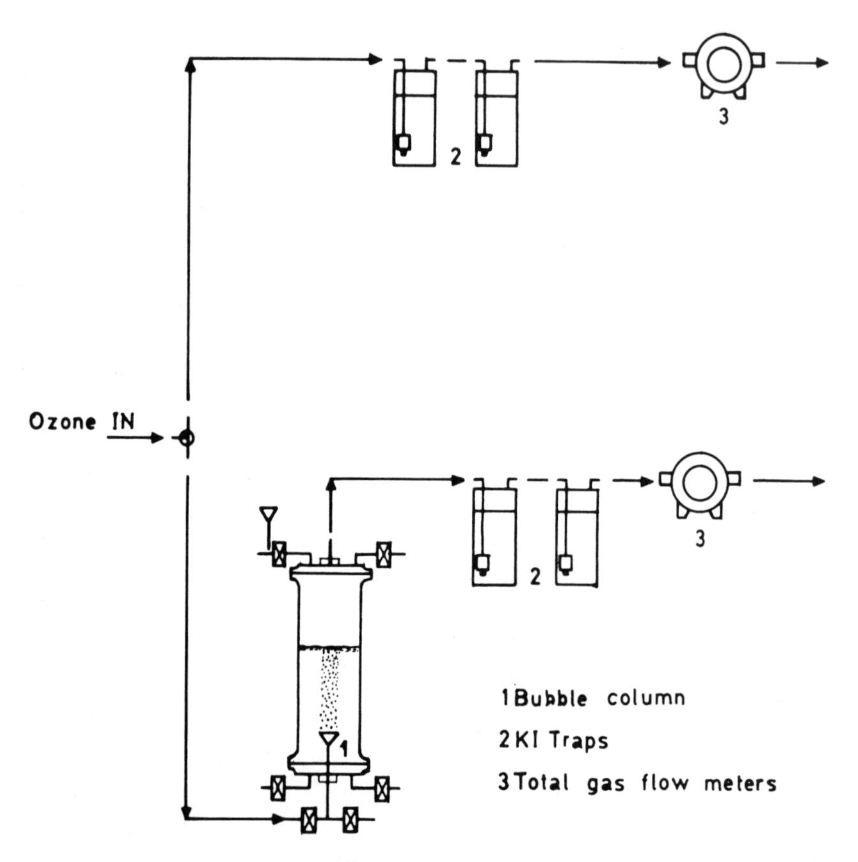

Figure 16. Scheme for testing ozone decomposition by bubbling through sintered glass.

culated through the system and partly destroyed without passing through the porous diffusing elements. This is due to several factors, which include the influence of light and turbulence in the gas flow. Therefore the concentrations taken into consideration for comparison must be rigorously those measured at the exit after or without passage through the diffusers.

Ozone concentrations are expressed at constant in- and outflow which is measured with a volumetric apparatus. It appears that diffusion through porous borosilicate glass has no measurable effect on ozone decomposition when carried out under conditions free of ozone demand and with negligible delays with respect to thermal decomposition of ozone. In the literature, part of the discussions of this subject probably may be related to the lower exchange capacities of rigid gas bubbles produced by fine diffusers even in analytical determinations.

Table II. Ozone Concentrations (g/Nm^3 gas) for Fritted Diffuser

Diffuser Pore Size (μm)	Water Column Height			
	0 mm		300 mm	
	Ozone in	Ozone out	Ozone in	Ozone out
150–200	2.48	2.50	2.47	2.49
40–90	2.52	2.53	2.52	2.53
10–20	2.47	2.42	2.47	2.46

PRACTICAL OZONE CONTACTING SYSTEMS: GUIDELINES FOR DESIGN AND OPERATION

Ozone Bubbling Through Porous Pipes

This method still remains the most widespread contacting system for water ozonation, especially in the treatment of clear waters. The dispersing elements generally are porous ceramic pipes; however stainless steel floors or plastic dispensers are also available.

The diffusers are installed at the bottom of the injection or contact chambers in which a sufficient reaction time must be maintained, e.g., up to 20 min average residence time of the water. Typical arrangements are a sequence of four to six baffled flow-through chambers as illustrated in Figures 17 and 18.

The philosophy of the process is based on the plugflow reactor principle, approaching "batch-type" reaction kinetics. Therefore, the total ozone consumption can be divided into appropriate substreams in each consecutive chamber to maintain a more or less constant residual concentration of dissolved ozone. The purpose of the method is to destroy refractory compounds and to provide viricidal action in the final treatment. In most cases the first injection satisfies the immediate ozone demand and is designed to inject 50–70% of the total ozone. A residual ozone concentration of 0.4 mg/l usually is attained in this first contact chamber, and this concentration is maintained by injection of the balance of the ozone.

The diffusers should produce bubbles with average effective radii of about 2 mm. In practice, diffusers with a pore size of 50–100 μm are installed at the base of a water column 4–6 m in height. The head loss of the immersed porous diffusers must be maintained at 300–500 mm water column. Average gas flow in each contacting column usually remains below 10% of the water flow.

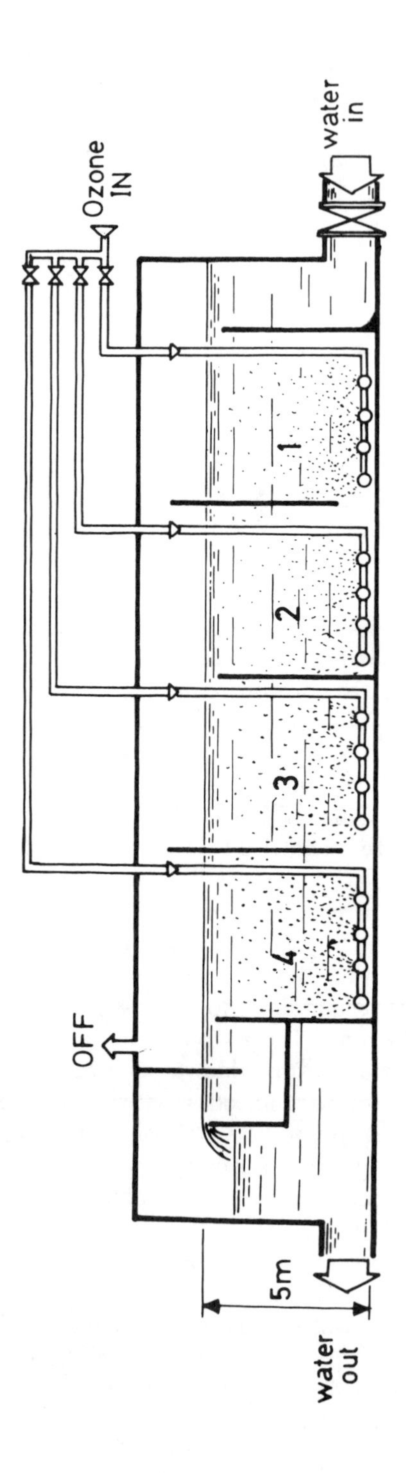

Figure 17. Multistage contacting chamber equipped with porous diffusers.

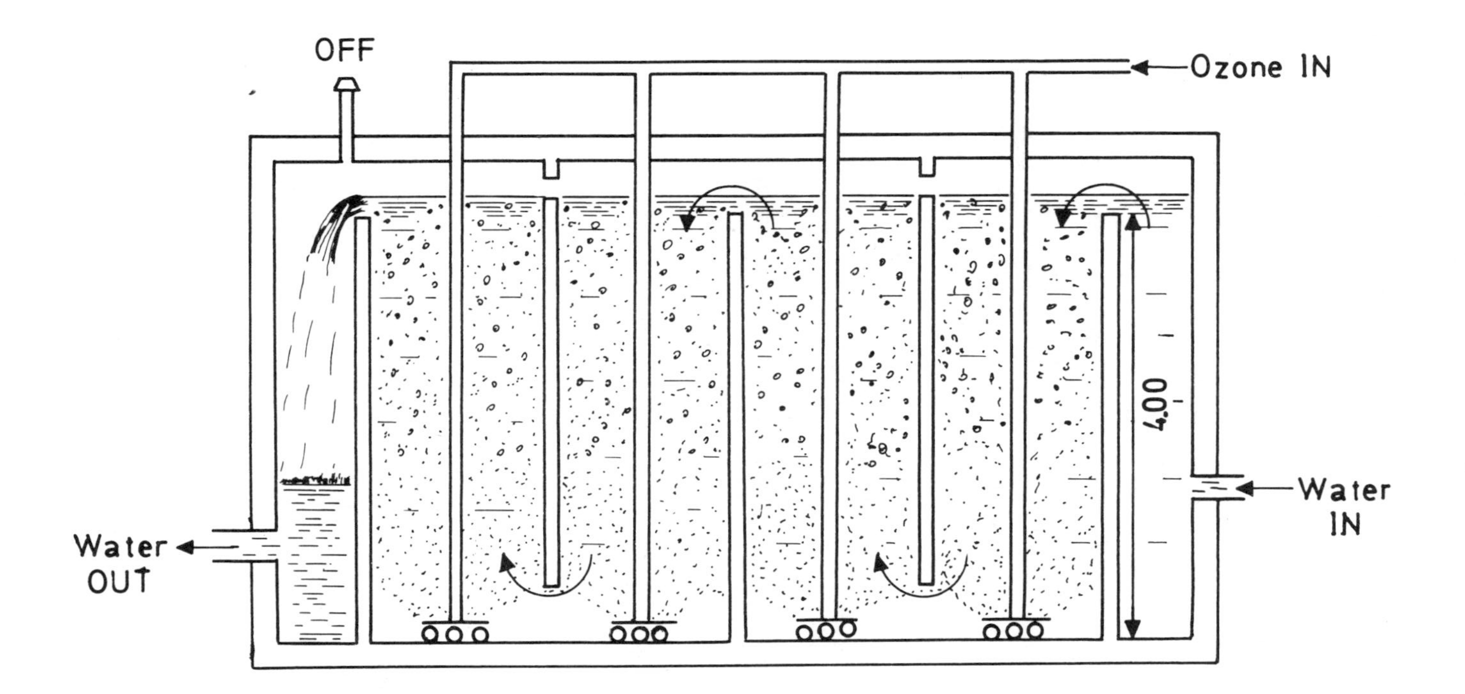

Figure 18. Multistage contacting chamber equipped with porous diffusers.

Based on bubbles with r = 2 mm, the total available surface for gas-liquid exchange per m^3 water is then approximated by the surface of the gas bubbles, that is $\simeq 0.150\ m^2/m^3$ water.

In conventional columns the average downflow velocity of the water is about 4–5 cm/sec. This has a limited effect on the velocity of bubble rise. The present tendency is to increase the velocity of the water up to 10–15 cm/sec. This method requires the injection of ozonated air at a pressure sufficient to overcome the head losses of the water column and the dispersing system, e.g., to 0.7 bar. The transit velocity within the baffles should be limited to a value below 30 cm/sec in order to prevent parasitic sucking of the bubbles into the adjacent chamber. To be able to use them as a prevention device, the different baffled chambers must be interconnected in the gas phase. As a general rule, which has sometimes been overlooked, the injection chambers should be designed in every possible circumstance to avoid short-circuiting and consequent over- or underdosing zones (Figure 19). Losses inherent to the experimental conditions can vary between 5 and 20% of the incoming gas concentration [18].

The advantages of the diffusion process are:

1. static operation without troublesome mechanical or electrical maintenance;
2. simple operation for maintaining a residual concentration of ozone by repeated injections; and
3. capability to double the air-injection capacity of given equipment.

The drawbacks of the method are:

1. tendency of vertical "channeling" of the air bubbles without intimate mixing of the air with the water; less ozone bubble contact results;
2. extreme importance of the porosity distribution in the injection pipes to ensure homogeneous injection. On leakages or breakage of a pipe, "torches" of ozonated air form, shortening the time available for transfer;
3. oxides of iron and manganese have a tendency to precipitate on the surface of the porous pipes at the sites where ozonated air emerges at high concentrations; porosity changes result; and
4. on rising, the bubbles expand because of diminishing hydrostatic pressure; with decreasing concentration and decreasing pressure, the tendency to dissolution diminishes.

The energy necessary to ensure sufficient air pressure is 2–3 W-hr/g ozone to be injected.

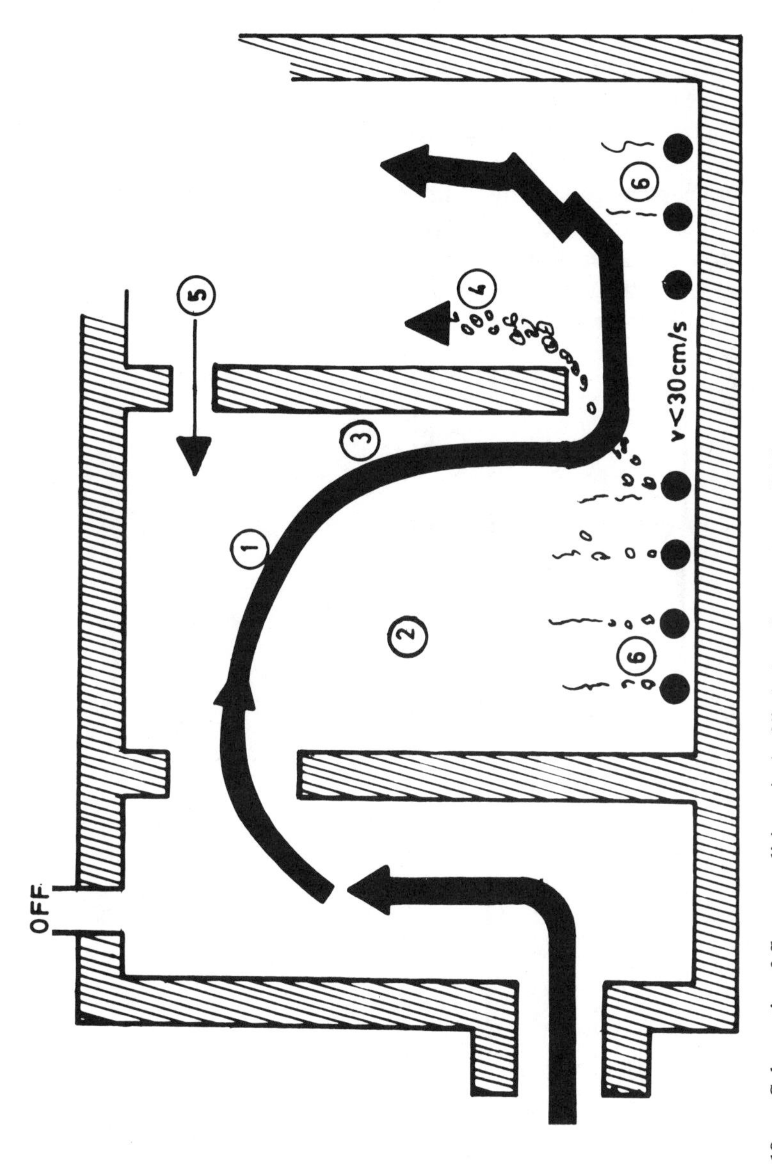

Figure 19. Schematic of flow conditions in baffled chambers. (1) possibility of (hydraulically) privileged flow; (2) possible zone of overozonation; (3) possible zone of underozonation; (4) parasitic flow of gas (to be avoided); (5) auxiliary exit of gas; (6) diffusers.

Turbine Modified Diffusers (Obermauch, Düren, Federal Republic of Germany)

A very elaborate tower system to prevent short-circuiting in ozone contacting has been designed at the city of Düren, in the Federal Republic of Germany, and operated with liquid circulating turbines. The concepts illustrated in Figure 20 are just as applicable to dynamic dispersing systems as to static contacting with diffusers. In this design the net contact time in the transfer zone ranges to about 150 seconds. The basic design of the contact chamber as illustrated here is to treat 600 m^3/hr. Operated with a gas-liquid circulating turbine [19], the losses are below 5% even at a residual concentration at the water outlet of 1–1.2 g/m^3. The success of the method is for the most part attributed to the symmetrical construction avoiding short-circuiting.

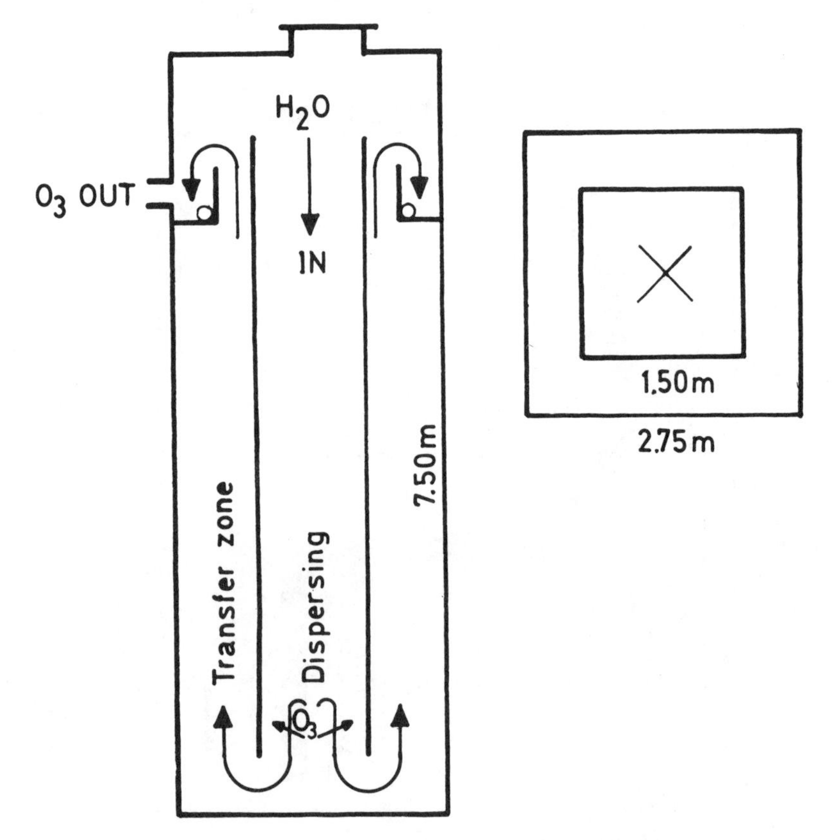

Figure 20. Design principle of ozone contacting chambers at Obermauch (Düren, Federal Republic of Germany).

Gagnaux Multiple Diffuser

A particular design for repeated injection in a single vertically superimposed contact chamber with repeated injection, as illustrated schematically in Figure 21, has been established by Gagnaux of the Swiss firm Sauter [20].

The technique is founded on the idea of transporting the gas by the horizontal flow of the liquid, enabling use of partially exhausted process gas in a first preozonation. What is particularly interesting is that no extra energy, other than that causing the water flow, is required. Furthermore, a longer contact time can be achieved with a comparatively lower construction cost than for double diffusion chambers discussed earlier. The system is proposed with static mixers as diffusers, hence the operating cost is comparable to that of the diffusers ($\simeq$2–3 W-hr/g ozone to be injected). As the achievability of the process depends mainly on the relative velocities of bubble rise and horizontal water flow, its flexibility is low. In this double-stage process, total ozone losses can decrease to less than 5%.

Van der Made and Welsbach Diffusers [21]

Other, more classical systems for diffusing ozone into water with porous elements are the systems of Van der Made (Figure 22A) (cocurrent) and of Welsbach (Figure 22B) (countercurrent). Ozone losses in both of these systems currently reach 20–30% for a specific contacting energy consumption of 2–3 W-hr/g.

Torricelli Contactor [21]

Another ozone contacting system is the Torricelli contacting chamber in which the ozone in the offgas is rediffused into the downstream current of incoming water (Figure 23). The basic diffusion of ozone-containing gas is achieved with porous diffusers in baffled chambers 2 m high and with inlet and outlet pipes 10–12 m high. Hydraulic compression of the process gas to 0.3–0.5 bar after the first injection is a typical feature of the system. Therefore leakages of the gas phase are one of the major operational problems of the system. In the downflow preozonation column the water velocity is below 150 mm/sec.

Losses in this system are below 5% of the injected ozone at conven-

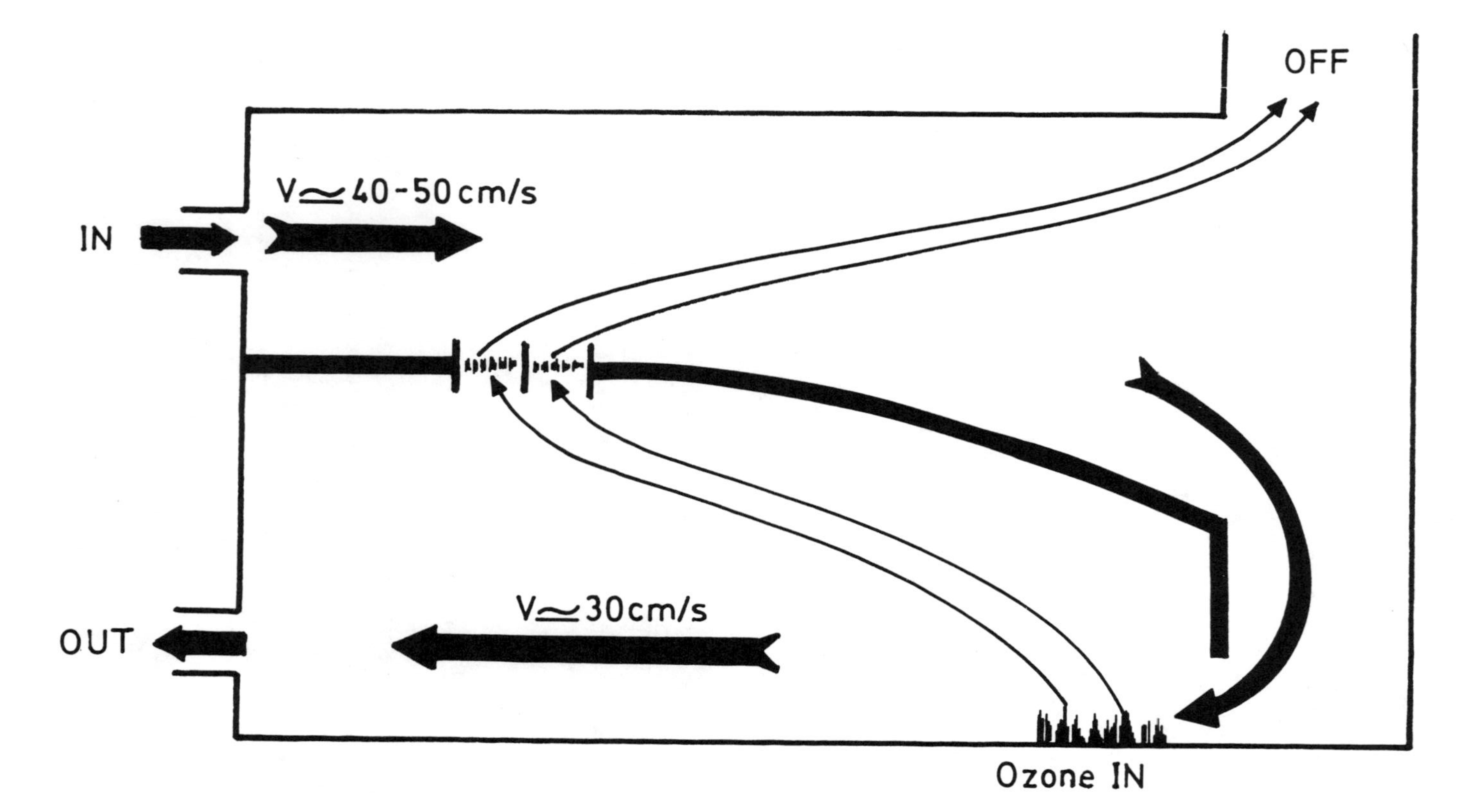

Figure 21. Gagnaux diffusion contactor [20]. Repeated injection through water flow.

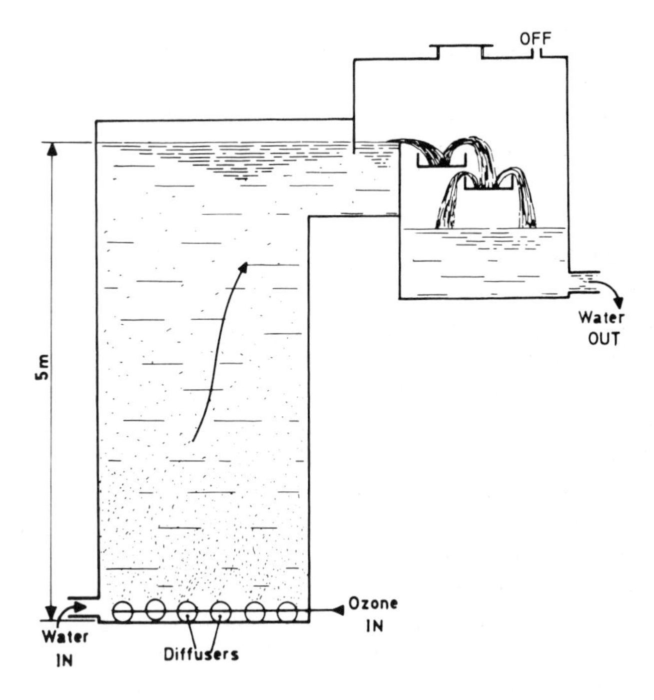

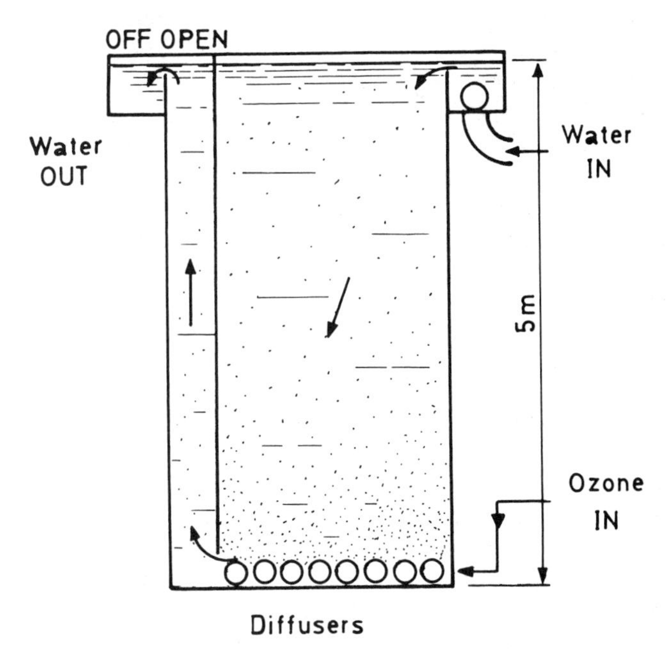

Figure 22. Van der Made (top) and Welsbach (bottom) contacting systems [21].

tional residual concentrations of 0.2–0.4 g/m^3 in the retention chambers. The operational cost may be estimated at 4–5 W-hr/g ozone produced for pressurizing the process gas, and at least 1 W-hr/g ozone contacted for water head loss (>1 m).

Static and Sonic Mixers

Most recently developed static mixers could replace the porous dispersers. Except perhaps for that of the VAR-mixer (Sauter), the use of these systems is not yet widespread. Other similar equipment includes the Kenics, Koch and Komax mixers (Figure 24A), the Ross ISG and the Yonkers sonic mixer (Figure 24B) [22].

These systems appear less suitable for ozone injection with maintenance of a residual concentration than for the treatment of rapidly reacting compounds, including immediate bactericidal action. Exact ozone losses in the absence of compounds reacting with ozone are unknown, but are estimated to be 10–20%. The flexibility of the air/water ratio for operation of the systems is a basic criterion for choice of specific equipment. Operating costs for full-scale use are unknown, but may be estimated at 4–5 W-hr/g ozone contacted: 2–3 W-hr for gas conditioning and 2 W-hr for accurate water flow. In the case of sonic mixing, the liquid introduced into the high energy field is dispersed in the form of droplets. By reversing the gas and liquid flows, the gas is dispersed into the liquid "as fine bubbles." Under conditions of high liquid flowrate and low gas flow (about 1%), the sonic mixer produces an extremely fine bubble pattern. At a more conventional operation of the system, e.g., one volume of gas per four volumes of liquid, the yields are similar to those of conventional systems. There exists no real cost evaluation as yet for pilot- or large-scale application of the ultrasonic mixing systems.

Contacting with Injectors

Historically, ozone dissolution using an injector is the first system used on large-scale since the Otto process was applied at Nice in 1906 (Figure 25). More recent versions are the partial injectors (contacting ozone with a side stream), leading also to direct "pipe injection" and sophisticated repeated injections, such as "slow downflow injection" including "deep-shaft" or "multisubmerse" methods. The Grace cocurrent contactor

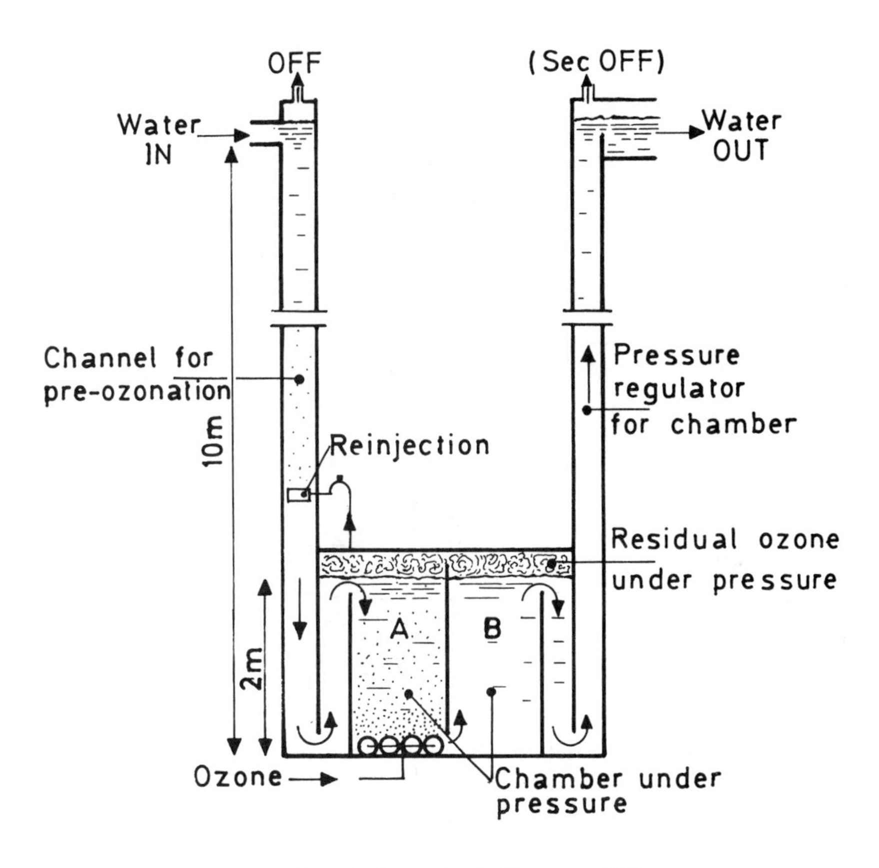

Figure 23. Torricelli system [22].

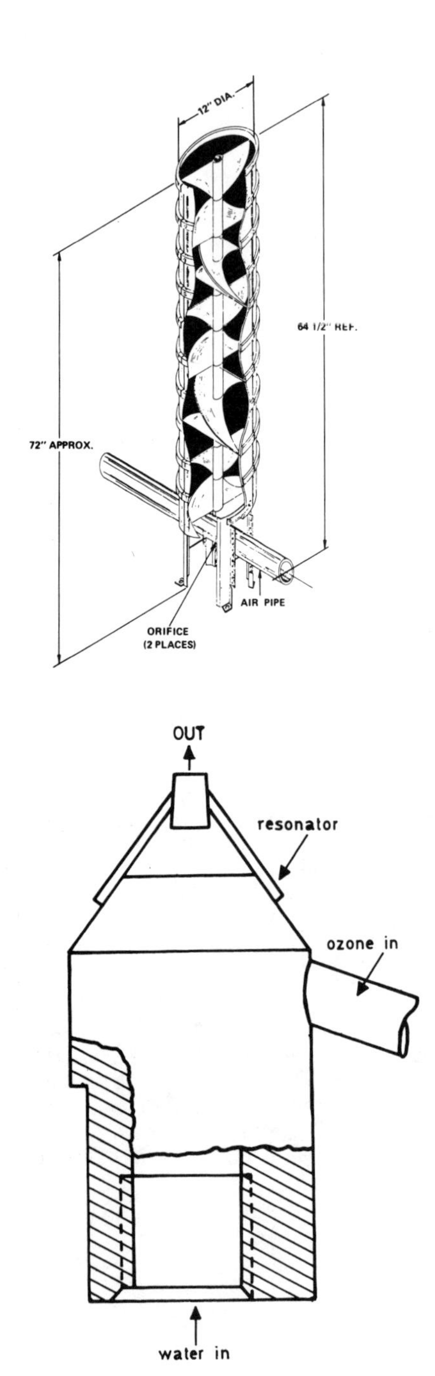

Figure 24. (A) Cutaway view of Kenics aerator; (B) Yonkers Sonic mixer.

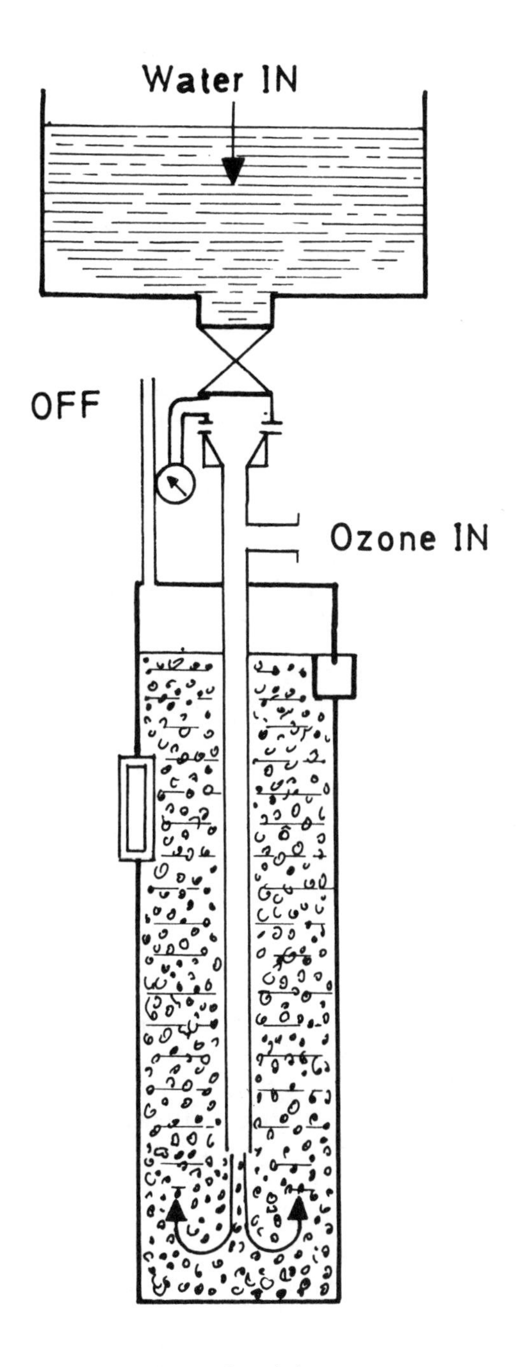

Figure 25. Otto injector.

(now called the Union Carbide Partial Pressure Injector) is based on similar principles.

In the Otto total injector system, (Venturi injector), ozonated air is sucked into a contacting column 5–6 m deep by the flow of the water system. By using a pressureless process gas, one must maintain a minimum downflow velocity in the expanding zone of the ejector at 40 cm/sec. Per phase of injection, the "minimum" head loss of the water is about 2 m; however, in practice, head losses of up to 20 m frequently are encountered. This head loss is necessary to prevent sudden variations in the offgas flow.

In this case the average energy required for ozone contacting approaches 15–20 W-hr/g ozone (costs up to 31 W-hr/g ozone are possible).

For a single injection the losses often are as high as 30–40% of the ozone introduced; hence the system is less favorable except perhaps for certain preozonation systems. Seeing that the process gas can be pressureless, the suction of offgases to a separate contactor appears as one of the attractive possibilities of this equipment. The water/air flow ratio often is critical, but this difficulty can be obviated by operating several injectors in parallel.

Direct injection of the ozonated gas to a transport pipe would be an attractive possibility to develop an ozone contacting system. The process is entitled "hydrokinetic injection" and has been attempted (Figure 26) [23]. The injection tube for the substream containing dissolved ozone is oriented in the opposite direction of the main flow. The relative velocities are given in Figure 26. The water flow of the substream is about 1/10 to 1/15 of the total flow. There is an increase in total pressure in the substream, which on introduction to the principal water flow brings about a certain expansion with bubble formation. Good results are obtained for bactericidal effects not requiring dissolved ozone residuals for long periods.

No practical data have been published on the dissolution yield or the percentage of ozone losses released at the vent. The major problem of this technique is associated with the gas purge after injection. This is necessary for venting of the nitrogen stripped, but the offgases may contain ozone in amounts of an unpredictable variability. Special attention must be given to the problem of corrosion, as the method is not suitable for use with standard materials.

At low ozonation rates, injection costs with this hydrokinetic method can decrease to as low as 0.5 W-hr/g ozone if the water pressure of the main flow is 5 kg/cm^2. The method could be suitable for partial preozonation of raw water to promote the micellization effect.

Multistage total injector systems have been described in the literature

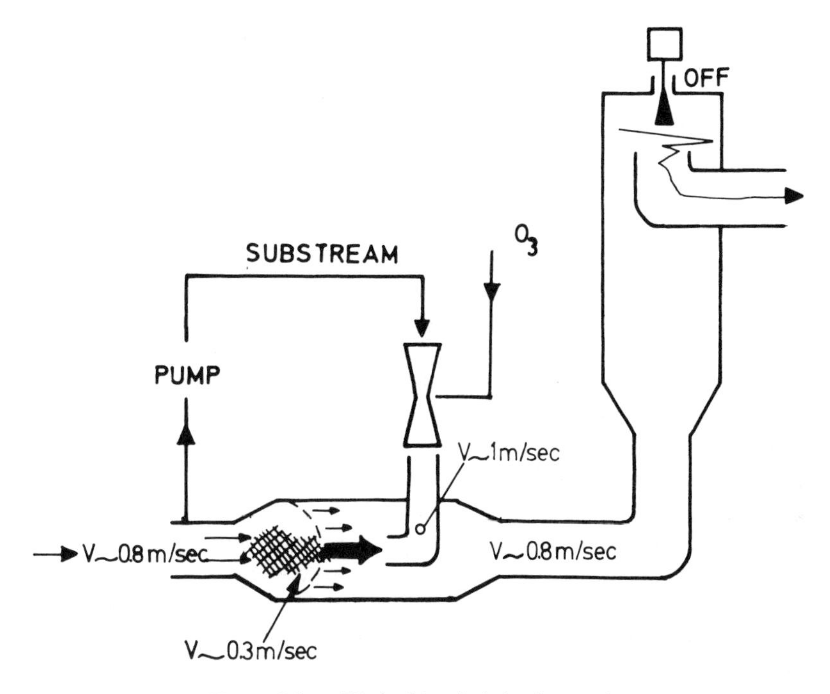

Figure 26. Hydrokinetic injection system.

[24] to improve the overall efficiency of ozone used. To limit the costs of the "injector system" the method of partial injecting has been proposed, in which ozonated air is sucked into and mixed with part of the water. The mixture then is injected at the bottom of a contact chamber in which the water to be treated is circulated. Obviously, from the standpoint of reaction mechanisms and kinetics, the system involves an excessive exposure of part of the water while another portion, the most significant, is not treated directly with ozonated air at high concentration. As a general rule, the proportional volume of air/volume of water operated in injectors depends on:

1. a decreasing function of the immersion depth of the injector shaft into the liquid;
2. an increasing function of water pressure; and
3. a maximum is reached at sufficient water pressure for a given injector.

In ideal operation, ozone losses can be below 15%, but the flexibility being poor (less than 20% in variation of vol air/vol water is admitted), the system is often difficult to operate in unsuited circumstances and therefore does not meet practical needs. In this case the losses increase significantly.

Injection of the substream can be performed at low counterpressure (CEO France or Düsseldorf system) or at high counterpressure (Chlorator system) (Figure 27). These systems are suitable for the injection of pressureless ozonated air. The injection depth is generally limited to less than 5 m.

The substream represents 5–10% of the total flow. The low-counterpressure system, when ideally run, enables ozone contacting with a cost of 4 W-hr/g ozone (e.g., Düsseldorf system) [25,26]. Average costs for the high-counterpressure method may rise to 10–20 W-hr/g ozone, but practical costs up to about 30–45 W-hr/ozone have been reported (e.g., Königsberg and Salzburg) [27–29].

Above all, one must consider the process as being surpassed by newer methods. It should only be emphasized in this present development when water pressure is available, for instance through gravitational head and in certain conditions for reinjection as a preozonation stage. The lack of flexibility can lead to complex systems in which several injectors of different nominal flows are incorporated to enable more flexibility in operation [30].

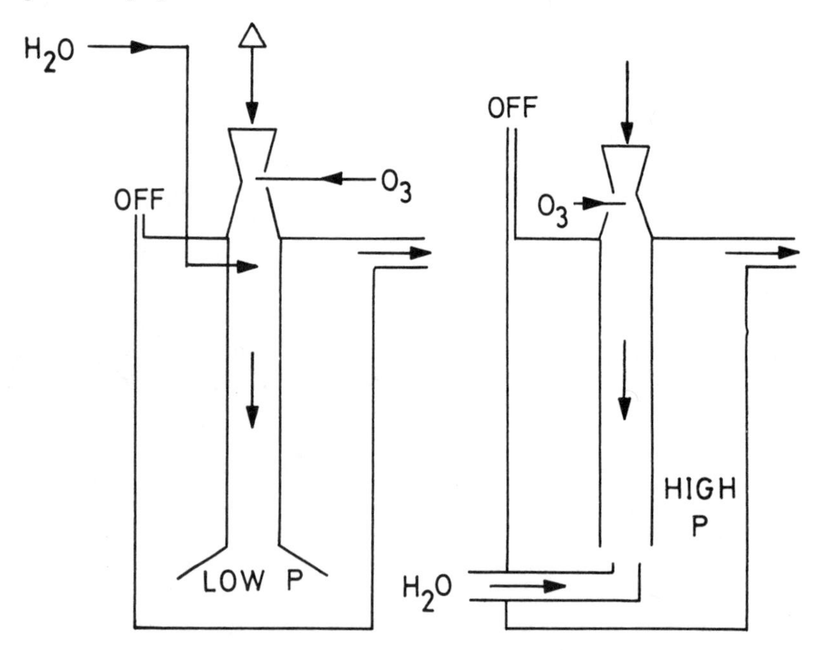

Figure 27. Scheme for low- and high-counterpressure injectors.

Ozone transfer in the injector system is promoted by the high turbulence in the injector shaft, in which the pressure increases gradually to reach a maximum. The downflow velocity in the injector shaft is >2 m/sec.

A better knowledge of the bubble migration velocity and ozone gas transfer has resulted in diminishing the downflow velocity of the water to be treated (Figure 28) [12,18]. The advantages of the process are:

1. Operation without mechanical equipment for agitation is possible.
2. The process involves a gradual increase of the hydrostatic pressure on the air bubbles when the ozone concentration in the air gradually decreases; this promotes dissolution.

The drawbacks of the method are:

1. There is a tendency for vertical travel of the air bubbles, with only partial direct ozone-bubble-to-water contact.
2. Critical importance of the air to water ratio results in less flexibility of operation of a given equipment; the process is characterized by practical operation conditions of 5–10% air flow compared with the water flow.
3. High reliability of the safety equipment is required to avoid ozone leakages when the flow of water is insufficient.
4. In the contact column, there is some tendency to strip off the residual ozone on release of the exhausted air.

To favor the dispersion of the ozonated air, the gas is best admitted to the system at a pressure of at least 0.5 bar. The resulting specific energy consumption is about 2–3 W-hr/g ozone. The input point of the ozonated air must be located at least 30–50 cm above the outflow level of the water. The head loss of the water, resulting from the critical downflow velocity, is 80–150 cm. The total necessary charge available ranges from 120 to 2 m water column. The additional resulting cost depends on the concept of the entirety of water flow. The energy required is generally about 2 W-hr/g ozone. The total energy required for the injection of 1 g O_3/hr ranges from 4 to 5 W-hr. With less flexibility of operation, the total injection technique presents comparable performance to the porous pipe injection systems.

The principle has lead to commercial systems such as the Submers reactor from Waagner-Biro (Vienna) and, to some extent the Grace cocurrent contactor (now the Union Carbide Positive Pressure Injector).

In the Submers reactor (Figure 29) an underwater pump recirculates process water at a head necessary to operate the system. Basic design

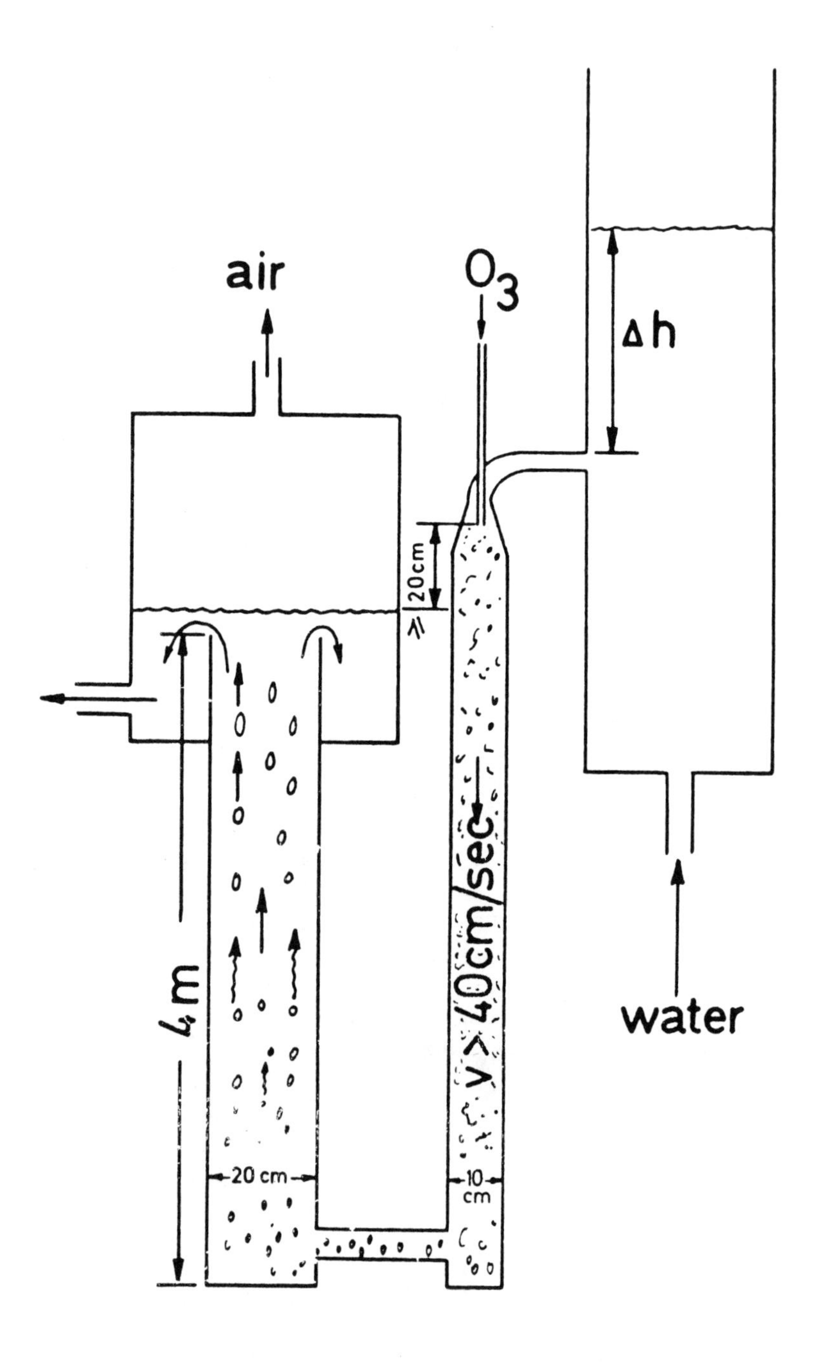

Figure 28. Experimental slow-downflow injector.

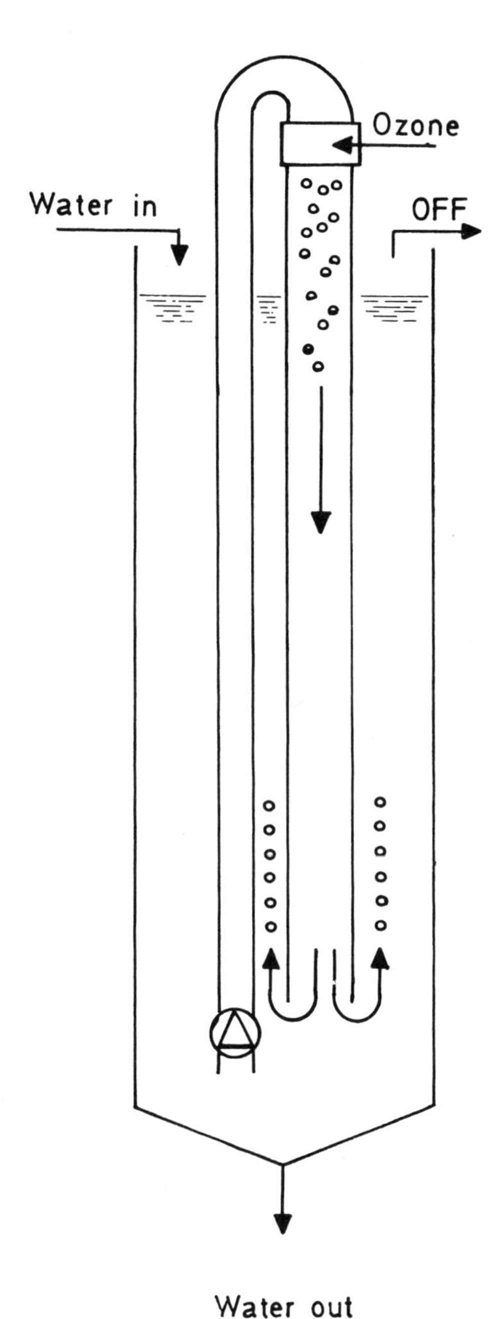

Figure 29. The Submers unit.

values for ozonation are a downflow velocity in the shaft of ≥ 30 cm/sec and a water head of about 1.0–1.20 m in the recirculating tube. Shaft depths can vary from 4 to 10 m, according to the overall design.

The basic design is conceived to recirculate 10–20% of the total water flow with an air-to-water volume ratio of 1/5. In the total contact basin as many Submers units as necessary can be installed so as to meet the demand. Hence, in principle the process is as suitable for low ozone demands, e.g., in swimming pool water or disinfection of groundwater, as for the treatment of highly polluted water, by installing the necessary quantity of units.

To disperse the incoming gas, ozonation is best performed at a gas pressure of 0.5 bar, hence 2–3 W-hr/g ozone are necessary for this purpose. The additional costs for water lifting range to 2 W-hr/g ozone per unit. Thus, the total cost ranges to 4 W-hr/g ozone for a single unit using air as the process gas. Similar dispersion with rapid submersed pump-ejector systems is more expensive and the dissolution yield can be lower (e.g., the Penberthy injector).

In the Grace cocurrent contactor (now the Union Carbide Positive Pressure Injector), the basic options for design are those of the slow downflow injector (Figure 30). This system involves bubble forming by direct contacting of the influent gas with the influent water, which is assumed to be a high ozone-consuming water, such as wastewater. Hence, the process is more or less related to the ICI deep-shaft technique for oxygen saturation. Basically the system is proposed for the use of oxygen-rich process gases. Consequently, it is suitable for extended oxygenation and ozonation in a single operation. The costs for contacting are those of slow downflow injecting systems to which must be added the additional head for benched overflow in consecutive baffles. The contacting system must be compared to combined ozonation-oxygenation (aeration) processes, rather than to ozonation alone.

Ozone Contacting by Liquid Dispersion into the Gas Phase

Spray contacting devices generally represent a lower investment cost than gas bubble to liquid transfer equipment. However, high gas solubilities are required to limit the number of stages necessary. Hence, direct spray towers, by which the liquid phase is dispersed in the form of droplets into the gas stream, are not used for ozonation except in cases of fast-reacting systems. Even cyclonic and high-turbulence spray units [31] are not suitable for maintaining a high level of dissolved residual ozone. Therefore the system probably will find only limited use in ozonation. Losses often range to 30–40% of the incoming ozone.

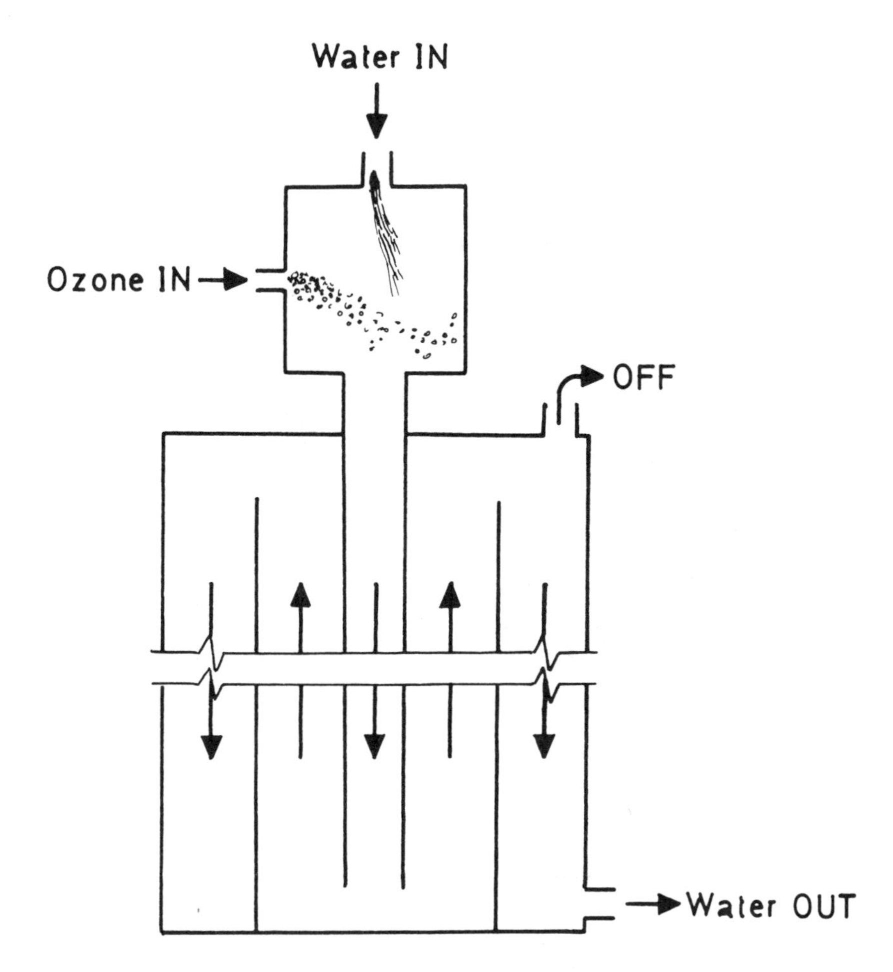

Figure 30. The Grace (Union Carbide Positive Pressure Injector) cocurrent contactor.

Packed Towers and Plate Columns

Packed contacting towers may sometimes appear interesting, as do plate columns. In these systems the wet contacting surface is increased by the packing material or plates. Ozonated air (Figures 31 and 32) is introduced at the base of a contacting column filled with liquid-surface generating material, e.g., Raschig rings or Berl saddles.

The wet surface can attain 200–250 m^2/m^3, thus increasing the exchange velocity. Computed on this basis and compared to the exchange surface of free gas bubbles, the theoretical equivalent height (HET) for

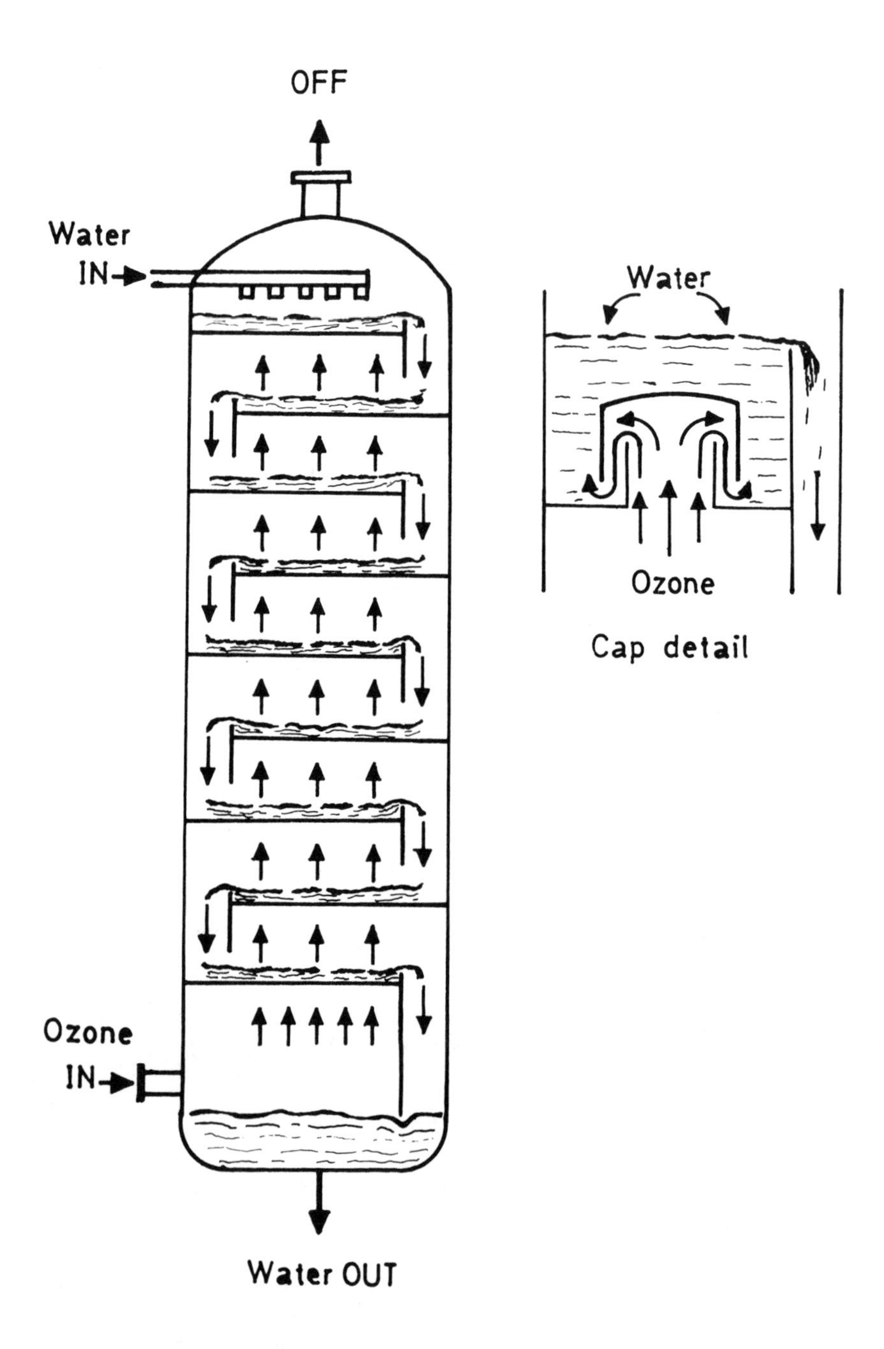

Figure 31. Schematic representation of plate contacting columns.

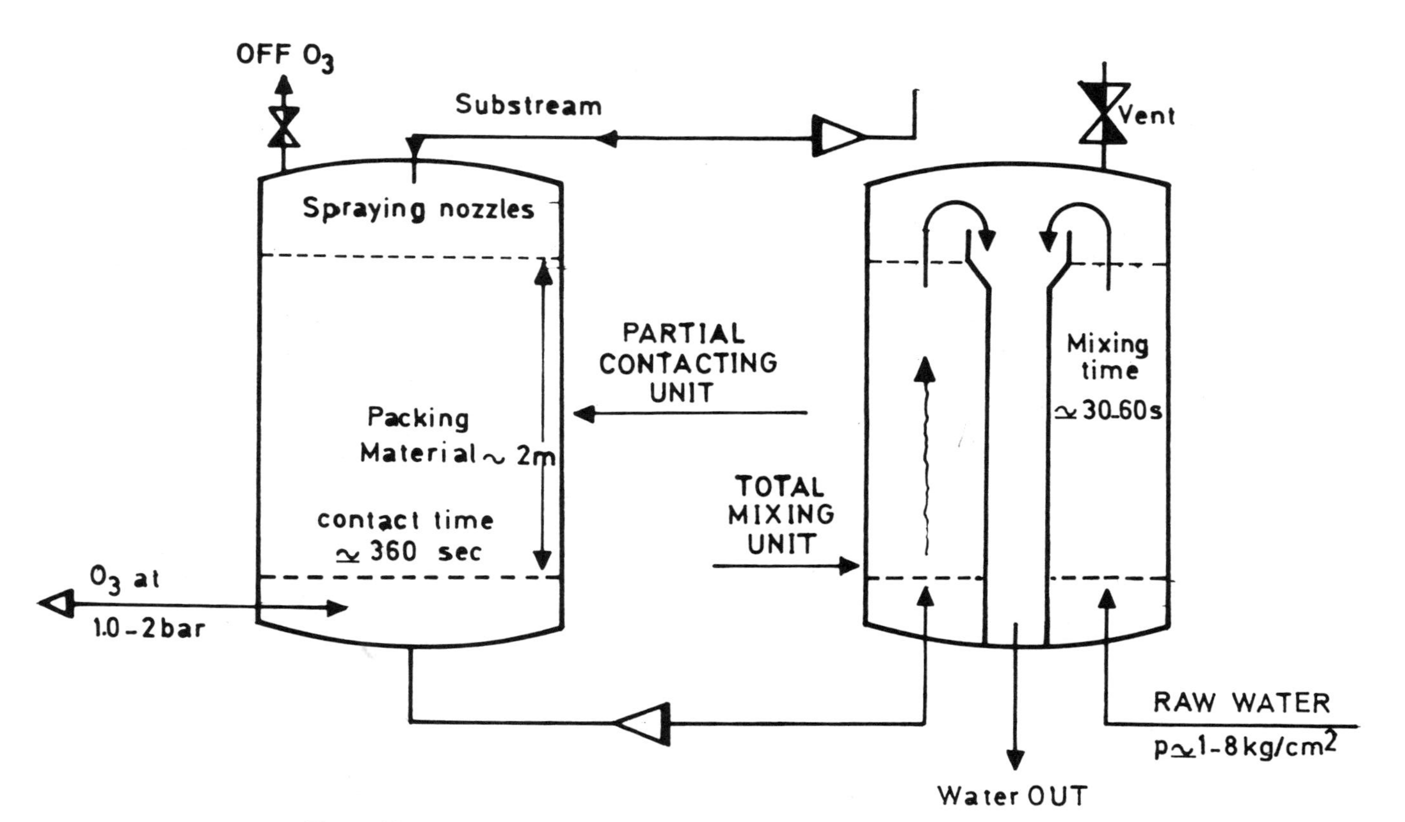

Figure 32. Schematic of packed tower for ozone contacting.

gas-to-liquid exchange is about three times less in packed towers than in bubble columns. Therefore, in this case the construction may be lower, e.g., 2 m. Plate diffusion columns with suitable bubble caps may have performances similar to those of packed columns. Normally, packed systems are operated so that the liquid forms a thin layer on the packing material. The film then is in contact with a relatively continuous gas flow.

The design of the spraying nozzles is largely similar to those for oxygenation. Ozonated gas is introduced under the contact elements where, in normal operation, a gas layer is formed. In varying conditions the column can also act as a "flooded contactor" or, at extreme water flow, as a bubble column. In normal operation the yield ranges to 95–98%.

The packed columns are often interesting, as they enable reaching higher concentrations in ozone (concentrations approaching saturation); however, under normal conditions of water treatment, this is not required. Furthermore, pressure conditions make the process more expensive: the minimum head loss for a 2-m contactor is about 4 m water column. For both of these reasons the process more often is designed to contact a substream of water to be treated with the total quantity of ozonated gas, and subsequently to mix this partial volume with the total flow of raw water.

Because ozonation, even as a final treatment step, may cause traces of iron, manganese and other deposits to precipitate on the packings, the substream more often is clear water. The relative flow of the substream to the raw water is 8–15%. The absolute pressure in the substream can attain 2 kg/cm^2. In this case the ozonated gas must attain at least the same pressure. A corrosion-free (AISI 316 or 318) water-ring compressor is the sole reliable equipment for this purpose known to be in practice. Operation is expensive. It may be estimated at 20 W-hr/g ozone (with air as process gas). The water substream in the contact column must have a pressure of 5–8 m water column. The cost, therefore, depends on the local circumstances, but may be estimated at 5–10 W-hr/g contacted ozone. For subsequent contact of the superozonated substream with raw water, it is necessary to reach a pressure of at least 1 kg/cm^2 more than that of the raw water flow. This may require a specific consumption of 6–40 W-hr/g ozone, depending on the pressure of the water to be treated.

Moreover, little is known about decomposition of dissolved ozone under compression resulting from pumping of the substream. Thus, the total costs of the system as a whole are likely to be high and dependent on local parameters such as pressure, ozonation rate, etc.

From the above comments one can deduce the energy costs to be 20–50 W-hr/g ozone. In the Sipplingen low-pressure process [32], the operating costs for the total flow contacting may be less. In this case the construction can be carried out in concrete. Estimated costs for the low-pressure packed columns are 15–20 W-hr/g ozone. In the high-pressure substream process, e.g., in Duisburg, Federal Republic of Germany, or in similar processes, e.g., the WABAG system, the construction must be airtight in steel resistant to moist ozone. In the latter system, the substream contacting and mixing towers are combined in a single pressure vessel.

Ozone Contacting with a Curved-blade Radial Gas Impeller

At the Tailfer station on the river Meuse (Brussels, Belgium), some full-scale experiments have been run with an air dispersing impeller. Injection of ozone into the water is performed during flow through baffled chambers (Figure 33). The average residence time in each injection chamber is 2 min. The residual concentration of ozone dissolved in the water is that measured after a 6-min total contact time, e.g., 2-min injection time and 4-min residual action time.

Each injection chamber, whether ozonation or preozonation, is composed of a square, 3- by 3-m compartment. The impeller is placed 5 m under the water level in the ozonation compartment. In a recirculation or reinjection compartment, the ozonated air is dispersed under a 3-m water column. This preozonation involves a compressor for recirculation of the air from the main ozonation chamber to the preozonation chamber. The ozonated air disperser is a commercial radial impeller with a backward-curved blade turbine. To favor lateral dispersion of the air admitted to the turbine when rotating at 2840 rpm, sophisticated deflectors are placed on a horizontal plate. Only movement resulting from dispersion is transmitted to the liquid by the turbine. The general construction is illustrated in Figure 34. Operational costs average 2–3 W-hr/g ozone per injection phase.

To improve the ozone contacting with gas impellers, partial recirculation of the liquid, e.g., water, has been applied. Although still basically a gas dispersing unit, the KERAG turbine enables partial liquid circulation (Figure 35).

Units of this type are capable of dispersing 25–1000 Nm^3/hr at normal water pressure, and 15–500 Nm^3/hr at 2 m immersion. The capacity of the equipment drops drastically on immersion, e.g., 7–250 Nm^3 at 0.5 bar are the comparative limits for this equipment. Hence, the system is essentially a surface gas disperser capable of partial liquid recirculation. It is most often set up just under the surface of the water to be treated. Its quintessence is the instant interaction of the concentrated ozonated air

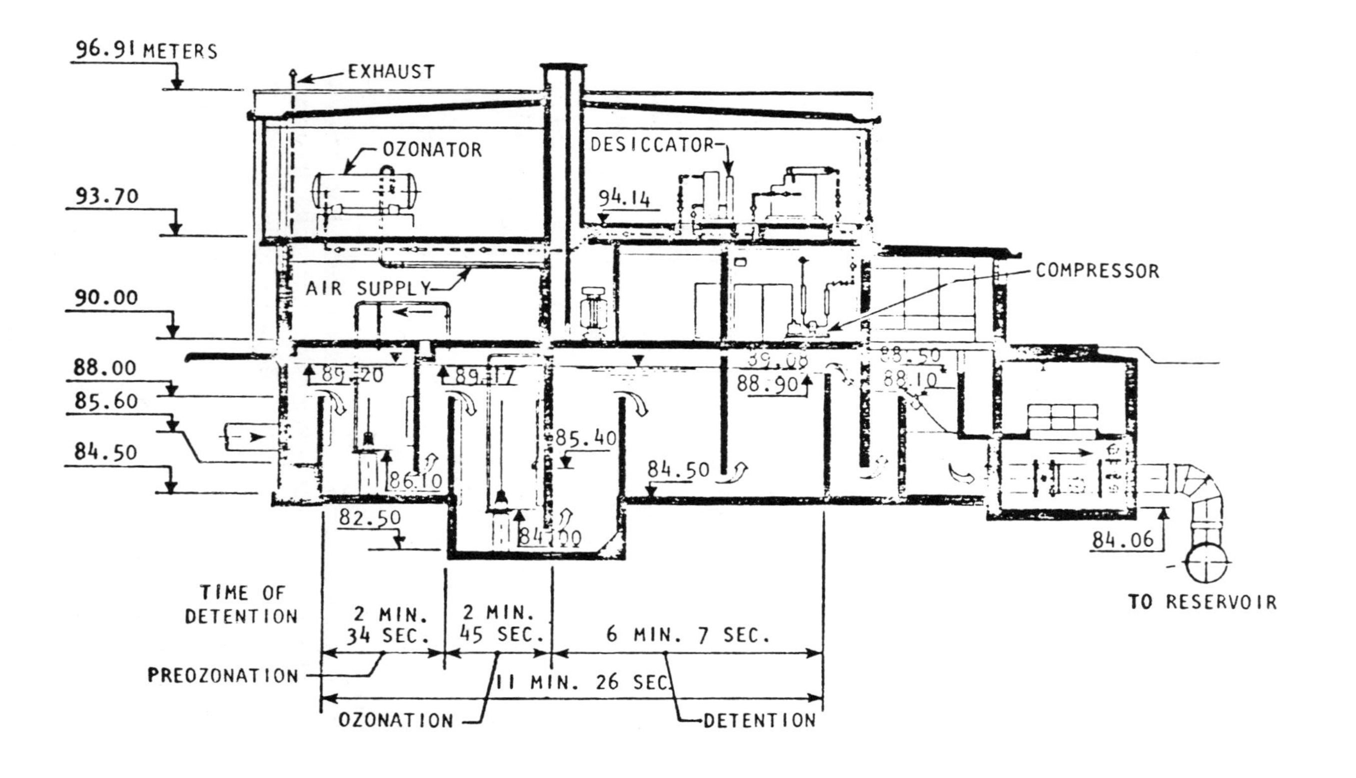

Figure 33. Contacting chamber at the Tailfer plant (Brussels, Belgium).

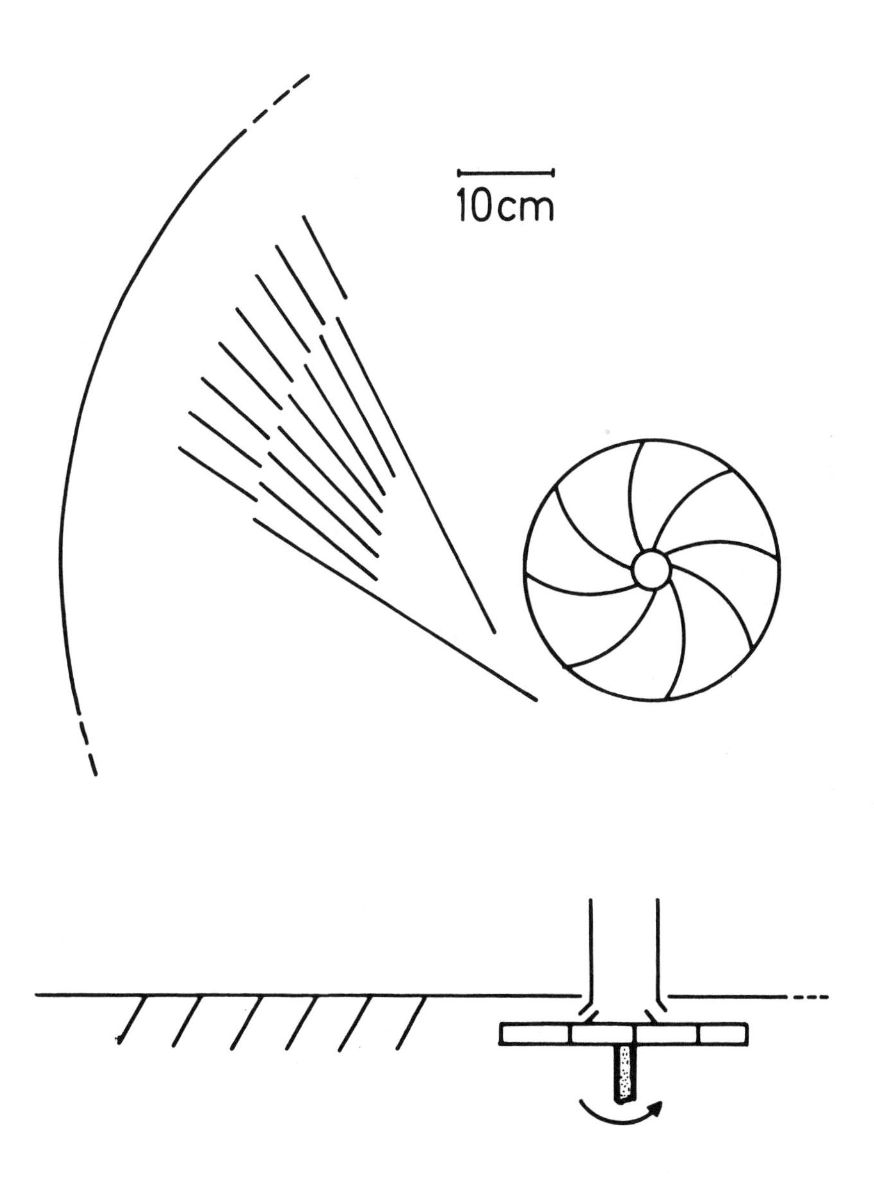

Figure 34. Gas impeller turbine (Rotoxyde type).

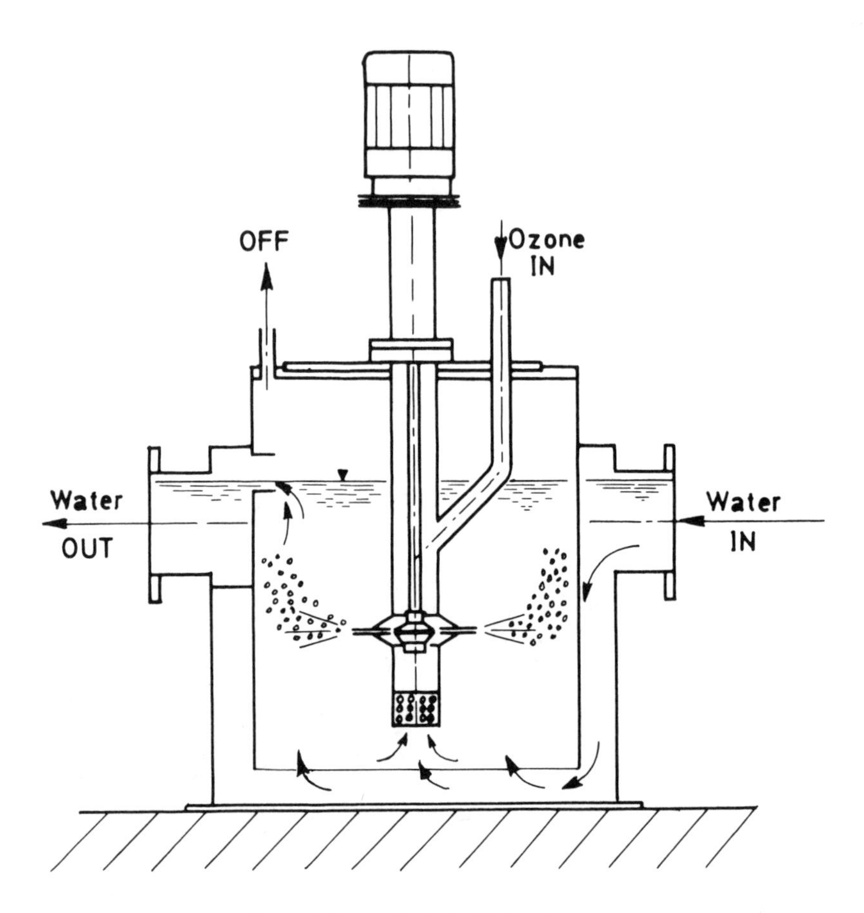

Figure 35. The KERAG unit.

bubbles with the water rather than the delayed action of a residual ozone concentration. According to pilot investigations, the loss of ozone at the exit represents 20–30%, and to provide a sufficient suction effect, the disperser requires about 7–10 W-hr/g ozone. The method appears less suitable for the principal ozonation, but can be adapted advantageously to preozonation. The process is advanced and produces very fine bubbles; therefore it is somewhat in contrast to the double-layer transfer theory. To limit the energy costs, it is convenient to apply a substream technique as, for instance, in the Rotterdam Waterworks, Kralingen (Figure 36) also repeated contacting can be favorable to the system [33].

Although this does not follow the historical development of mechanically agitated contactors, presently accepted concepts are that liquid circulating devices are more efficient than gas dispersing turbines. More-

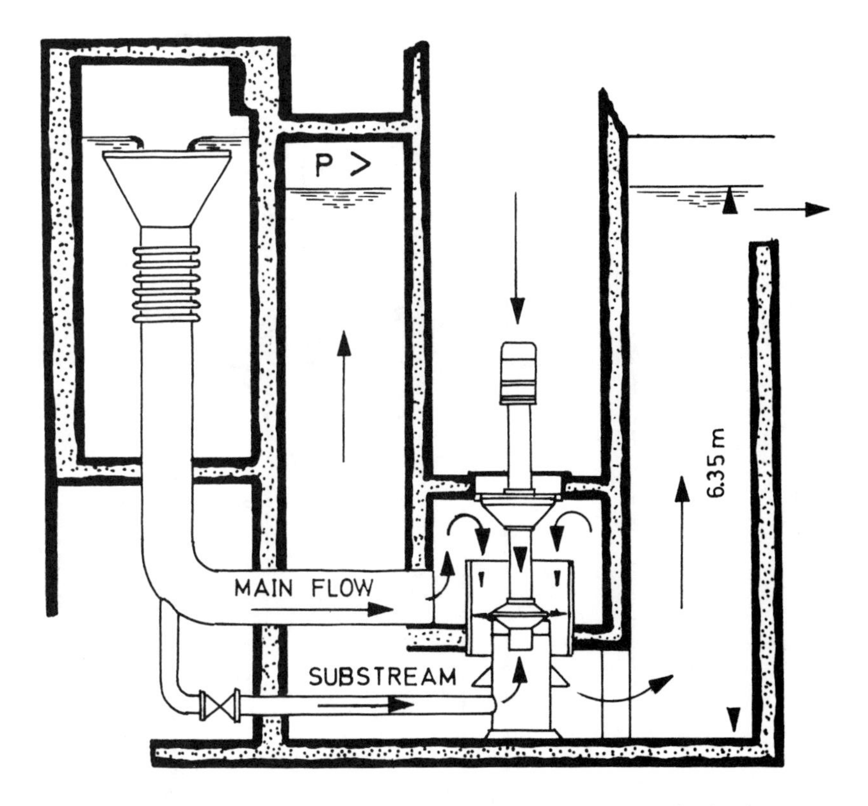

Figure 36. Substream ozonation in Kralingen, the Netherlands.

over, partial ozone decomposition has been considered possible in gas dispersing systems involving compression-decompression in the absence of a circulating liquid. The KERAG turbine is an approach to these concepts of gas-liquid circulating.

Several "reduced-scale" and pilot systems based on this gas-liquid circulation concept have given satisfactory results and increased efficiency of bactericidal action [12].

When immersing a helix propeller to about one-third of the vessel's depth, one can create a vortex by rotating the propeller at an appropriate velocity. The movement imparted to the liquid enables a direct dissolution of ozone by swirling the ozonated air at 0.5 bar into the vortex zone (Figure 37). This process enables good solubilization rates under proper operational conditions, and ozone losses amount to less than 5–10% of the quantity introduced.

The total contact time plays an important part in ozone dissolution and greatly influences treatment efficiency in liquid circulating contact-

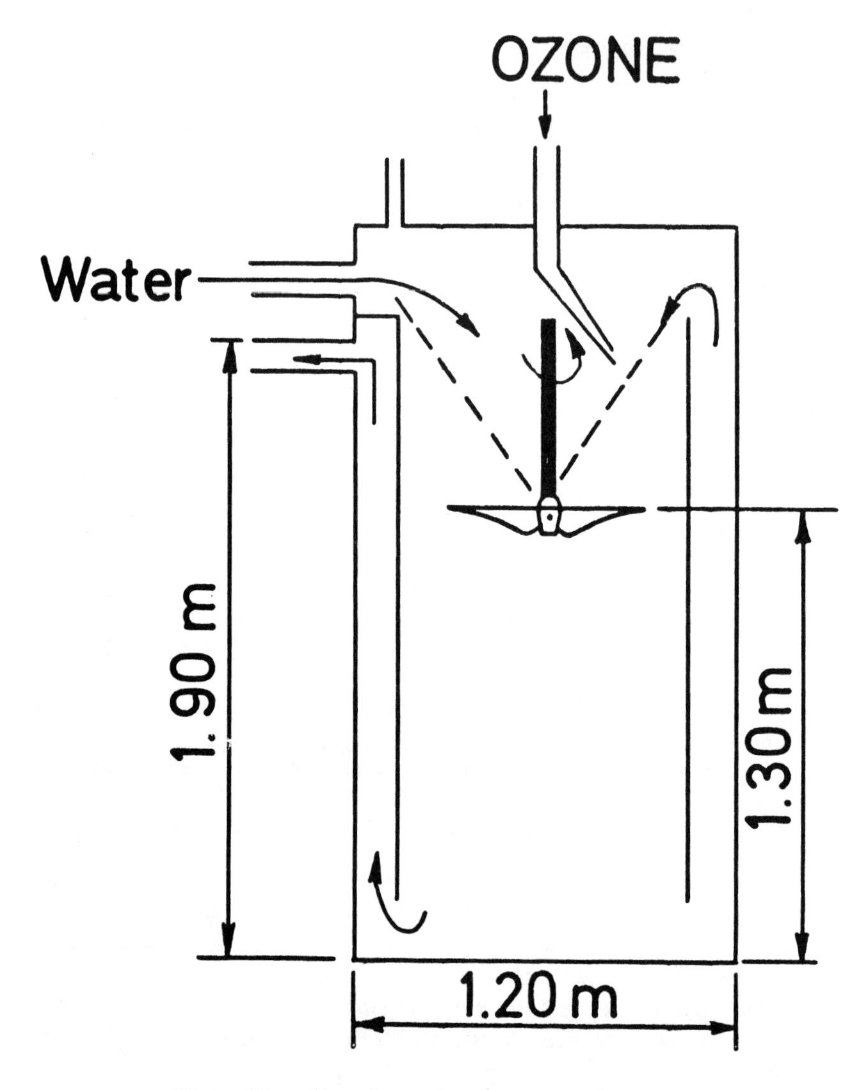

Figure 37. Experimental vortex contacting system.

ing. Therefore a liquid circulating impeller was tested with and without crown-dispersion of the incoming gas [12].

Advantages of the process are:

1. Intimate contact is guaranteed of each portion of the water with the incoming concentrated ozonated air.

2. The pressure of the ozonated air can be limited to 0.15–0.2 bar, since the injection point does not need to be immersed more than 1.5 m.
3. The dissolution yield is good, and the recirculating process promotes the "concentrated ozone bubble contact," with instant bacterial killing.

Calculation of the recirculating characteristics of the propeller can be performed on the basis of continuous mixing. The process suggests a preferable circular construction of the injection chimney in which the downflow velocity of the water is best maintained at 15–20 cm/sec. This velocity contributes to a ballasting of the air bubbles.

Operation costs for the process are 0.5–1 W-hr/g ozone for the air compression and about 3 W-hr/g ozone for the propeller enabling a single injection (Figure 38).

After preliminary full-scale experiments at the Tailfer plant, the latter has been equipped for twofold injection systems using liquid-circulating aeration turbines (Figure 39). The turbine consists of a water-circulating millwheel that sucks and mixes the ozonated air into the circulated water. The unit is powered by a 1450-rpm motor (resistant to significant overloading) for a turbine of an injection capacity of 100 Nm^3/hr. During normal operation, equal volumes of water and air are circulated, e.g., 100 Nm^3 air/hr into 100 m^3 water/hr for a motor power of 5.5 kWh. The device permits normal operation ranging from 50 to 170% of its total operating capacity. If less air is admitted to the system, more water is circulated. The circulated air/water mixture is dispersed into the bulk of the water to be treated through a series of outlet pipes that are placed in a crown form between the deflecting blades. The turbine acts like a series of injectors in a crown position between horizontal deflecting blades.

An obvious advantage of the system is the excellent dissolution obtained even when high residual ozone concentrations are maintained in the water. Dissolution yield increases when less air and more water are circulated. The turbine is self-aspirating, thus the ozonated air does not necessarily need to be pressurized before injection. Such a turbine therefore is particularly suitable for preozonation techniques for which no intermediate compressor is required.

Energy consumption of the system, expressed on a comparative basis with the air disperser, was 5.7 kWh, for a nominal injection capacity of 100 Nm^3/hr. These figures represent approximately 5–6 W-hr/g ozone injected. The cost of the air compression (2–3 W-hr/g ozone) eventually must be added to these values if the ozone is generated in pressurized air. The turbines finally installed at the Tailfer plant for full production

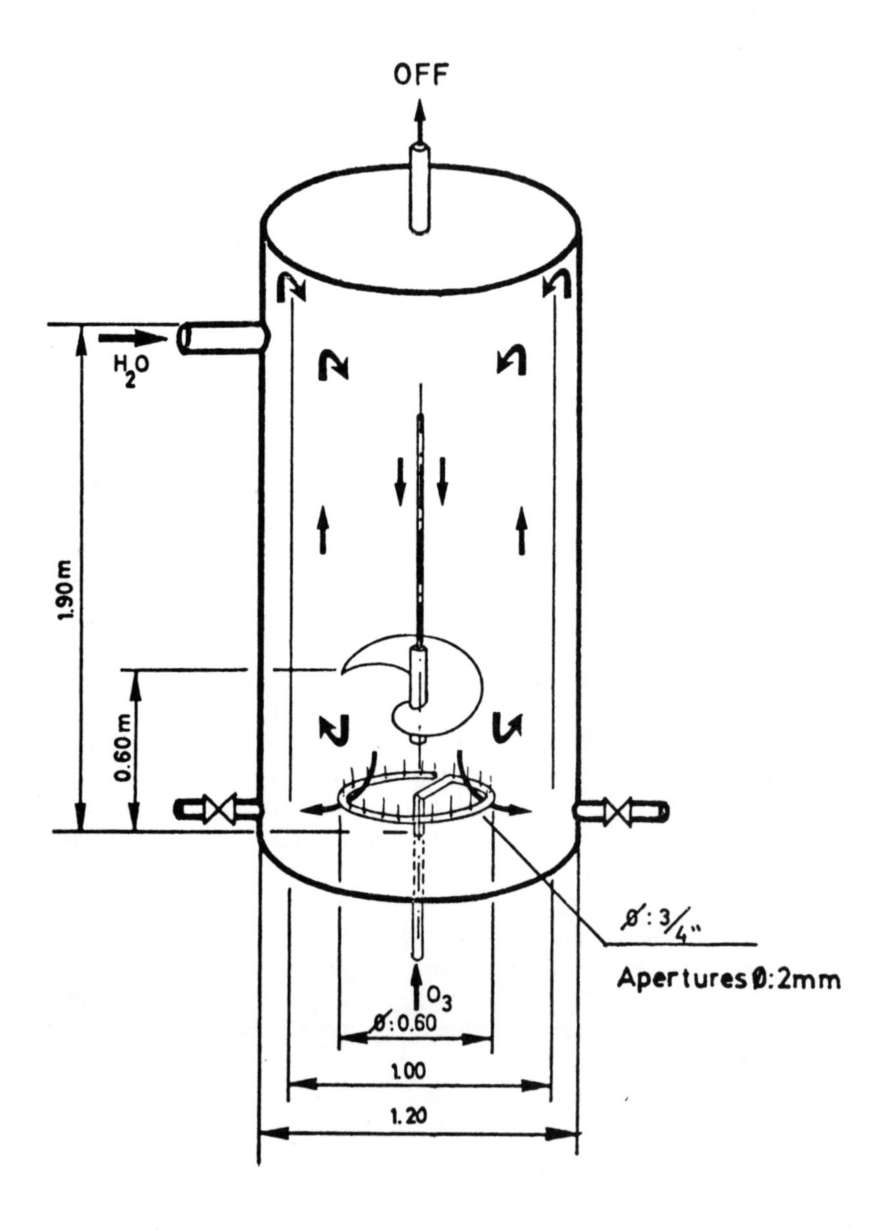

Figure 38. Water-recirculating propellor.

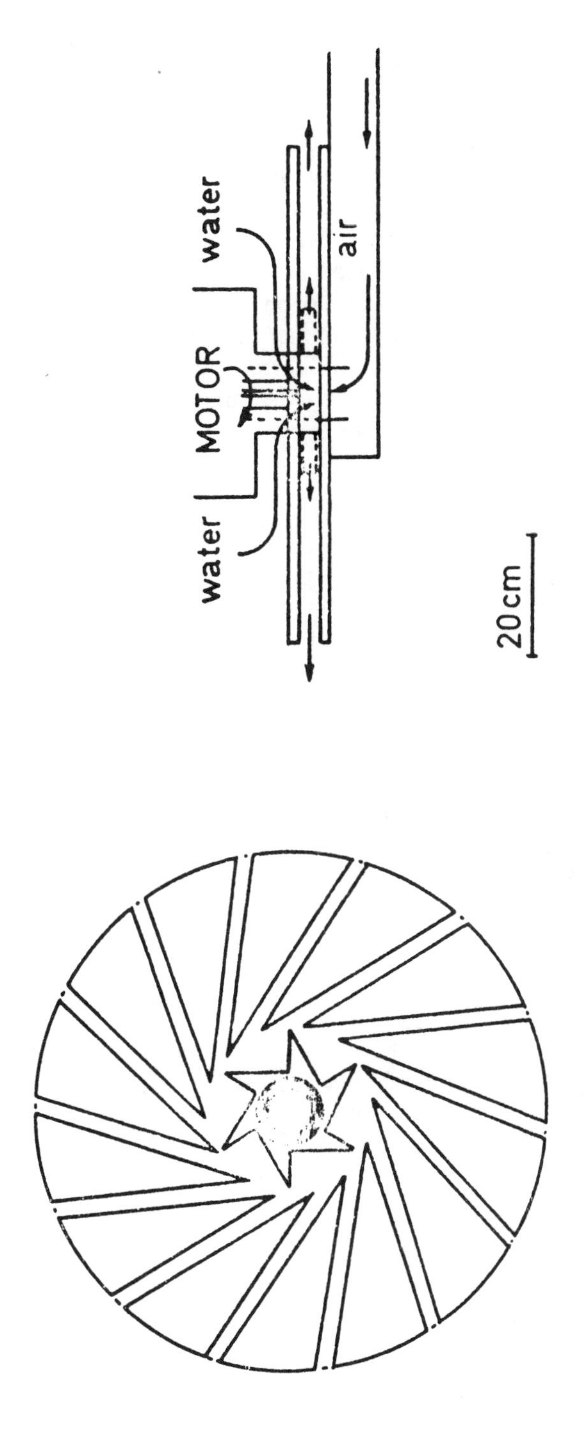

Figure 39. Liquid-circulating aeration turbine.

capacity have a higher injection rate (Frings type 600 T-VAS for ozonation and type 900 T-VAS for preozonation phases, respectively).

The nominal capacity for injection in the ozonation step is 200 Nm^3/hr, but flow changes between 150 and 300 Nm^3/hr are possible. In the Tailfer plant, process gas aspirated by the turbine is available with a pressure of 0.5 bar and the turbine circulates between 110 and 215 m^3/hr effective gas volumes with a 1.4-m water head for aspiration. This aspiration level equals the difference in levels between the dispersing deflectors and the exit of the guard tube. In the preozonation phase, the flow aspirated by the turbine is 320 m^3/hr at an immersion level of 3.1 m. The minimum aspiration level of the turbine for correct operation is 1 m water.

Results obtained on full scale are the same as those obtained during the preliminary investigation of the system.

Conclusion

In selecting ozone/water contacting systems for any particular case, the following fundamental and engineering criteria must be taken into consideration on a comparative basis:

1. average vent losses at a given residual ozone concentration in water equivalent to the practical dissolution yield attainable;
2. overall costs of operating the systems reported on a reliable basis of ozone consumption;
3. flexibility of air/liquid contacting ratio;
4. combined action of instant bubble contact and prolonged residual dissolved ozone;
5. the possibility of combination of repeated or multiple contacting units;
6. approach of bubble size distribution to the theoretical effectiveness of the contacting process;
7. simplicity in prolonged operation vis-a-vis being complicated by moving parts or clogging of static elements;
8. operation in pressurized or vacuum systems related to the possible escape of offgases;
9. necessity of protection of contacting basins against corrosion or alterations; and
10. the possibility of up-rating flow conditions in the contacting structures.

Table III shows comparisons of the salient features of the various types of ozone contactors discussed above.

Table III. Comparison of Contacting Systems

System: Example	Major Advantage	Major Disadvantage	Average Estimated Operating Cost (W-hr per g O_3)
Dispersion with Porous Elements: Obermauch, Düren	Static	Clogging and Channeling	2–3
Repeated Static Injection: Sauter	Static	Critically Flow-dependent	2–3
Static Mixing: VAR-mixer; Kenics; Ross ISB	Static	Losses	4–5
Sonic Mixing: Yonkers	Static	Costs	Unknown
Total Injectors (rapid): Otto; PPI	High Contact Rate	High Losses	15–20
Pipe Injection (Hydrokinetic)	Low Cost	Bumping	0–5
Partial Injection			
High Counterpressure: Chlorator	High Turbulence;	Partial Over- and Under-	10–45
Low Counterpressure: CEO France; Düsseldorf	Dissolution	Ozonation	4
"Slow Downflow" Injection: Submers, Waagner-Biro	No Moving Parts	Channeling	4–5
Spray Towers	Low Investments	High Losses	Unknown
Packed or Plate Column	High Yields	Clogging and Pressure Dependence	15–40
Gas Impeller: Tailfer, Brussels, Belgium	Breakages	Less Ozone-Bubble Contact (losses)	2–3
Surface Turbine: KERAG	Accessibility	Losses	7–10
Vortex System	Simplicity	Experimental; Instability	4–6
Recirculating Propeller	High Bubble Contacting	Moving Turbines	5–7

TREATMENT OF OFFGASES OF OZONE CONTACTING SYSTEMS

Background

When injection of ozone into water is accomplished with ozonated air, the initial concentration of the process gas is on the order of 20 g O_3/Nm^3. Assuming a 90% practical yield for an average contacting system, ozone concentrations in the offgas may attain 2 g/Nm^3. On repeated injection or preozonation, the final losses in the offgas can be on the order of 0.2–0.5 g/Nm^3.

The threshold limit value (TLV) for ozone is 0.1 ppm (vol) or about 0.2 mg/Nm^3, while a concentration of 0.6 mg/Nm^3 (0.3 ppm vol) is allowable for short residence times (less than 30 min). The toxicity of zone is reviewed elsewhere [9,34,35], particularly for ambient concentrations higher than the TLV or maximum allowable concentration (MAC) values.

Another aspect of the treatment of offgases containing ozone is related to the possible impact of ozone in the atmosphere. In most areas, the "natural" atmospheric ozone concentration varies with altitude from 0.04 to 4 ppm. Average ground-level concentrations are on the order of 0.02 ppm (vol). Seasonal changes have an important impact on these concentrations. Daily factors also influence ground-level concentrations of ozone; the highest concentrations normally are observed at the end of the day in highly populated areas.

According to the most recent studies on the stability of ozone in the ambient air, the half-life of ozone in "fog" is about 4–5 hr. The problem is related to complaints originating from the proximity of ozonation plants to houses.

The detection limit of ozone odor is in the range of 0.02–0.05 ppm (vol). Working capacity of humans is not inhibited at concentrations up to 0.3 ppm; at this level, the complaints remain subjective. No adverse effect on intelligence has been reported after exposure to 0.2–0.3 ppm of ozone. The limiting concentration of ozone that does not damage plant growth is 1 ppm at ground level.

Preventive measurements for the safety of working areas and of equipment must also be considered, and have been reviewed in the literature [35]. These include appropriate venting of the rooms, hydraulic joints to tighten the contacting chambers, and special devices to isolate possible leakages from cable and pipe joints into concrete contacting chambers. Special stress must be placed on the fact that taking appropriate measures for the destruction of ozone in the offgases and the venting of the working areas also is necessary to protect the electronic parts of monitoring equipment.

Presently available methods for elimination or destruction of ozone in contactor offgases are:

1. preozonation;
2. recycling of the offgases (to the ozone generator);
3. dilution (before venting);
4. washing and/or chemical scrubbing;
5. thermal decomposition;
6. adsorption and reaction on combustible support;
7. catalytic decomposition; and
8. adsorption accompanied and/or followed by decomposition.

Some of the techniques are still experimental in nature. At present, thermal decomposition is the most widely used method, but the high energy costs involved stimulate further research on the subject.

Preozonation

In preozonation, the air escaping from the contacting chamber is reinjected into unozonated water (see Figure 33). In the case of ozonation used as a finishing step in drinking water treatment, the yield of ozone absorption in the preozonation phase is once again on the order of 90%. The problem remains; however, the ozone concentration is now lowered by a factor of 10, e.g., 0.1 g/Nm^3 instead of 1 g/Nm^3.

By preozonating raw water containing rapidly reacting dissolved substances and bulk material, ozone in the offgases can be destroyed quantitatively. However, most existing treatment plants have not been designed for this purpose. Thus the raw water intake is often far from the ozonation plant. Also the basins and work areas where the raw-water flow is accessible for ozone contacting usually are not foreseen in the initial construction, with the securities required for ozone contacting. In new designs the preozonation of raw water may receive more attention when establishing designs for the future.

The preozonation system needs either a self-aspirating device, e.g., an injector of water-circulating turbine, or a compressing stage with an appropriate stainless steel water-ring compressor. Energy costs to operate the systems range in the following order:

- injector: 200 (max. 800) W-hr/Nm^3;
- turbine: 100 (max. 200) W-hr/Nm^3; and
- compressor: 80 (max. 150) W-hr/Nm^3;

Part of these costs may be deduced as being due to the use of ozone in the preozonation phase: ±40 W-hr/Nm^3 offgas recontacted.

Recycling

Recycling of the offgases to the ozone generator normally is practiced when an oxygen-rich gas is used for ozone generation. This technique is based on economical reuse of oxygen, which is the determining factor. The method has been attempted by the city of Paris at the St. Maur plant. To be successful, the offgas must be either pressurized or sucked through the gas conditioning system of the ozone production chain, as indicated in Figure 40.

However, gradual enrichment of nitrogen and carbon dioxide contents in the recycled gas [6] is inherent to the process even if air only is used. Therefore a purge and a supplement of fresh gas are necessary to prevent a drop in the ozone production yield. Efficient trapping of trace organics also is necessary, to avoid their gradual accumulation on the adsorbent in the drying tower. Corrosion-free materials or materials resistant to moist ozone also are necessary at certain points in the recycling chain.

Ozone in the contactor offgases gives no real increase in ozone concentration at the exit of the ozone generator. This corresponds to the concept of the ozone generator being a chemical reactor operating at equilibrium [7].

The extra costs for treating the offgases associated with the process are essentially those of compressing: 80–100 W-hr/Nm^3. The extra cost for special corrosion-resistant materials used in the gas preparation and recycling chains is dependent on the plant. It may vary between 5 and 10% of the costs of the ozone producing and contacting system.

Dilution

Dilution of the offgas containing the lost ozone by fresh air in the venting system is sometimes a practical method. However, the dilution ratio needed to reach the safety objective of 0.1 ppm ozone in the vented offgas directly could be extremely high, e.g., between 5000 and 10,000. Therefore, the method is only practicable after extensive use of the residual ozone, e.g., through preozonation, and by ensuring an appropriate rate of atmospheric dilution, e.g., 8–10, by adapting exhaust stacks. Under these conditions, a dilution ratio by forced venting of 100–120 is sufficient. With a lowering of 10 mm water column at the point of aspiration, the operating costs amount to 8–10 W-hr/Nm^3 offgas.

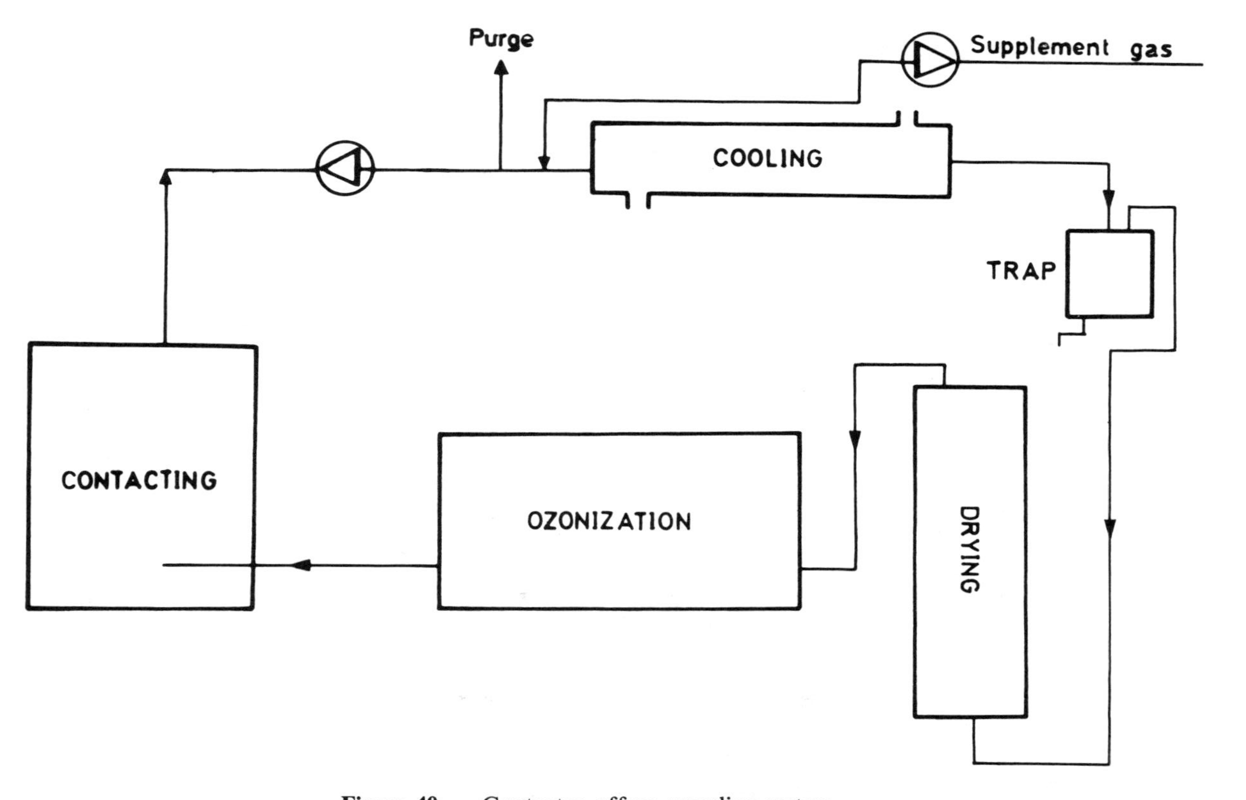

Figure 40. Contactor offgas recycling system.

Despite these very favorable operating costs, the technique remains rarely used. Major problems are the noise produced by the enormous centrifugal ventilators, which exceeds the permitted limit of 60 dBA and gas flow regulation for the different conditions of production, which is of very small variability and may interfere with the progress of ozone contacting. Practical designs for application of the dilution method are the use of air ejectors placed in a noise-absorbing space and thus aspirating the offgas (Figure 41). By this technique, less regulating equipment is necessary.

In the case of the Notmeir plant in Belgium, where the capacity of ozonation is not high (max. 6 kg O_3/hr), offgases from ozonation can be mixed with the exhaust gas of diesel engines or pump motors. In the latter case, the residual ozone reacts with impurities of the exhaust gas. The necessary dilution ratio thus can be lowered to 35 (max) or even, under extreme conditions, to 10.

Washing

Washing the offgas with water in spraying towers is not an efficient method for removal of ozone from offgases, even when the contact tower is packed with Raschig rings [35]. At higher concentrations than the usual, ozone abatement can reach 50%. Actually no studies have yet

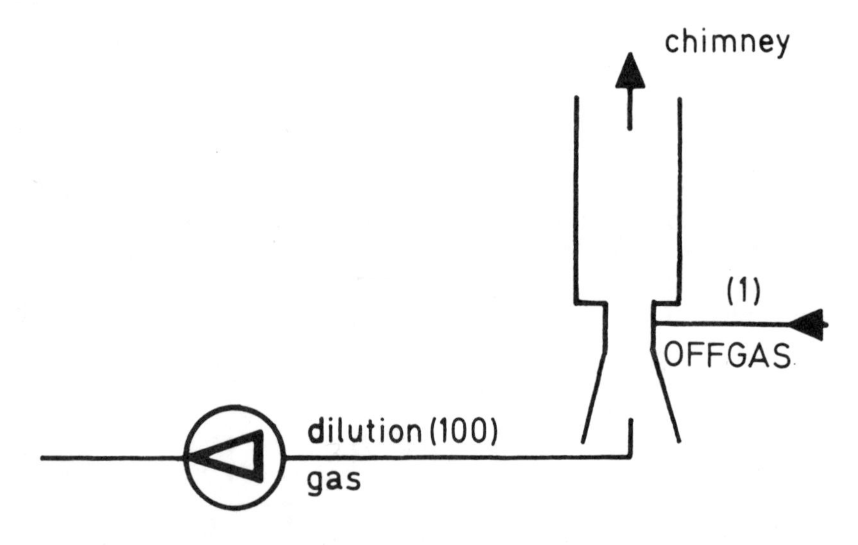

Figure 41. Treatment of offgases by air ejection.

been reported that use ozone-reducing products. Operating costs are limited practically to those of ventilation, and are about 5 W-hr/Nm^3.

Elimination of residual ozone by scrubber-ventilation equipment using an appropriate reductant at first sight may appear to be an attractive possibility. In this case, one must add to the ventilation costs (5–6 W-hr/Nm^3) those for the reductant which are estimated at 20–50 W-hr/Nm^3 and are dependent on the ozone concentration in the offgas. Several different reductants have been studied in the scrubbing system. The most important include solutions of ferrous sulfate and/or sodium chlorite. From preliminary investigations, this technique seems less suitable, since the reaction rates are insufficient to remove the ozone to appropriate levels [35].

Thermal Decomposition

Thermal decomposition is at present the most widespread technique for elimination of ozone in the offgases of ozonation plants. Three principal techniques are available:

1. heating in a single-passage electrical resistance;
2. heating through a thermal exchange; and
3. heating and combustion by burning.

The respective investment costs are 1, 2.5 and 1.3, respectively, for the three techniques.

Ozone in air is more stable than in water; its half-life in the gas phase may vary from 4 to 12 hours at ambient temperature.

Thermal decomposition of ozone in air starts as early as 30°C and is significant at 40–50°C. At 200°C the destruction rate is about 70% and at 230°C, 92–95% within 1 min. At 300°C and higher, 100% decomposition is achieved within a reaction time of 1–2 sec (Figure 42).

Heating in a single-passage electrical resistance is a simple flow-through process with large capabilities for easy automation. Head losses are on the order of 20–30 mm water column. The released gases attain temperatures of 250–300°C, and require the use of refractory materials in the construction of the exhaust chimneys. Moreover, overdimensioning of the exit pipes is required to enable localization of the heating units. A gas flow of 300 ± 100 Nm^3/hr demands a section of 0.6×0.6 m to be treated. Operating costs of the system vary from 130 to 170 W-hr/Nm^3 offgas.

Heating the offgas in a heat exchanger enables recovery of part of the energy by preheating of incoming gas. The total construction is larger

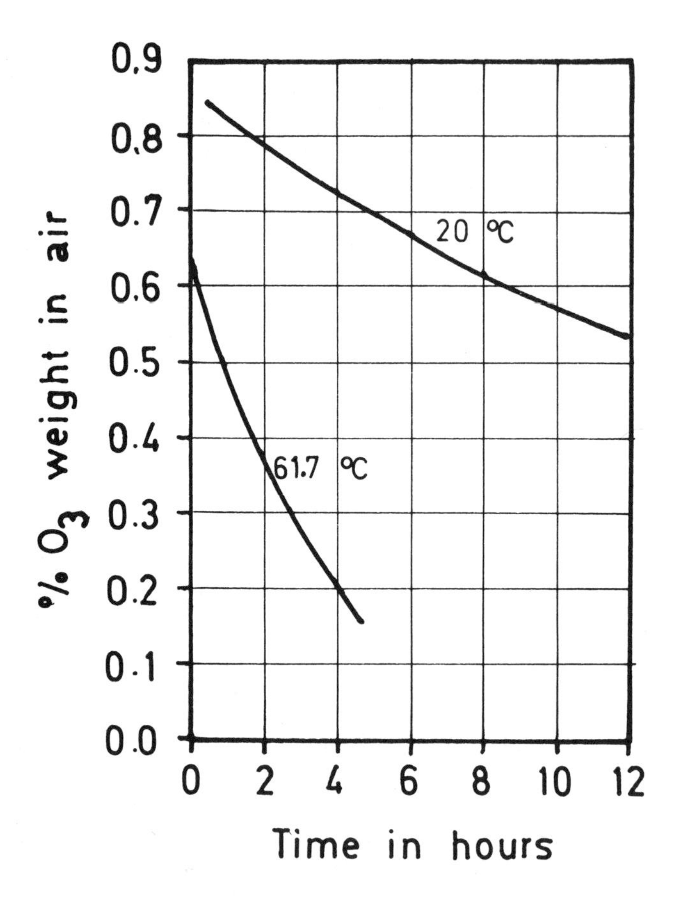

Figure 42. Ozone half-life in air as a function of temperature.

than for resistance heating and is illustrated in Figure 43. Operating costs may be evaluated on the basis of practical existing systems at 85 W-hr/Nm^3. Because of the exchange process, the final temperature of the exit gas attains 90–100°C, thus enabling the use of conventional materials for construction of conduits.

Head losses in the systems can amount to 1 m water column. This high value renders automation difficult. Furthermore, centrifugal ventilators are hardly resistant to corrosion caused by moist ozonated gases. Therefore, they are best placed after the destruction units to evacuate the gases by aspiration and blowing. For heat exchangers and blowers placed upstream, the exchanger must be built of corrosion-resistant materials, e.g., stainless steel AISI 316 or 318. In this sequence, the blowers must be

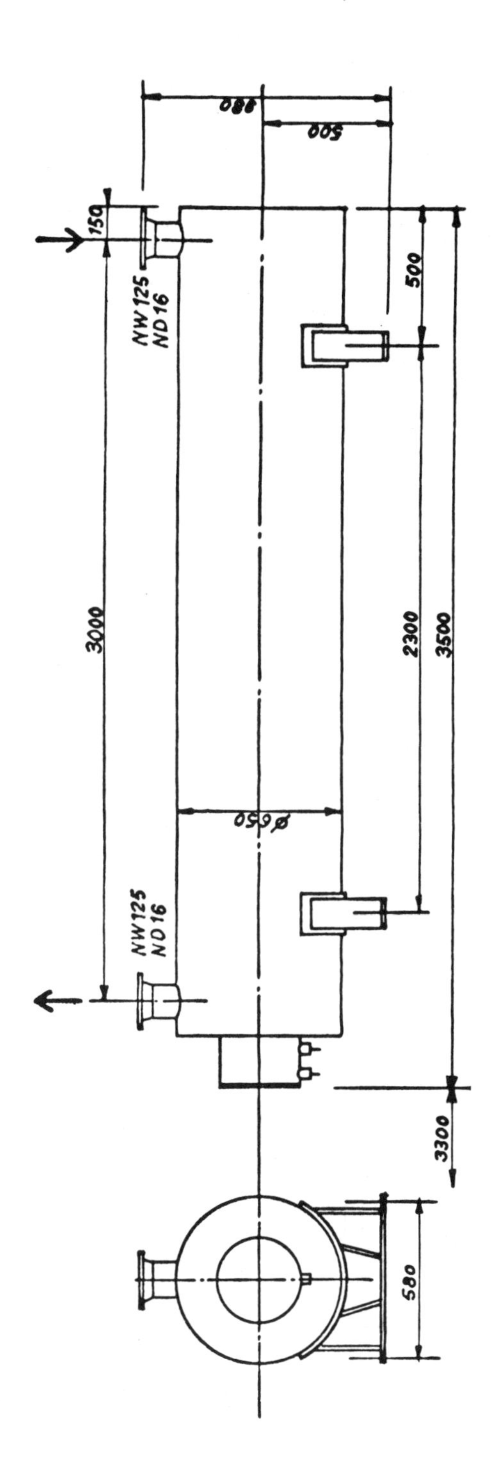

Figure 43. Space requirements for a heat exchanger.

of the water-ring type, and operation costs are prohibitive. Therefore, aspirating equipment is best placed at the exit of the destruction unit. In this case, conventional constructions with corrosion-resistant epoxy coatings are sufficient. With oxygen-rich process gases the equipment should correspond to the applicable safety requirements.

Besides the directly heated thermal exchanger, destruction can be achieved in an indirect exchanger, e.g., of the Frölich type (Figure 44). The system is installed in the Amsterdam works at Weesperkapsel. In this system, the heated gases are exchanged with the inflowing offgases that reach 200°C at the exit. The exchange yield is 60–70% and the preheated gas is then directed to an oven equipped with a fuel-oil burner operated to heat the offgases at 300°C. The exit gas of the oven is then directed to the Frölich exchanger to preheat the incoming offgases. The exchanger is constructed of stainless steel and equipped with borosilicate glass pipes through which the heated gas is circulated.

The flexibility of the total system is based on the large flow fluctuations admitted by the burner system, e.g., between 5 and 100% of the

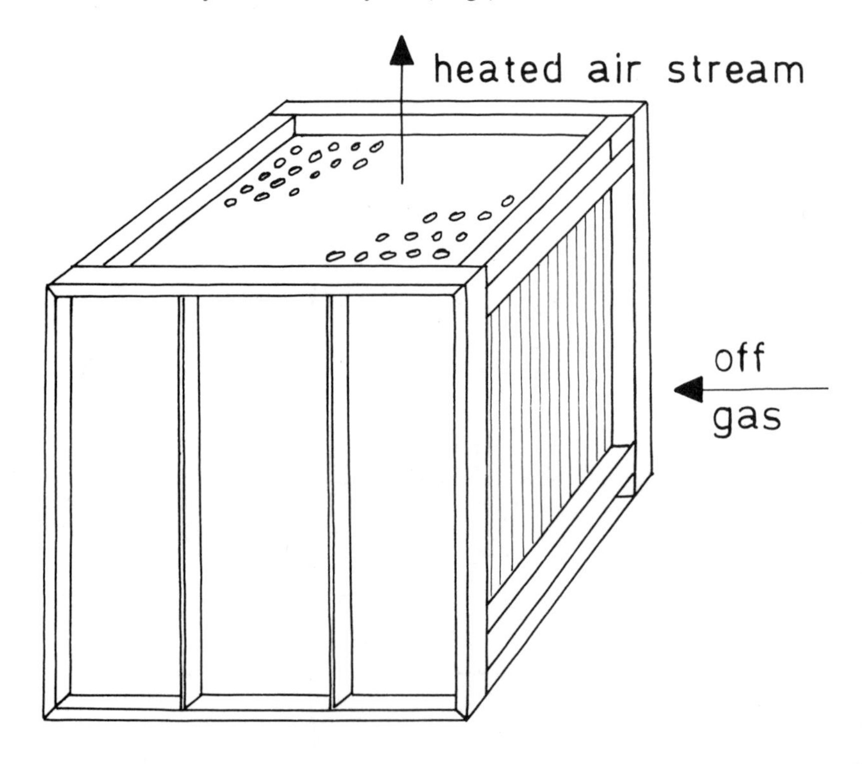

Figure 44. Frölich exchanger.

design capacity. Operation costs require 10 ml fuel oil/Nm^3 offgas and 10 W-hr/Nm^3 for the auxiliaries.

Preheating the offgases by a suitable exchange process not only enables economizing part of the operation costs, but also enables lowering of the process temperature in the combustion zone. At a 120-sec residence time of the offgas in the heated oven, 350°C is necessary to achieve complete destruction. The design also must ensure sufficient mixing of the gases in the oven. The total equipment occupies considerable space (Figure 45).

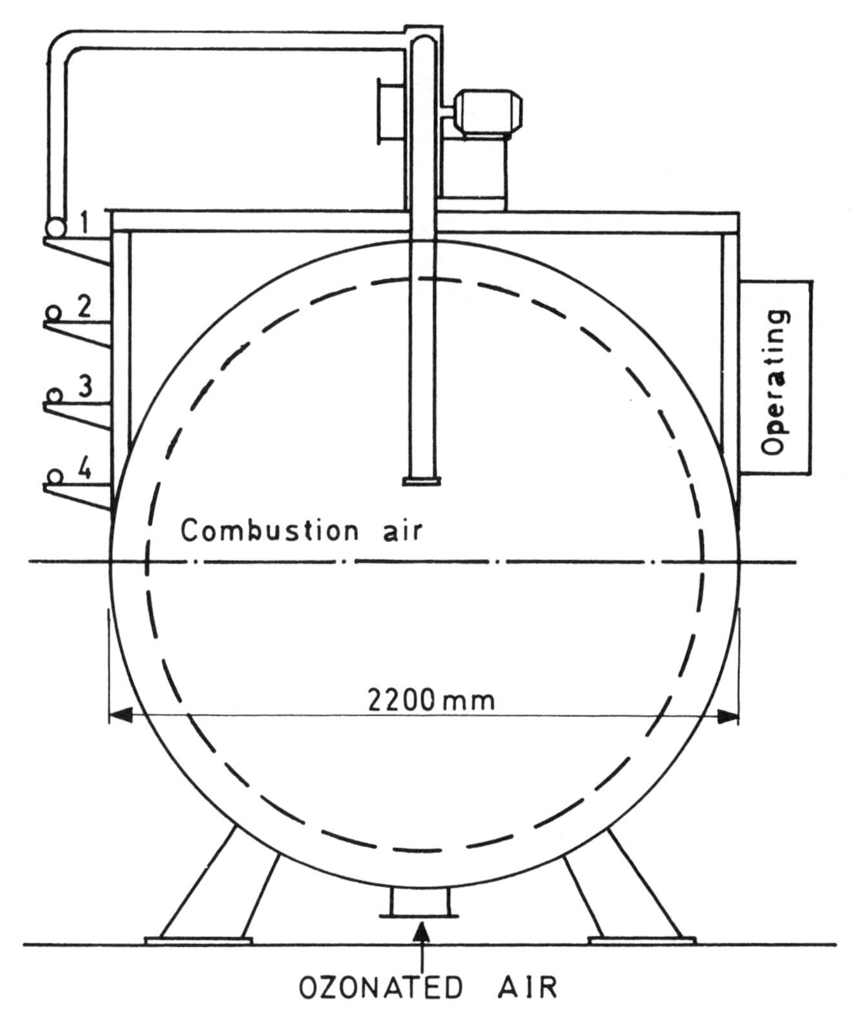

Figure 45. Scheme of a combustion oven for ozone destruction.

The approximative dimensions of the oven illustrated in Figure 45, designed to treat a flow of 400 Nm^3/hr of offgas, are 2.2 m in diameter and 10 m in length; hence, an interior volume of about 10 m^3 is required. The total gas flow must be regulated as well as the combustion rate to maintain the desired temperature. Without secondary heat exchangers, the operational costs require 30 ml fuel oil/Nm^3 offgas to be treated, while 10 W-hr/Nm^3 is necessary for auxiliary equipment, blowing and regulation of the burner.

Adsorption

Destruction of ozone by adsorption on a combustible support consists in practice of the use of an upflow filter filled with an activated carbon layer. The ozone consumes the carbon by slow-rate combustion.

Fundamental design parameters are to use 2 liters (~1 kg) of activated carbon to treat 1 Nm^3/hr offgas and to construct the filter with a carbon layer thickness of 1.2 m. The head loss which results is 0.2–0.3 bar. To obtain a complete reaction, the filter layer is best heated to 60–80°C. This warmup is best obtained by circulating boiling water into a peripheral exchanger surrounding the filter.

The process can sometimes evolve to unsafe conditions in which severe explosions can occur. These are caused by accumulation of unstable reaction products of ozonation, e.g., hydroperoxides. CO radicals also can be formed, inducing high energy-releasing conversions of carbon oxides. These risks are avoided by appropriate water-spraying over the carbon layer. The system is also forbidden in using oxygen-rich gas for generation of ozone.

Its low cost of operation is an advantage of the process, as it requires only 12 W-hr/Nm^3 offgas to be treated.

Catalytic Decomposition

Catalytic decomposition of residual ozone in the offgas enables a faster decomposition of ozone than with activated carbon. Most currently available catalysts are based on palladium, but other metal oxides such as manganese and nickel oxides also can be active. Sometimes the active catalyst is coated on a support to enable easier operation, e.g., palladium-based coating on aluminum granules. The exact formulation of commercially available catalysts always remains proprietary to the manufacturers. Furthermore, present knowledge in this field is only preliminary [35–37]. The most widespread catalysts for use in ozone destruction are COO37 and E 221P, obtained from Degussa, and the Harsaw-MnO-201T catalyst. All of these catalysts are rapidly exhausted when moisture is present. As this is always the case in ozone contactor offgases, permanent heating of the mass is necessary.

For the COO37 (Degussa) catalyst the optimum operating temperature is between 70 and 80°C, while during regeneration periods the temperature must be raised to 120°C, but must not exceed 130°C. Acid oxides of boron, nitrogen oxides and most chlorine compounds may deactivate the catalyst irreversibly. The minimum contact time necessary to obtain a significant destruction rate is on the order of 0.4 sec at an ozone concentration of 10 g/Nm^3.

The velocity number (hr^{-1}) is (Table IV):

$$VN = (hr^{-1}) = \frac{\text{vol offgas treated } (Nm^3)}{\text{vol catalyst } (m^3) \times \text{time of 1 cycle}}$$

This velocity number is a direct function of the ozone concentration in the gas under consideration (Figure 46).

Experiments on dry ozonized gases treated at 15°C have confirmed the basic relationships of Figure 46 [37]. At ambient temperatures, the same quantitative decomposition of ozone is obtained with a moist offgas; however, the length of the process cycle is shortened.

According to our experiences in Brussels, the optimum operating temperature of the catalyst bed can be between 30 and 40°C [37], when

Table IV. Ozone Decomposition Rates over COO37 Catalyst as a Function of Catalyst Bed Temperature (≥99% Destruction)[a]

Temperature (°C)	Dew Point (°C)	Gas/Catalyst Volume Ratio (Nm^3/m^3)
15	20	4,800
26	20	13,300
31	23	280,000
42	20	590,000
46	20	107,000
60	5	67,000

[a]Empty bed contact time ≃ 1 sec.

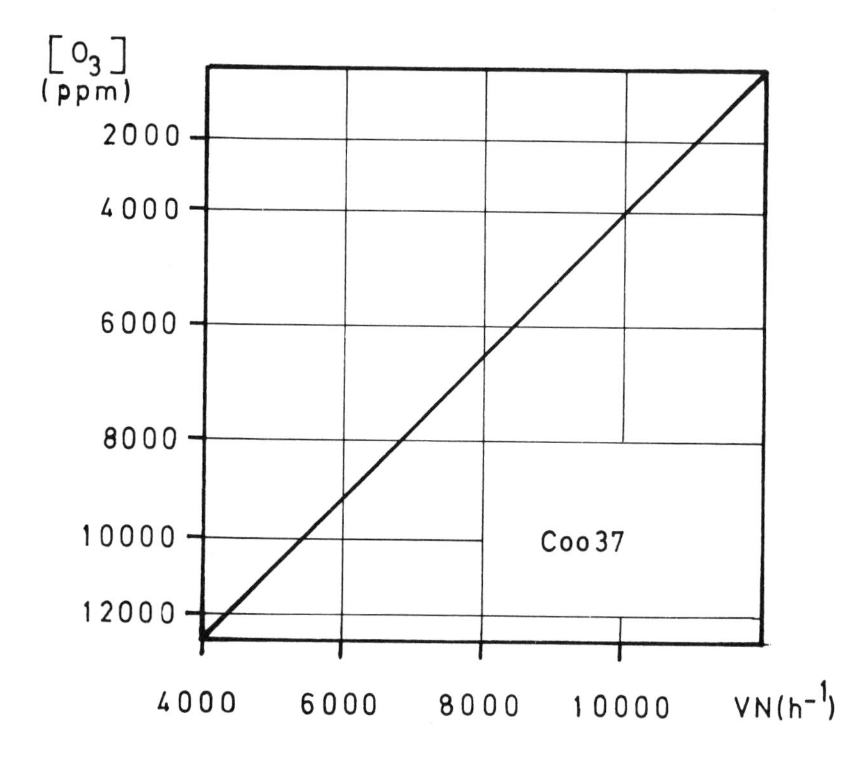

Figure 46. Correlation of velocity number and ozone concentration for COO37 catalyst.

the destruction yield is considered with the costs associated with the heating intensity.

The E 221P catalyst is a palladium catalyst described to be able to support a more intensive regeneration than COO37. Regeneration temperatures up to 520°C during 8 hours have been advanced. Also the inactivation by nitrogen oxides and chlorinated products is reversible on intensive thermal regeneration. Sulfur compounds poison this catalyst.

On the same basis as the data for the COO37 catalyst, the volume ratios for at least 99% decomposition (at the initial concentration of 3 ± 1.5 g O_3/Nm^3 are established for the E 221P catalyst as shown in Table V. The Harsaw MnO-201 T (1/8 in.) catalyst provides equal performances at ambient temperature when applying a dry process gas. With moist offgases, however, the catalyst gives a lower yield, even at higher temperatures of the contact bed (Table VI). These manganese-containing catalysts appear to be less efficient for ozone destruction in an offgas saturated with water.

Table V. Ozone Decomposition Rates over E 221P Catalyst[a]

Catalyst Temperature (°C)	Gas/Catalyst Volume Ratio (Nm^3/m^3)
15	36,000
25	13,000
28.5	27,000
30	400,000
37	1,070,000
43	560,000

[a]Based on a theoretical contact time of 1 sec. Dew point is 20°C.

Table VI. Ozone Decomposition Rates over MnO-201 T Catalyst[a]

Catalyst Temperature (°C)	Gas/Catalyst Volume Ratio (Nm^3/m^3)
40	32,000
50	67,000
60	93,000

[a]Theoretical contact time = 1 sec. Dew point = 20°C.

At present, further investigation is necessary of both the costs and operational characteristics of the use of catalysts for ozone destruction. The available data reported here enable a first approximation of direct operation costs of about 5 W-hr/Nm^3 offgas to be treated. These include heating of the contacting layer containing the catalyst. Poisoning frequency and the price of the catalyst are other important economic aspects inherent to the process.

Adsorption/Decomposition

Adsorption and decomposition is another technique for possible auxiliary destruction of ozone, e.g., in the gas recycling processes.

Silica gel has the property, when freshly activated, to fix the ozone from a transient gas. The data are summarized in Figure 47. The contact time theoretically amounts to 8–10 sec and the mass is gradually

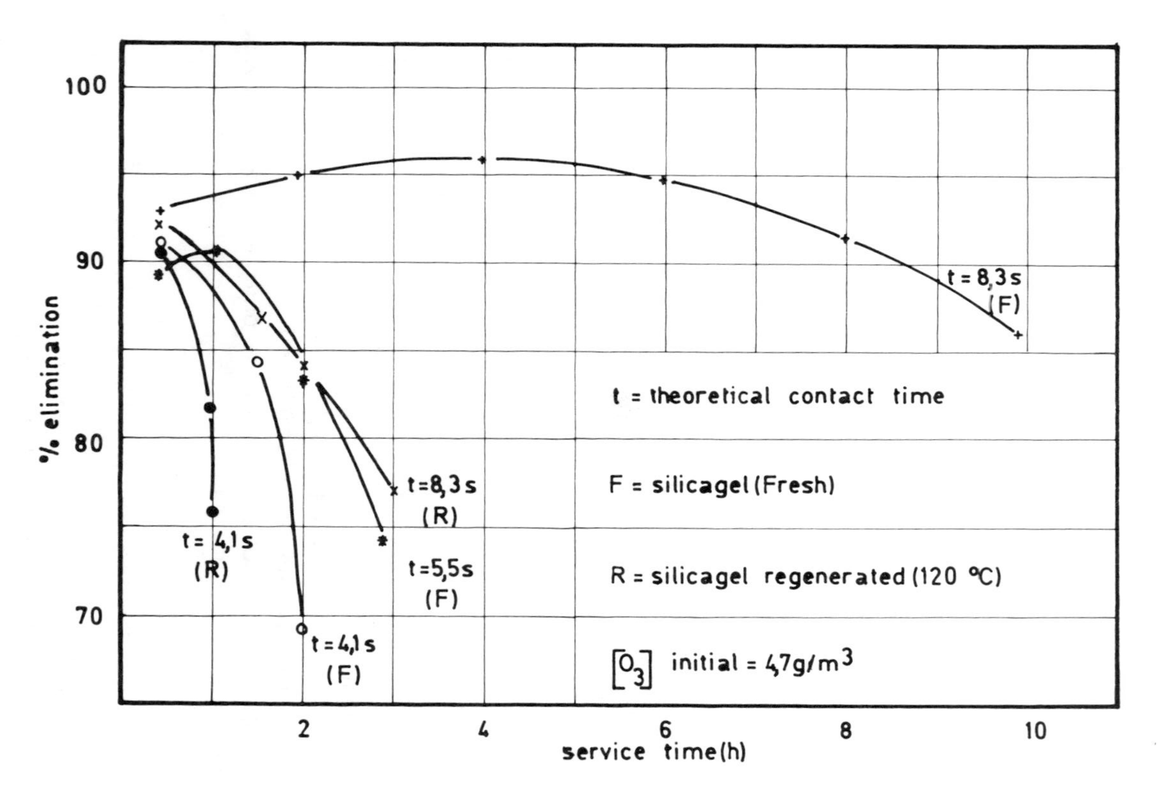

Figure 47. Adsorption of ozone on silica gel as a function of residence time in the adsorption layer.

exhausted. After several thermal regenerations, the ozone decomposition properties of the mass are lowered. Zeolite-containing components behave similarly to silica gel but exhaust slower than does silica gel. Moreover, the deactivation of active sites is not irreversible after a prolonged operation time.

The exact conditions for the operation of the adsorption-decomposition technique, involving thermal decomposition of the adsorbed ozone during the regeneration, still require further investigation. These are more particularly concerned with the abrasion and degradation of the adsorbing material on repeated thermal regeneration.

The basic idea of these processes is to find a contact material capable of concentrating ozone by adsorption from the offgases and to enable thermal and/or catalytic destruction by heating a reduced volume compared to that of the original flow of the offgas. Data on the subject have already been published [35], but further research is still under way. The most promising materials are adsorbing zeolites.

Some manufacturers of ozonation equipment, however, are concerned that adsorption of ozone onto solid surfaces may pose a safety problem. This is because volatile organic materials, as well as ozone, may also concentrate onto the adsorbent surfaces. If the concentrations of ozone and organics become high enough, oxidation of the adsorbed organics (along with decomposition of the adsorbed ozone) may occur violently. Because of these possibilities, the manufacturers recommend that ozone in contactor offgases be destroyed before the treated gases are recycled through adsorbents.

The possible operating costs of the adsorbent process are on the order of 4–6 W-hr/Nm^3, necessary for thermal regeneration of the adsorbing mass.

As a conclusion to this discussion of the principles of the treatment of ozone contacting offgases, a comparative table of operation costs of the different systems may be illustrative (Table VII).

ACKNOWLEDGMENTS

The preparation of this contribution has required many investigations, much study and many developments in which several collaborators have been involved. The preparation of the manuscript was carried out with the invaluable assistance of Mrs. Edith Rogge and Mr. André Henin. To my friend Dr. Rip G. Rice I am grateful for providing the opportunity to write this contribution to an important undertaking so as to provide reliable and broad international exchange of information on ozone.

Table VII. Comparative Operational Costs of Ozone Offgas Treatment Systems

System	Operation Cost (W-hr/Nm³)	Major Advantage	Major Disadvantage
Preozonation	80–150	Ozone is Used	Partial Destruction Only
Recycling	80–100	No Effluent	Corrosion Risks
Dilution	8–10	Easy Operation	Noise Problem
Scrubbing	25–60	Safe Equipment	Partial Reaction Only
Heating			
Single	130–170	Easy Monitoring	Hot Offgases
Exchanger	85	Good Yield	Difficult to Automate
Combustion	≅150	Total Destruction	Large Equipment
Adsorption on Activated Carbon	10–15	Static Operation	Explosion Danger
Catalysts	5	Small Equipment	Catalyst Poisoning
Adsorption on Silica Gel	2	Partial Yield	Irreversibility on Regeneration
Adsorption on Zeolites	1–2	Long Cycle Periods	Still Experimental

REFERENCES

1. Murray, C. N., and J. P. Riley. "The Solubility of Gases in Distilled Water and Sea Water (Oxygen), *Deep-Sea Res.* 16:311 (1969).
2. Murray, C. N., J. P. Riley and T. R. S. Wilson. "The Solubility of Gases in Distilled Water and Sea Water (Nitrogen)," *Deep-Sea Res.* 16:297 (1969).
3. Carpenter, J. H. "New Measurements of Oxygen Solubility," *Limnol. Oceanog.* 11:264 (1966).
4. Landine, R. C. "A Note on the Solubility of Oxygen in Water," *Water Sew. Works* (August 1971), p. 242.
5. Rawson, A. E. "Studies in Ozonization. Part 2. Solubility of Ozone in Water," *Water Water Eng.* (March 1953), p. 102.
6. Masschelein, W., G. Fransolet, J. Genot and R. Goossens. "Perspectives de l'ozonization de l'eau au départ d'air enrichi en oxygène," *T.S.M.-L'Eau* 71:385 (1976).
7. Masschelein, W. J. "L'optimalisation de la capacité de production d'ozone dans le traitement des eaux," *T.S.M.-L.'Eau* 72:177 (1977).
8. Simon, M., and H. Scheidtmann. "Die neue Ozonanlage der Stadtwerke Duisburg," *Gas Wasser Forsch.* 32:877 (1968).
9. Hoigné, J. "Mechanisms, Rates and Selectivities of Oxidations of Organic Compounds Initiated by Ozonation of Water," Chapter 12, this volume.
10. Gomella, C. "Diffusion de l'ozone dans l'eau," *Houille blanche* 4:439 (1967).

11. Guillerd, J. R. "L'évolution dans le traitement des eaux par l'ozone aux cours des 15 dernières années," *T.S.M.-L'Eau* 63:279 (1968).
12. Mignot, J. "Pratique de la mise en contact de l'ozone avec le liquide à traiter," in *Proceedings of the Second International Symposium on Ozone Technology,* R. G. Rice, P. Pichet and M.-A. Vincent, Eds. (Vienna, VA: International Ozone Association, 1976), pp. 15–46.
13. Masschelein, W. J. "Techniques de dispersion et de diffusion d'ozone dans l'eau," in *International Symposium, Ozon und Wasser, Wasser Berlin – '77* (Berlin, Federal Republic of Germany: AMK Berlin, 1978), pp. 118–139.
14. Carré, J. "Détermination simultanée de l'asymptate et de l'exposant sur l'exemple de la dissolution de l'oxygène dans l'eau," *T.S.M.-L'Eau* 71:35 (1976).
15. Malleviale, J., M. Roustan and H. Roques. "Détermination expérimentale des coefficients de transfert de l'ozone dans l'eau," *Trib. Cebedeau* 377:175 (1975).
16. Danckwerts, P. V. *Gas-Liquid Reactions* (New York: McGraw-Hill Book Company, 1970).
17. Masschelein, W. J., R. Goossens and C. Houbrechts (in preparation).
18. Masschelein, W. J., G. Fransolet and J. Genot. "Techniques for Dispersing and Dissolving Ozone in Water," *Water Sew. Works* 122:12–57 (1975).
19. Bredtmann, M. "Ozonausuntzung durch Anwendung des Belüfterprincips bei der Trinkwasseraufbereitung," *Gas Wasser Forsch.* 115:326 (1974).
20. Gagnaux, A. "Gas-Liquid Contacting Apparatus," German Patent 2,253,396 (May 10, 1973).
21. Miller, G. W., R. G. Rice, C. M. Robson, R. Scullin, W. Kühn and H. Wolf. "An Assessment of Ozone and Chlorine Dioxide Technologies for Treatment of Municipal Water Supplies," U.S. EPA Report No. 600/2-78-147 (Washington, DC: U.S. Government Printing Office, 1978), pp. 115–21.
22. Nebel, C., P. C. Unangst and R. D. Gottschling. "An Evaluation of Various Mixing Devices for Dispersing Ozone in Water," *Water Sew. Works* Ref. No. R-6 (1973).
23. Kurzmann, G. E. "Das indirekte Ozon-Druckverfahren," *Wasser Luft Betrieb.* 7:11 (1963).
24. O'Donovan, D. C. "Treatment with Ozone," *J. Am. Water Works Assoc.* 57:1167 (1965).
25. Hopf, W. "Zur Wasseraufbereitung mit Ozon and Aktivkohle," *Gas Wasser Forsch.* 111:83,156 (1970).
26. Schenk, P. "Die Wasseraufbereitungsanlage des Wasserwerkes Düsseldorf 'Am Staad'," *Gas Wasser Forsch.* 103:791 (1962).
27. Berger, K. "Über den Betrieb der Ozonanlage Königsberg der Wasserversorgung Bern," *Gas Wasser Forsch.* 105:1338 (1964).
28. Kolin, L. "Vorteil und Kostungliederung der Ozonisierung von Trinkwasser," *Gas Wasser Wärme* 19:79 (1965).
29. Kopecky, J. "Entkeimungsprobleme des Salzburges Wasserwerke," *Österreich. Wasserwirts.* 20:198 (1968).
30. "Loch Turret Water Scheme," *Water Water Eng.* 869:263 (1968).
31. Stahl, D. E. "Ozone Contacting Systems," in *First International Symposium on Ozone for Water and Wastewater Treatment,* R. G. Rice and

M. E. Browning, Eds. (Vienna, VA: International Ozone Association, 1975), pp. 40–55.

32. Meyer-König, W. C. E., and E. Carl. "Die Mikrosieb- und Ozonanlage der Bodensee Wasserversorgung," *Gas Wasser Forsch.* 113:25 (1972).
33. Grombach, P. "Ozon als Oxidationsmittel bei der Seewasser-Aufbereitung," *Gas Wasser-Abwasser* 55:533 (1975).
34. Masschelein, W. J. *Ozone in Water Treatment* (Brussels, Belgium: Ecochem, 1975).
35. Masschelein, W. J., G. Fransolet and B. Avalosse. "Destruction de l'ozone. Coût et sécurité dans les installations," in *Applications de l'Ozone au Traitement des Eaux* (Paris: International Ozone Association, 1979), pp. 469–514.
36. Weissenhorn, F. J. "Führung von Ozon im Kreislauf und Umwandlung," in *International Symposium, Ozon und Wasser, Wasser Berlin—'77* (Berlin, Federal Republic of Germany: AMK Berlin, 1978), pp. 140–147.
37. Masschelein, W. J., G. Fransolet and B. Avalosse (in preparation).

SECTION 3

MATERIALS OF CONSTRUCTION

CHAPTER 7

RECENT EXPERIMENTAL STUDIES DEALING WITH CORROSION AND DEGRADATION OF MATERIALS IN OZONE-CONTAINING ENVIRONMENTS

R. Zawierucha and H. Charleson

Materials Engineering Laboratory
Union Carbide Corporation
Linde Division
Tonawanda, New York

A key aspect of any system for generating or contacting ozone in liquid or gaseous environments involves the materials of construction. This chapter acquaints the reader with recent experimental studies dealing with this aspect. The experimental studies were conducted by the Linde Division of Union Carbide Corporation at its Tonawanda, NY, site.

BACKGROUND

Producers of air separation plants and industrial gas distribution systems use components handling dry oxygen that require the selection of materials to minimize oxygen compatibility problems, e.g., combustibility. Through years of testing and plant operation the material database is well established.

Introduction of the oxygen activated sludge UNOX® system (Union Carbide Corp., New York, NY) resulted in requirements for material suitability data in high-humidity gas- and liquid-phase applications. To obtain these data, extensive environmental testing was done using specimen racks in UNOX pilot plants and a number of early commercial plants. Hundreds of UNOX plants, in operation and under construction, testify to the feasibility of this approach.

Of particular interest in this chapter is the use of high-purity oxygen to generate ozone. Two basic systems for using oxygen-based ozonation systems are (1) the recycle system used for municipal and industrial water and wastewater treatment, and industrial process applications, and (2) the once-through system integrated with an oxygen-activated sludge wastewater treatment system for wastewater disinfection. The authors' laboratory participated in a program to generate corrosion data applicable to these requirements. Candidate materials were tested in high-humidity ozone-bearing gaseous environments and in aqueous phases containing high levels of dissolved oxygen and ozone. The results of these tests will be discussed in this chapter.

EXPERIMENTAL

The testing of materials for use in ozone contactors has been conducted in a fashion similar to the previously discussed UNOX systems materials testing. All of the testing has been done at Tonawanda in the course of ongoing development activities. This is because of the still-limited number of operating ozone water and wastewater treatment facilities in this country, and the lack of demand for onsite pilot-plant testing of the type done in the early days of oxygen activated sludge design.

Alloy selection included a number of materials that were desirable because they had worked well in oxygen systems, and materials that were reputed to work well in ozone environments. Resistance to combustion in oxygen environments, corrosion resistance, economic considerations and fabricability were the major factors considered in choosing materials for testing.

The sample racks were prepared according to generally accepted techniques used in field or laboratory corrosion tests. The coupons were normally several inches square with a mounting hole in the center. The samples were dielectrically isolated from each other and the mounting structure, to eliminate galvanic interaction between samples. Samples and spacers were clamped together with a threaded rod. The area of sample under the spacers, therefore, provided an area for evaluation of the material's susceptibility to crevice corrosion. The samples were weighed and identified before being installed in the contactor.

When the racks were removed, they were disassembled, and all of the samples were examined individually before and after cleaning. Chemical cleaning via inhibited agents was utilized, when required, on metallic samples to remove corrosion products. Nonmetallic samples were cleaned using only a soft brush, soap and water. After the postcleaning observations were made, all of the samples were weighed. The weight changes of metallic samples were used as the basis for calculating general corrosion rates.

Metal samples were examined further for evidence of general corrosion, pitting, crevice and edge attack. As a final step, the maximum penetrations due to pitting or crevice attack were measured and recorded. Due to the localized nature of pitting and its tendency to proceed at erratic rates, general corrosion rates determined by weight loss data can be used only as rough guides to serviceability where pitting has occurred.

In contrast to metals, permeability and chemical aging considerations may affect the compatibility of plastic and rubber materials with an environment. Therefore, for the nonmetallic materials, it was necessary to analyze not only kind and rate of material removed but also the change in overall properties. For this reason, the nonmetallics in most cases were tested as-received for hardness and ultimate tensile strength, as well as for their post-test hardness and strength. The range of variation in these properties is often much higher in these materials than in metals, particularly after environmental exposure. This is especially true of materials that are composite in nature, where selective attack may leach the strength-producing fibers and/or remove the material around them. Also, the number and direction of fibers in any given sample may vary greatly. The values presented in this report are average values.

The results of this testing are presented in Tables I through IV. It may be noted that ozone and dissolved ozone concentrations were higher during these experiments than normally might be expected in an ozone contactor. This was done intentionally to accelerate the results as much as possible, since the exposure periods, ranging between 86 and 187 days, were relatively short.

OBSERVATIONS/RESULTS

While Tables I through IV summarize the test data, selected photomacrographs will be presented to enable the reader to interpret the tabulated numerical results. These are particularly illustrative when effects of crevice corrosion and pitting were significant (with metallic samples) and when other forms of deterioration were noted (with the nonmetallic samples).

Figure 1 shows a good example of a specimen that suffered severe local pitting and crevice corrosion, and was perforated at location P. The specimen shown in Figure 1 was 6061-T6 aluminum, exposed to a liquid-phase environment.

Figures 2 and 3 show Monel® K-500 samples (Huntington Alloys, Inc., Huntington, WV) exposed to vapor- and liquid-phase attack, respectively. Both samples suffered heavy crevice attack, but the vapor phase sample suffered a much heavier pitting attack.

For comparative purposes, several nonmetallic and coated samples are of interest. Extensive surface checking is visible in a vapor-phase Hypalon® sample (E. I. du Pont de Nemours & Co., Inc., Wilmington, DE) shown in Figure 4, while spalling and checking damage were sustained by a vapor phase Ceilcote® 640 sample (General Signal Company, Cleveland, OH) shown in Figure 5.

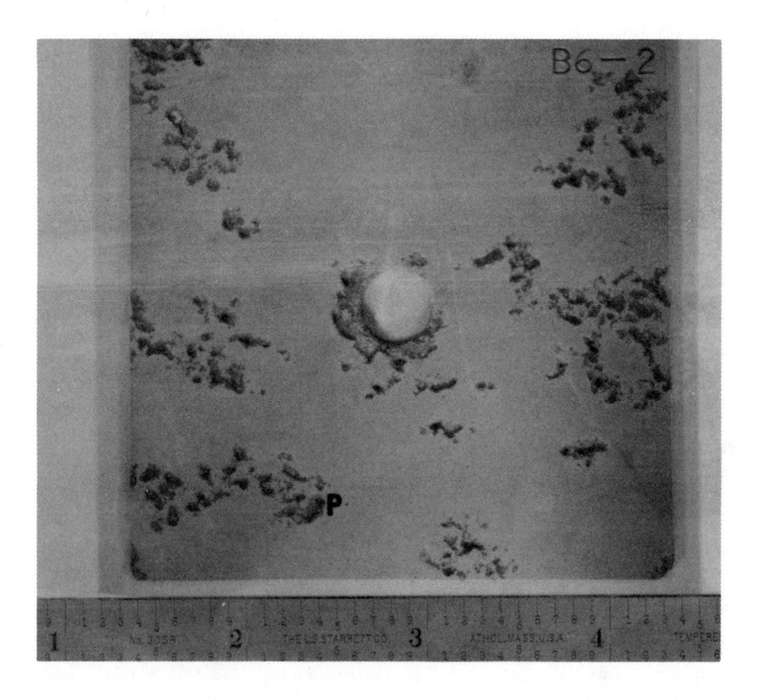

Figure 1. Photomacrograph of 6061-T6 aluminum after testing in a liquid-phase environment containing ozone.

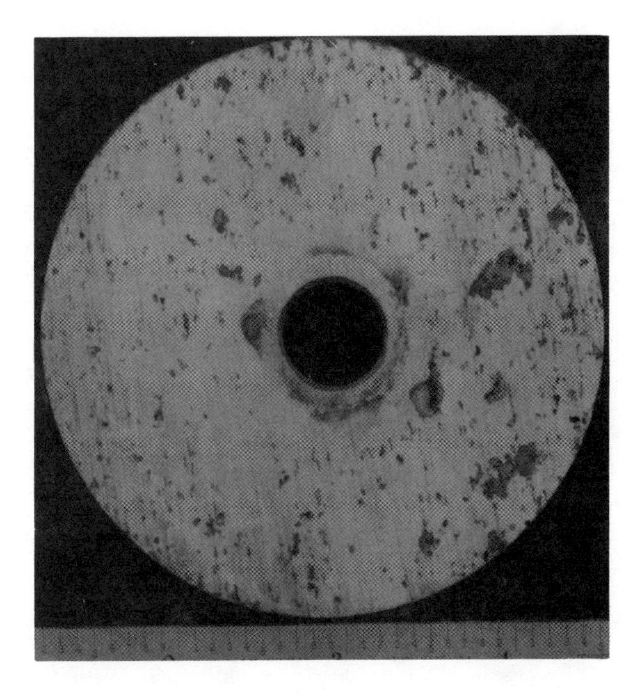

Figure 2. Photomacrograph of Monel K-500 after testing in a vapor-phase environment containing ozone.

Metallic Materials

Stainless Steels

The subcategories of stainless steels tested were the following:

1. austenitic stainless steel, e.g., 302, 303, 304 and 316;
2. martensitic stainless steel, e.g., 410;
3. ferritic stainless steel, e.g., 444L; and
4. precipitation hardening stainless steel, e.g., 17-4PH.

Subcategories 1 and 2 are hardenable via heat treatment, whereas the others are not.

Tables I and II show that 44L stainless steel, a ferritic alloy of the 18Cr-2Mo type, was the best performer of the alloys tested in both the vapor and liquid phases. This assessment included both parent metal and simulated weld (sensitized) exposures. Sensitization is the term used to

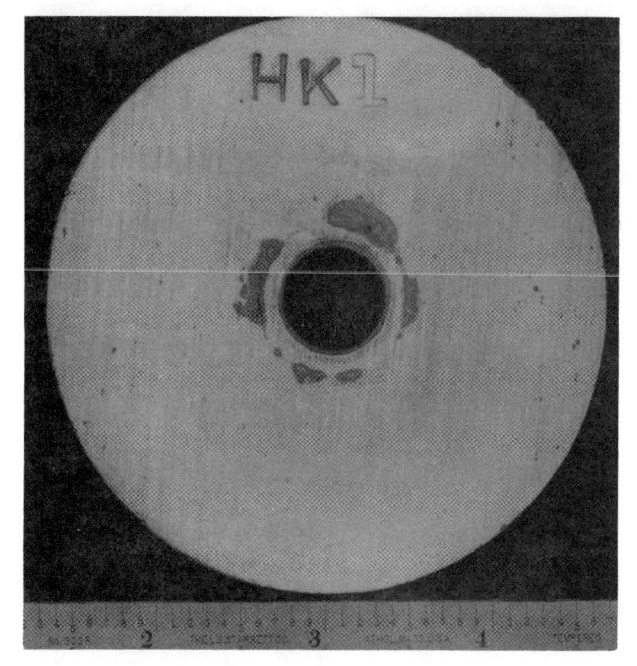

Figure 3. Photomacrograph of Monel K-500 after testing in a liquid-phase environment containing ozone.

describe thermally induced carbide precipitation, which may have a deleterious effect on the corrosion resistance of many stainless steels.

In general, the performance of the stainless steels was better in the vapor phase than in the liquid phase. Sensitization heat treatments designed to simulate weld heat-affected zones resulted in a marked increase of the susceptibility to corrosion damage of the 304 and 316 stainless steels, particularly in the liquid-phase tests. Chromium levels seemed to be important; the 410 stainless steel, the only stainless alloy with a nominal 12% chromium, was markedly inferior to the other stainless alloys with nominal chromium levels of at least 17%. Molybdenum steels gave mixed results. The 444L and 316 stainless steels both contain approximately 2% molybdenum, but the latter alloy was markedly inferior to the former. Alloys 17-4PH and 410, which are hardenable via heat treatment, were more susceptible to corrosion than the nonheat-treatable stainless steels.

Aluminum Alloys

As with the stainless steel samples, it is obvious that corrosion was more severe in the liquid phase than in the vapor phase. Examination of

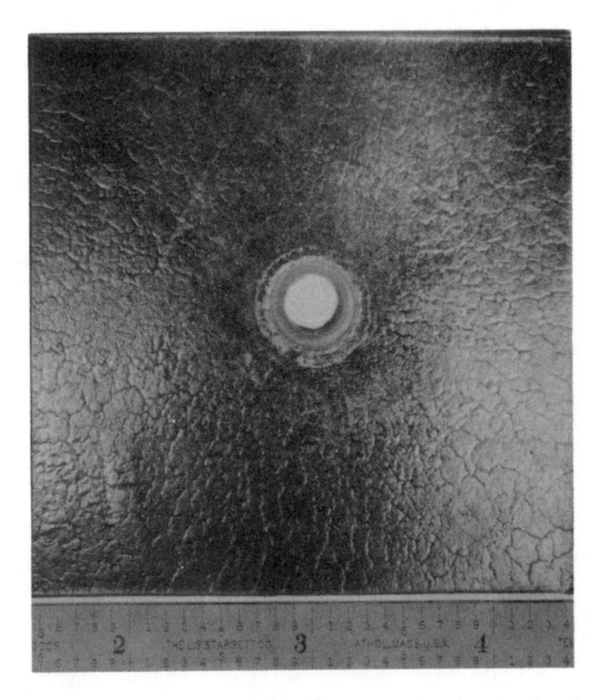

Figure 4. Photomacrograph of a Hypalon sample after testing in a vapor-phase environment containing ozone.

the specimens revealed that local corrosion mechanisms, e.g., pitting and crevice corrosion, were much more significant than general corrosion. In fact, perforation via pitting can occur even in the presence of a low general corrosion rate.

Copper and Nickel Alloys

In contrast to the previous two groups, these materials exhibited more serious corrosion damage in the vapor phase than in the liquid phase, particularly with respect to general corrosion characteristics. This is considered somewhat unusual and unfortunate, since the selection of cuprous and nickel alloys presumably would be made on the basis of combustion resistance in oxygen- or ozone-rich gas. Of these alloys, the free machining brass appeared to be the most promising.

Nickel Superalloys and Plating

The nickel superalloys, represented by a number of Inconel® alloys (Huntington Alloys, Inc., Huntington, WV), performed very well in both

Figure 5. Photomacrograph of a Ceilcote 640 sample after testing in a vapor-phase environment containing ozone.

the liquid and vapor phases. Generally, the corrosion behavior of these materials was similar to that of austenitic stainless steels. Inconel 625, which, on the basis of its molybdenum content would be expected to be the most resistant to pitting, was not tested. The Inconels generally show oxygen compatibility, i.e., resistance to combustion, that is better than average. However, these materials are not as resistant to combustion in oxygen as the cuprous alloys, although they are superior in this regard to stainless steels.

Only one plated composition was tested: Kanigen® nickel (Chemplate Corp., Los Angeles, CA), on a substrate of CF-8M (a cast equivalent of 316 stainless steel). The performance of these samples was considered to be excellent; this material is worthy of further consideration for use in ozone environments.

Babbitt

Babbitt alloys are alloys of tin that may also contain some copper, antimony and, possibly, lead. It is clear that the #1 Babbitt (which con-

Table I. Performance of Metallic Materials in Ozone Contactor, Vapor Phase[a]

Material	Test Duration (days)	General Corrosion Rate[b] (mil/yr)	Max. Pit Depth (in.)		Visual Observation/Comments
			Outside Crevice	Inside Crevice	
304 Stainless Steel	105	0.10	c	Nil	A few small pits scattered over sample
316 Stainless Steel	105	0.10	Nil	Nil	No apparent attack other than slight weight loss
444L Stainless Steel	105	0.10	Nil	Nil	No apparent attack other than slight weight loss
17-4PH Stainless Steel	105	0.10	Nil	Nil	No apparent attack other than slight weight loss
304 Stainless Steel	105	0.10	d	d	Simulated weld, slight crevice attack typical of non-304L sample
316 Stainless Steel	105	0.10	Nil	Nil	Simulated weld, no apparent attack other than slight weight loss
444L Stainless Steel	105	0.10	Nil	Nil	Simulated weld, no apparent attack other than slight weight loss
410 Stainless Steel	105				Moderate crevice attack; heavy edge attack; scattered small pitting
410 Stainless Steel	187	0.26	e	d	Slight crevice attack; heavy edge attack, scattered small pitting
302 Stainless Steel	86	0.10	Nil	Nil	No apparent attack other than slight weight loss
303 Stainless Steel	86	0.10	Nil	Nil	No apparent attack other than slight weight loss
Aluminum, 3003	105	0.50	d	f	Severe crevice attack; scattered local attack; slight edge attack
Aluminum, 6061-T6 Sheet	105				Heavy crevice attack; remainder of sample sustained high density of moderate to heavy local attack
Aluminum, 6061-T6 Tubing	105				Scattered areas of shallow localized attack, with areas of heavy attack at tube ends
Aluminum, 6061-T6 Sheet	187	0.49	f	f	Severe crevice attack; high density of intense local attack
Aluminum, 6061-T6 Tubing	187				Slight pitting attack
Aluminum, Anodized Coupling and Nuts	105				No apparent attack on anodized coupling; both nuts received heavy surface attack
Aluminum, Anodized Fittings	187				Some slight general attack was noted
Aluminum Bronze	105	11.12	f	f	Heavy crevice and edge attack; high density of pits over entire sample

Table I, continued

Material	Test Duration (days)	General Corrosion Rate[b] (mil/yr)	Max. Pit Depth (in.)		Visual Observation/Comments
			Outside Crevice	Inside Crevice	
Phosphor Bronze	105	9.34	f	e	Heavy crevice and edge attack; high density of pits over entire sample
Free Machining Brass	105	1.14	d	e	Concentrated heavy crevice attack; scattered small pitting outside crevice region
Monel 400, Sheet	105	5.74	f	f	Heavy crevice attack, high density of broad deep pits outside crevice, heavy edge attack
Nickel 200, Casting	86	10.27	f	f	Heavy crevice attack; severe pitting attack, possible intergranular attack outside crevice
Monel K-500	86	3.57	f	f	Heavy crevice attack; heavy edge attack, high density of deep pits elsewhere
Monel 400, Casting	86	4.68	f	f	Heavy crevice attack; severe edge attack; high density pitting elsewhere
Monel 400, Bar	86	2.78	f	e	Moderate to heavy crevice attack; high density pitting elsewhere
Nickel, ASTM 296, P-108	86	5.98	f	f	Moderate to heavy crevice attack; high density of small pits and possible intergranular attack outside crevice
Inconel 600	105				Moderate to heavy edge attack; scattered small pits outside crevice region
Inconel 600	187	0.21	e	e	Moderate crevice attack; high degree of pitting; heavy edge attack
Inconel 601	105				No apparent attack; slight staining
Inconel 601	187	0.10	Nil	Nil	Slight edge attack
Inconel 610	86	0.10	e	f	Moderate to heavy crevice and edge attack; scattered shallow pitting outside crevice
Inconel X-750	105				Slight crevice attack; remainder of sample in excellent condition
Inconel X-750	187	0.10	Nil	e	Moderate to heavy crevice attack; incipient pitting outside crevice

Kanigen Nickel Plating on CF-8M Substrate	86				Metallographic examination of coating and substrate showed no deterioration
#1 Babbitt	105	0.10		f	Heavy crevice attack; some scattered pitting outside crevice
#4 Babbitt	105	28.62			Heavy pitting over entire sample, geometry made measurement impossible
Bare Carbon Steel	105	14.49	f	f	Heavily encased in rust, severe crevice attack; heavy pitting attack, scattered edge attack
Galvanized Steel	105	1.11	d	d	Light film of residue, one small pit in crevice and outside, uniform attack on galvanized surfaces; attack on uncoated edges
D_2M Austenitic Iron	105	10.07	e	f	Heavy crevice attack; general attack accompanied by pitting; heavy edge attack

[a]Ozone concentrations = 11–34 mg/1.
[b]General corrosion rate is based on weight loss data and should be used only as a guide where pitting is noted.
[c]Less than 0.001 in.
[d]0.001–0.005 in.
[e]0.005–0.010 in.
[f]Greater than 0.010 in.

Table II. Performance of Metallic Materials in Ozone Contactor, Liquid Phase[a]

Material	Test Duration (days)	General Corrosion Rate (mil/yr)	Max. Pit Depth (in.)		Visual Observation/Comments
			Outside Crevice	Inside Crevice	
304 Stainless Steel	105	0.10	d	c	Very slight crevice attack; 1–2 small pits per side outside crevice area
316 Stainless Steel	105	0.10	d	c	Small pits scattered over the sample, both in and out of the crevice area
444L Stainless Steel	105	0.10	Nil	Nil	No apparent attack other than slight weight loss
304 Stainless Steel	105	0.38	f	f	Simulated weld; severe crevice attack; moderate to heavy edge attack, scattered tunnel pits outside crevice, possible intergranular attack
316 Stainless Steel	105	0.66	f	f	Simulated weld; heavy crevice attack, possibly intergranular, moderate edge attack; scattered pitting
444L Stainless Steel	105	0.10	Incip.	Incip.	Incipient crevice and pitting attack
17-4PH Stainless Steel	105	1.05	Nil	f	Severe crevice attack with tunneling inside mounting hole, no other attack
410 Stainless Steel	105				Heavy rust buildup over ⅓ of one side; severe crevice attack; a few pits outside crevice
410 Stainless Steel	187	1.92	Nil	f	Severe crevice attack with a maximum depth of 0.162 in.; slight edge attack
303 Stainless Steel	86	0.10	Nil	Nil	No apparent attack other than slight weight loss
302 Stainless Steel	86	0.10	Nil	Nil	No apparent attack other than slight weight loss
Aluminum 3003	105	2.39	f	f	Severe crevice attack; numerous areas of severe local attack; heavy edge attack
Aluminum 6061-T6, Sheet	105				Sample covered with tenacious brown stain, severe crevice attack; scattered severe local attack
Aluminum 6061-T6, Tubing	105				Tenacious brown stain over entire sample; severe local attack at tube ends; minor pitting over the rest of the sample

Aluminum 6061-T6, Sheet	187	2.28	Perforated	f	Severe crevice attack; high incidence of severe local attack outside crevice with one perforation
Aluminum 6061-T6, Tubing	187				Severe end grain; severe pitting and crevice attacks
Aluminum Anodized Fittings	105				No apparent attack on coupling; both nuts received heavy attack
Aluminum Bronze	105	2.28	f	f	Moderate to heavy crevice attack; clustered pitting outside crevice; heavy edge attack
Phosphor Bronze	105	0.58	d	d	Moderate crevice attack; scattered small pitting
Free Machining Brass	105	1.12	d	c	Slight crevice attack; scattered small pitting outside crevice
Monel 400, Sheet	105	1.46	f	f	Severe crevice and edge attack; scattered broad, deep pits outside crevice area
Monel 400, Casting	86	0.98	e	f	Heavy crevice attack; scattered isolated pits; slight edge attack
Monel 400, Bar	86	1.05	f	e	Heavy crevice attack; high density of elongated pits on round surfaces; deep pitting on flat surfaces
Monel K-500, Casting	86	1.05	e	f	Heavy crevice attack; moderate density of small, pinhole pits outside crevice area
Nickel 200	86	0.88	f	f	Heavy crevice attack; scattered deep pitting on round and flat surfaces outside crevice region
Nickel, ASTM 296,P108	86	0.47	f	f	Heavy crevice attack; scattered deep pits on round and flat surfaces outside crevice area
Inconel 600	105				Slight crevice attack; remainder of sample appeared unattacked
Inconel 600	187	0.10	Nil	Nil	No apparent attack other than slight weight loss
Inconel 600	105				Crevice attack on inside wall of mounting hole, remainder of sample unimpaired
Inconel 601	187	0.10	Nil	Nil	Apparent intergranular and grain attack within mounting hole; very slight edge attack
Inconel 610	86	0.10	d	d	Moderate crevice attack; slight edge attack; a few small, scattered pits outside
Inconel X-750	105				Moderate to heavy crevice attack; no other visible damage
Inconel X-750	187	0.10	Nil	d	Large area of shallow crevice attack
Kanigen Nickel Plating On CF-8M Substrate	86				Metallographic examination revealed no attack on substrate, coating performed satisfactorily

Table II, continued

Material	Test Duration (days)	General Corrosion Rate (mil/yr)	Max. Pit Depth (in.)		Visual Observation/Comments
			Outside Crevice	Inside Crevice	
#1 Babbitt	105	8.56	f	f	Heavy crevice attack; general and pitting corrosion outside crevice; heavy edge attack; difficult to assess because of geometry
#4 Babbitt	105	15.69	f	f	Severe crevice attack; general and pitting corrosion outside crevice; moderate to heavy edge attack; difficult to assess because of geometry
Bare Carbon Steel	105	1.69	f	f	Heavily encased in rust after exposure; severe crevice attack; general corrosion and pitting; one side more heavily attacked
Galvanized Steel	105	5.39	d	Nil	Dark residue on surface; two small pits in crevice; uniform attack on galvanized surfaces; attack on uncoated edges
D_2M Austenitic Iron	105	3.87	f	f	Heavy crevice and edge attack; heavy pitting on one side, moderate on the other

[a] Dissolved ozone concentration = 2.6–8.2 mg/1.
[b] General corrosion rate is based on weight loss data and should be used only as guide where pitting is noted.
[c] Less than 0.001 in.
[d] 0.001–0.005 in.
[e] 0.005–0.010 in.
[f] Greater than 0.010 in.

tains no lead) was much superior to the #4 Babbitt. Although the #4 Babbitt exhibited the highest general corrosion rate of any material tested in both liquid and vapor, the #1 Babbitt was susceptible to crevice corrosion in both vapor and liquid phases.

Carbon Steel and Austenitic Iron

Bare carbon steel and austenitic iron samples performed poorly in both liquid and vapor phases. In addition to exhibiting relatively high general corrosion rates, they proved to be highly susceptible to pitting and crevice corrosion.

The effect of galvanizing was to reduce the general corrosion rate of carbon steel in both the liquid and vapor phases and to decrease susceptibility to pitting and crevice corrosion. It should be noted that disappearance of the galvanized layer would result in an increase in corrosion rate of the carbon steel which would be equivalent to that of bare carbon steel. For short-duration service requirements, galvanized steel surfaces could be acceptable.

Nonmetallic Materials

Tables III and IV tabulate the test results obtained with nonmetallic materials. In general, there was an increase in the hardness of the materials with exposure that generally accompanied a decrease in tensile strength, although there were several exceptions to this observation. This effect probably was due to changes in the internal structure of the material during exposure, similar to those which occur during the aging of plastics and rubbers. In some of the tests, the sample racks were located in areas where they were subjected to splashing when exposed to the vapor phase, or to relatively high liquid velocities when exposed to the liquid phase. This might have accounted for the weight losses observed in some materials, while similar materials showed a weight gain in other tests. Hypalon compared to Hypalon 9102 is a good example of this. Weight gains are evidence of permeation of the materials by the environment.

Most of the plastics and rubbers tested did not perform satisfactorily in the tests. Only the Teflon® (E. I. Du Pont de Nemours & Co., Inc., Wilmington, DE) and unplasticized polyvinyl chloride (UPVC) samples performed in a satisfactory manner in both phases. Type F polyethylene and the Viton® (E. I. Du Pont de Nemours and Co., Inc., Wilmington, DE) sample also appeared to perform well in the vapor phase, if the sur-

Table III. Performance of Nonmetallic Materials in Ozone Contactor, Vapor Phase[a]

Material		As-Received		Post-test				
Sample Identification	Supplementary Description	Avg. UTS[b] (psi)	Durometer Hardness	Avg. UTS (psi)	Durometer Hardness	Test Duration (days)	Weight Change (%)	Observations/Comments
Teflon, 1/16" Thick		1,849	58D	2,338	55D	187	−0.05	No apparent attack
Teflon, 1/8" Thick		2,054	59D	2,894	59.5D	187	−0.02	No apparent attack
Silicone Rubber, 1/8" Thick	Vinyl methyl polysiloxane rubber	1,324	56A	388	60A	187	−0.36	Slight surface degradation
Silicone Rubber, 3/16" Thick	Vinyl methyl polysiloxane rubber	909	65A	256	76A	187	−0.51	Slight surface degradation; apparent loss of ductility
Silicone Rubber, K1448	Vinyl methyl polysiloxane rubber	810	82A	519	92A	105	−6.65	Scattered rust stains; corner spalled off
Silicone Rubber, UC-5	Vinyl methyl polysiloxane rubber	1,575	51A	675	66A	105	−0.42	Light rust stains on 50% of sample
Viton	Vinylidine fluoride and hexafluoropropylene copolymer	1,580	71A	1,360	72A	105	−1.95	Slight surface erosion and decomposition
Viton 985, 1/16" Thick	Vinylidine fluoride and hexafluoropropylene copolymer	1,416	68A	1,300	81A	187	+7.76	Sample distortion; some decomposition around mounting hole
Viton 985, 1/8" Thick	Vinylidine fluoride and hexafluoropropylene copolymer	730	53A	671	67.5A	187	+0.86	Sample distortion; decomposition and heavy attack 3/4 distance around mounting hole
CPE 9401	Chlorosulfonate polyester	1,886	60A	1,540	57.5A	187	−0.33	Apparent pitting type surface attack; sample very tacky to touch
CPE 9405	Chlorosulfonate polyester	2,072	51A	2,574	40.5A	187	+1.76	High degree of surface checking; sample tacky to touch
Hypalon	Chlorosulfonate polyester	2,095	74A	1,915	75A	105	−13.00	Severe surface erosion corrosion
Hypalon 9102	Chlorosulfonate polyester	1,598	50A	1,960	46A	187	+11.20	Apparent general surface attack and checking
Vinylex PVC	Polyvinyl chloride	1,685	74A	2,158	80A	86	−1.55	Slight surface abrasion
UPVC, Type I	Unplasticized polyvinyl chloride	9,475	90.0R^c_L	9,375	89.3R^c	105	−0.39	Color changed to gray white; dark spot adhering near crevice

UPVC, Type II	Unplasticized polyvinyl chloride	7,295	$69.5R^{c}_{L}$	6,930	$66.9R^{c}_{L}$	105	−2.25	Color faded, half gray, half white; heavy rust deposit on one side
UPVC, Reinforced	Unplasticized polyvinyl chloride	7,260	$67.2R^{c}_{L}$	5,840	$55.7R^{c}_{L}$	105	−5.08	Color changed to white; slight surface abrasion
UPVC, Solvent Welded Butt Joint	Unplasticized polyvinyl chloride	56[d]		70[d]		187	+0.10	Sample bleached; some apparent surface attack on butt joint
UPVC, Solvent Welded Lap Joint	Unplasticized polyvinyl chloride	115[d]		149[d]		187	+0.10	Sample bleached from exposure
Polyethylene, Type A		1,465	51D	1,505	54D	105	−6.90	Yellowish stain over 75% of sample; visible surface attack
Polyethylene, Type F		4,100	67D	4,100	69D	105	−2.09	Scattered rust stains over 10% of sample; slight surface abrasion
Polypropylene, Type O		5,090	74D	5,110	$95.4R^{c}_{R}$	105	−7.44	Slight surface abrasion; slight dark staining
EDPM 6550	Ethylene propylene terpolymer	2,140	70A	[e]	[e]	105	−93.36	Sample badly decomposed
Nordel®[f]	Ethylene propylene terpolymer	1,750	70A	1,565	74A	105	−27.54	Heavy surface erosion corrosion
Neoprene		2,770	69A	[e]	[e]	105	n/a	Sample decomposed and was lost in test
FRP, UTR	Fiberglass-reinforced polyester resin	13,425		13,165		105	−11.77	Surface delamination in evidence
FRP, TSF	Fiberglass-reinforced polyester resin	8,255		9,530		105	−6.69	Surface delamination in evidence
FRP, UTS	Fiberglass-reinforced polyester resin	10,635		10,760		105	−2.90	Surface delamination in evidence
Nylon		9,190	83D	8,865	85D	105	−21.44	Heavy surface attack except area under spacers
Sikaflex, T-68	Urethane diisocyanate	3,300	55A	864	52.5A	187	−41.20	Apparent heavy surface attack with white and brown coating; perforations near mounting hole

Table III, continued

Material		As-Received		Post-test				
Sample Identification	Supplementary Description	Avg. UTS[b] (psi)	Durometer Hardness	Avg. UTS (psi)	Durometer Hardness	Test Duration (days)	Weight Change (%)	Observations/Comments
Sikaflex-IA, Disc	Polyurethane					187	+2.1	Sample showed good ductility; chalking and cracking indicated decomposition; surface very sticky to touch
Ceilcote 640	Epoxy grout					187	−18.46	Sample catastrophically attacked
Ceilcote 648	Epoxy grout					187		Sample apparently destroyed during test

[a]Ozone concentration = 11–34 mg/l.
[b]UTS = ultimate tensile strength.
[c]Rockwell hardness test.
[d]Numbers indicate breaking load (lb).
[e]Not available.
[f]Registered trademark of E. I. du Pont de Nemours & Co., Inc., Wilmington, DE.

Table IV. Performance of Nonmetallic Materials in Ozone Contactor, Liquid Phase[a]

Material		As-Received		Post-test				
Sample Identification	Supplementary Description	Avg. UTS (psi)	Durometer Hardness	Avg. UTS (psi)	Durometer Hardness	Test Duration (days)	Weight Change (%)	Observations/Comments
Teflon, 1/16" Thick		1,849	58D	1,874	60D	187	−0.04	No apparent attack other than slight surface corrosion
Teflon, 1/8" Thick		2,054	59D	1,728	62D	187	−0.09	No apparent attack other than slight surface corrosion
Silicone Rubber, 1/8" Thick	Vinyl methyl polysiloxane rubber	1,324	56A	586	45A	187	−0.22	No apparent attack other than slight weight loss
Silicone Rubber, 3/16" Thick	Vinyl methyl polysiloxane rubber	909	65A	584	66.5A	187	−0.72	Apparent general surface attack
Silicone Rubber, Ki448	Vinyl methyl polysiloxane rubber	810	82A	370	85A	105	−1.66	Two corners spalled off; yellow staining in crevice area
Silicone Rubber, UC-5	Vinyl methyl polysiloxane rubber	1,575	51A	750	56A	105	−0.55	Light rust stains impregnated 50% of sample
Viton	Vinylidine fluoride and hexa-fluoropropylene copolymer	1,580	71A	1,375	72A	105	−2.38	Slight surface attack and blistering
Viton 985, 1/16" Thick	Vinylidine fluoride and hexa-fluoropropylene copolymer	1,416	68A	2,953	72A	187	+9.88	General surface attack
Viton 985, 1/8" Thick	Vinylidine fluoride and hexa-fluoropropylene copolymer	730	53A	714	66A	187	+2.15	General surface attack
CPE 9401	Chlorosulfonate polyester	1,886	60A	1,218	56A	187	+1.12	Surface attack on one side; apparent loss of ductility
CPE 9405	Chlorosulfonate polyester	2,072	51A	1,910	45A	187	+1.85	Apparent pitting, trenching and blistering
Hypalon	Chlorosulfonate polyester	2,095	74A	1,980	79A	105	−23.33	Heavy surface erosion corrosion
Hypalon 9102	Chlorosulfonate polyester	1,598	50A	1,106	41A	187	+16.84	Heavy surface attack and checking
Vinylex PVC	Polyvinyl chloride	1,685	74A	2,158	80A	86	−1.55	Slight surface attack
UPVC, Type I	Unplasticized polyvinyl chloride	9,475	90.0R_L[b]	8,845	87.9R_L[b]	105	−0.30	Black soot-like coating over entire sample; scattered rust-colored stains

Table IV, continued

Material		As-Received		Post-test		Test	Weight	
Sample Identification	Supplementary Description	Avg. UTS (psi)	Durometer Hardness	Avg. UTS (psi)	Durometer Hardness	Duration (days)	Change (%)	Observations/Comments
UPVC, Type II	Unplasticized polyvinyl chloride	7,295	$69.5R^{b}_{L}$	6,375	$66.5R^{b}_{L}$	105	−1.20	Heavy soot-like coating on one side; rust-colored deposits on the other side
UPVC, Reinforced	Unplasticized polyvinyl chloride	7,260	$67.2R^{b}_{L}$	6,490	$62.3R^{b}_{L}$	105	−11.13	Moderate attack; brown coating over entire sample
UPVC, Solvent Welded Butt Joint	Unplasticized polyvinyl chloride	56[c]		45[c]		187	−0.10	Similar to vapor phase specimen in appearance
UPVC, Solvent Welded Lap Joint	Unplasticized polyvinyl chloride	115[c]		146[c]		187	−0.10	Similar in appearance to vapor phase specimen
Polyethylene, Type A		1,465	51D	1,510	53D	105	−5.32	Scattered rust-colored deposits; slight blistering of surface on one side
Polyethylene, Type F		4,100	67D	4,165	77D	105	−2.89	Scattered rust-colored deposits
Polypropylene, Type O		5,090	74D	5,055	$93.7R^{b}_{L}$	105	−10.96	Scattered rust-colored deposits
EDPM 6550	Ethylene propylene terpolymer	2,140	70A	2,005	76A	105	−20.44	Heavy surface erosion corrosion
Nordel	Ethylene propylene terpolymer	1,750	70A	1,480	73A	105	−16.74	Heavy surface erosion corrosion
Neoprene		2,770	69A	N/A	69A	105	−35.58	Heavy surface attack pitting-like deterioration on one side
FRP, UTR	Fiberglass-reinforced polyester resin	13,425		16,255		105	−10.26	Surface delamination evident
FRP, TSF	Fiberglass-reinforced polyester resin	8,255		7,755		105	−8.35	Surface delamination evident
FRP, UTS	Fiberglass-reinforced polyester resin	10,635		9,070		105	−4.90	Surface delamination evident

Nylon		9,190	83D	10,980	82D	105	−75.59	Severe thinning by erosion; perforation at crevice edge
Sikaflex, T-68	Urethane diisocyanate polyurethane	3,300	55A	904	70.5A	187	−3.51	Heavy surface attack and apparent cracking
Sikaflex-IA, Disc	Urethane diisocyanate					187	+37.8	Sample showed good ductility; surface revealed no evidence of decomposition; large weight increase indicates permeation
Sikaflex-IA Joint	Urethane diisocyanate					187		Some surface decomposition products; sample showed good ductility
Ceilcote 640	Epoxy grout					187	−1.16	Heavy attack on glaze side; chalking and spalling on other side; heavy edge attack
Ceilcote 648	Epoxy grout					187	−0.076	Similar to 640 above; apparent permeation by some form of greasy substance.

[a] Dissolved ozone = 2.6–8.2 mg/l.
[b] Rockwell hardness test.
[c] Numbers indicate breaking load (lb).

face erosion and weight loss can be attributed to the abovementioned test conditions. Sikaflex-1A® (Sika Chemical Corp., Lyndhurst, NJ), an elastomeric joint compound, also showed good performance, although it was apparently quite permeable, judging from its weight gain. All other samples were unsatisfactory in performance, either by change in strength, ductility, weight or general evidence of attack.

[Editor's note: Data given in Tables III and IV for the performances of UPVC in ozonated atmospheres were gathered with maximum exposure times of six months. On the other hand, UPVC ozone-piping systems in a number of operational U.S. wastewater treatment facilities have manifested embrittlement and longitudinal cracking on long-term (>6-month) dry ozone service (Reid, Quebe, Allison, Wilcox & Associates, Indianapolis, IN, unpublished report, 1981). The recommendation of a leading ozone systems supplier (Cottier, D., Welsbach Corp., Philadelphia, PA, personal communication, 1981) to use stainless steel (304/304L above water; 316/316L below water) is a more conservative approach, based on the current knowledge of UPVC in ozone services. See Chapter 12 for a more detailed discussion of this topic.]

Limited testing of nonmetallic materials in dry ozone indicated considerable variability in the performance of materials belonging to a generic classification. It appears that nonmetallic materials used for ozone should be specified precisely in terms of proven chemical compositions, instead of by general material type. This in fact is no different from what is done when metallic alloys are specified for a given application. Future nonmetallic materials evaluation will involve evaluation of individual materials within generic classifications to define more precisely optimal materials for use in ozone generators.

DISCUSSION

Tables I and II presented the results of the ozone exposure tests in vapor and liquid for various metals. This information is compared with similar results obtained with material exposed to oxygenated wastewater environments (Tables V and VI). These tables contain a brief summary, which shows the range of data obtained in UNOX pilot plants situated at Wyandotte, MI; Philadelphia, PA; Tampa and Jacksonville, FL, and Tonawanda, NY. Wastewaters treated at these sites are classified as municipal, industrial and mixed industrial-municipal. Consequently, they showed varying degrees of corrosivity. While the tests are not exact duplicates of the ozone tests, it may be seen that with few exceptions the ozone data are similar to those obtained in the UNOX pilot plants.

Table V. Summary of Corrosion Data Obtained by Exposure of Metal Samples at Five UNOX Pilot Plants for up to 198 Days

Alloy	General Corrosion Rate (mil/yr)	Pit Depth (in.)	Pitting Density	Crevice Corrosion
Carbon Steel	0.57–10.80	0.003–0.035	Moderate to Heavy	Nil to Severe
3003 Aluminum	Nil–6.29	Nil to Sample Perforation (>0.051 in.)	Nil to Scattered	Nil to Severe
Monel 400	Nil–1.27	Nil–0.002	Nil to High	Nil to Moderate
202 Stainless Steel	<0.10	Nil–0.005	Nil to Moderate	Nil to Heavy
400 Stainless Steel	Nil–0.64	Nil to Perforation (>0.025 in.)	Nil to Heavy	Nil to Catastrophic
304 Stainless Steel	Nil–<0.10	Nil–0.005	Nil to Scattered	Nil to Moderate
439 Stainless Steel	Nil–<0.10	Nil–0.008	Nil to Light Scattering	Nil to Slight
444L Stainless Steel	Nil–<0.10	Nil–0.008	Nil to Light Scattering	Nil to Slight

Table VI. Summary of Corrosion Data Obtained via Exposure of Sensitized (Simulated Welds) Stainless Steel Samples at Five UNOX Pilot Plants for Periods up to 198 Days

Alloy	General Corrosion Rate (mil/yr)	Pit Depth (in.)	Pitting Density	Crevice Corrosion
202 Stainless Steel	Nil–0.37	Nil–0.007	Nil to Moderate	Nil to Intense
304 Stainless Steel	Nil–0.36	Nil–0.017	Nil to Moderate	Nil to Heavy
400 Stainless Steel	Nil–1.00	Nil–0.617	Nil to Heavy	Slight to Heavy
439 Stainless Steel	Nil–0.67	Nil–0.011	Nil to Heavy	Nil to Heavy
444L Stainless Steel	Nil–0.12	Nil–0.004	Nil to Moderate	Nil to Moderate

The best universal performer in the UNOX tests cited was 444L stainless steel, an 18Cr-2Mo material. Corrosion rates for carbon steel in ozone and UNOX environments are similar. The behavior of the as-received stainless steels and simulated welds are also similar. The performance of the Monel 400 was notably different in the UNOX and ozone environments. While the performance of Monel 400 in the UNOX environment was affected by the type of wastewater treated, i.e., municipal vs industrial, its behavior in ozone environments was much worse. Copper and nickel alloys, with the exception of nickel superalloys, simply do not perform well in ozonated environments.

The test results obtained in this study are limited when compared to the vast range of potential of ozone requirements expected to be encountered in the future. Fruitful areas for future work have been identified, however. Further testing of materials should serve to reduce further the data gaps that remain and increase the latitude of the designer. While the testing results presented here are limited, they do provide sufficient information to enable the design of safe, reliable oxygen-based ozone systems. This is particularly true when these data are integrated with the vast materials database available in the air separation industry and more recent information obtained by operation and testing in UNOX plants. It can be concluded that materials that are unacceptable for oxygen service will also be unacceptable in ozone service. However, good performance in oxygen service does not guarantee good performance in ozone service.

SUMMARY

Compatibility data were generated for a number of materials in high-humidity ozone-bearing gaseous environments and in liquid phases with high levels of dissolved oxygen and ozone. With the exception of the copper and nickel-copper alloys, which performed poorly, the performance of the metallic materials in the ozone test program was very similar to what has been experienced in UNOX and air activated sludge systems. This finding tends to increase confidence in the safety of high-purity oxygen-based ozone systems.

Severe pitting and crevice corrosion may occur in wet ozone applications. Perforation may occur even when accompanied by low general corrosion rates. From the standpoint of pitting and crevice corrosion, the 444L stainless steel (18Cr-2Mo) proved to be a superior performer. While individual alloys exhibited markedly better performance in either the vapor phase or liquid phase, neither environment is generally more severe than the other.

Caution must be exercised in the specification and use of nonmetallic materials due to compositional variations within generic groupings. Consequently, significant performance differences in ozone service may occur. Further experimental work with nonmetallic materials on a generic basic in ozone service may be warranted, depending on specific requirements.

SECTION 4

ANALYTICAL ASPECTS

CHAPTER 8

ANALYSIS OF OZONE IN AQUEOUS SOLUTION

John Stanley and J. Donald Johnson
University of North Carolina at Chapel Hill
Chapel Hill, North Carolina

The application of ozone for water treatment processes usually involves onsite generation of an ozone-containing gas mixture, which is introduced into the aqueous solution using a gas-liquid contacting system. Control of this process is based on the measurement of the applied or utilized ozone dose, or the dissolved residual ozone. The applied ozone dose is determined by measuring the ozone concentration in the gas feed stream to the ozone contactor. The utilized, or absorbed, ozone dose generally is taken to be the difference between the concentrations of ozone in the contactor exhaust gases and in the contactor feed gas.

In applications where dissolved ozone concentrations can be maintained in solution, dissolved residual ozone measurements assist in control and evaluation of process performance. Ozone residuals are observed once the spontaneous ozone demand of the solution has been satisfied. Total ozone demand consists of three factors:

1. relatively slow, continuous decomposition of ozone to oxygen;
2. conversion of ozone molecules to hydroxyl radicals, which occurs rapidly only at high pH values; and
3. often fast, chemical reduction of the ozone molecule by reaction with oxidizable constituents of the solution.

These chemical reductions of the ozone molecule often produce new oxidants, which may mistakenly be measured as residual ozone. The use of residual ozone measurements to control ozonation processes is most common in the disinfection or oxidation of drinking water supplies. Here, in contrast to wastewater treatment, the ozone demand can be satisfied by applying excess ozone to obtain a residual ozone concentration and thus assure ozonation process completion as a measure of microbiological safety.

In comparison with aqueous analytical methods, gas-phase measurements of ozone can be performed with relative accuracy and freedom from interferences. Gas-phase ozone analysis and its application for ozonation process control purposes is an important subject, but is beyond the scope of this chapter [1]. On the other hand, the use of dissolved residual ozone measurements is not as well developed as is gas-phase analysis, in part because of the questionable sensitivity and lack of selectivity or accuracy due to interferences in present methods applied to measure and monitor dissolved ozone. Analytical procedures in solution are influenced by the chemical instability of the ozone molecule in water and often are subject to interferences from oxidizing compounds, either present in solution initially or formed during the ozonation step.

The present availability and use of analytical techniques for measurement of ozone in water are the subject of this review. It is intended to update earlier reviews [2,3] and to include recent analytical developments in this field. Currently available methods are examined and compared in light of potential interferences and limits to detection sensitivity. Developments in the application of electrochemical methods of analysis also will be reviewed. From this discussion, it is anticipated that the analyst will be better able to select the most appropriate analytical methodology with a full understanding of its advantages and limitations for use in a given specific test situation.

GENERAL CONSIDERATIONS IN SELECTING AN ANALYTICAL METHOD FOR OZONE

Although ozone is approximately 13 times more soluble in water than is oxygen at equilibrium at the same partial pressure of each gas [4], concentration levels observed in water during actual practice seldom exceed a few milligrams per liter, and are usually maintained below 1.0 mg/l by design [5]. Rapid depletion of the concentration of the ozone molecule is caused by autodecomposition of molecular ozone or by reduction to highly reactive intermediate free radicals (Figure 1) [6,7]. It is very important to note here that the rate of ozone decomposition depends on a number of factors, including the solution pH, degree of mixing, temperature, the type of organic and inorganic substrates present, and the

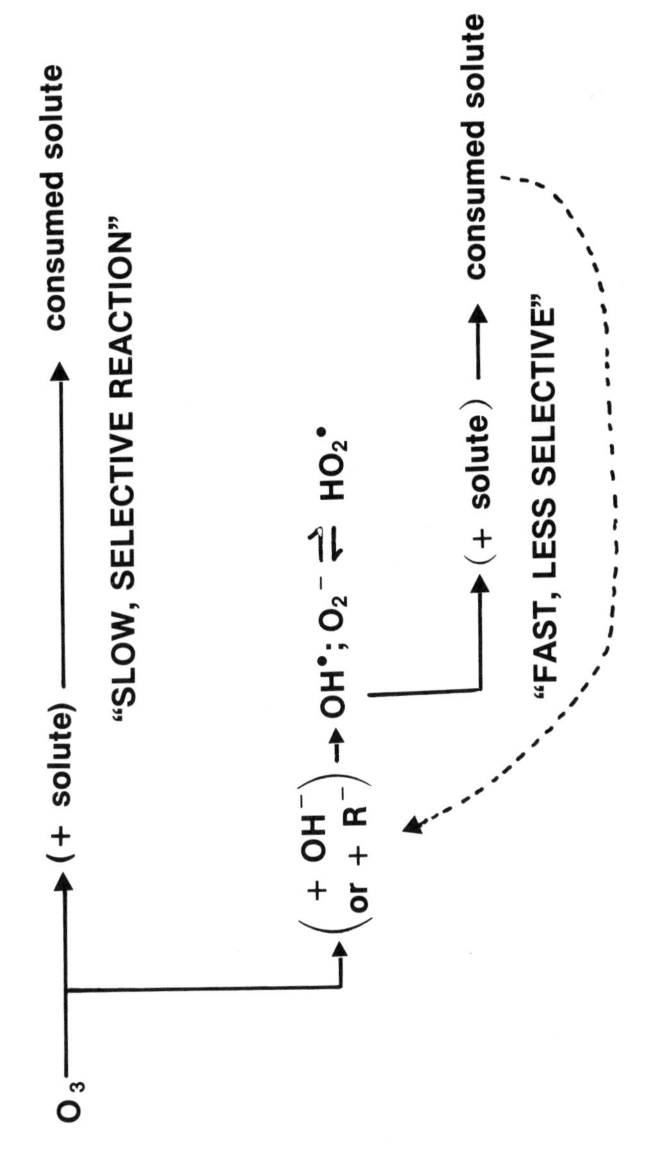

Figure 1. Direct slow reaction and fast radical intermediate reaction pathways [6].

relationship between the various reaction rates of competing chemical oxidation/reduction reactions.

Evidence for this can be found in the conflicting reports of numerous research efforts to define decomposition rates for ozone in water [8,9]. First-order reaction rate constants from 5 to 9000 hr^{-1} and reaction orders from 0 to 2 have been reported in various buffer media at different pH levels. In the most recent study [8], the decomposition of ozone in water was catalyzed by hydroxide ions to the 0.5 order and was second-order from pH 2.0 to 9.5. Direct reactions between the O_3 molecule and other compounds are selective and relatively slow. Indirect reactions through intermediates like hydroxyl radicals are rapid and nonselective. The rates of these latter reactions can be decreased by materials such as carbonate ions and aliphatic alcohols, which "scavenge" the reactive hydroxyl radical intermediates formed by decomposition of the ozone molecule. Both direct and indirect reactions of this type, in which the ozone molecule is chemically reduced, are responsible for the large differences between applied ozone doses and measured ozone residual concentrations. They make it important to measure and control ozonation processes on the basis of residual ozone concentration. These reactions, in general, do not constitute interferences to the analytical determination of ozone, unless they occur because of contaminated reagents or dilution water.

A secondary effect of ozone decomposition is the formation of products that possess chemical or physical characteristics similar to those of ozone. Consequently, some of these decomposition products interfere by causing a positive response during the measurement for ozone. The most frequently encountered situation is the case in which ozone forms new compounds that are strong oxidants and react with analytical reducing agents in the same manner as does the ozone molecule. One of the principal decomposition products of ozone, the hydroxyl radical ($OH^\cdot$), is a highly reactive oxidant [6,10,11]. At least two workers have observed a response on ozonation of phenol to iodometric test methods in the absence of molecular ozone [8,12]. It is thought to be a response to peroxides, like peroxyacids and hydrogen peroxide, formed as end-products of the radical processes. More common components of natural waters like manganous and ferrous ions also are oxidized in ozonated solutions to forms that can interfere significantly with certain test procedures. While not specifically identified, it is speculated that other classes of compounds, including organic acids, aminoorganic compounds and some unsaturated compounds, when present in sufficient quantities during the ozonation step, will form potential oxidant interfering materials, such as peroxy compounds.

Some estuarine waters contain bromide, so that the addition of ozone (or chlorine) produces elemental bromine and a number of bromoinorganic and bromoorganic species with oxidizing capabilities that will not be distinguished from ozone in some analytical test procedures [13,14]. Finally, in treatment processes where chlorine or chlorine dioxide is added in conjunction with or in sequence with ozone, it should be recognized that chlorine, chloramine compounds and chlorine dioxide all have been reported as potential interferents in analytical tests for ozone [15].

Obviously, knowledge of the chemical fate of or the chemical nature of by-products produced by the ozone molecule in any application is essential for the analyst to be able to select an accurate ozone measurement technique. As discussed above, analytical selectivity is a key parameter influenced by decomposition rates of the ozone molecule as well as the specific water treatment process design. The question of potential interferences is better understood when knowledge of the chemical constituents of the treated water is obtained. A most common occurrence during ozone analysis is failure to distinguish between the content of total oxidant and molecular ozone, O_3.

Sensitivity of ozone analytical measurements is also important. Detection below levels of 1.0 mg/l is essential and detection at the microgram-per-liter level is desirable. Finally, a third concern, often underemphasized, is adaptability of the analytical method. This consideration includes a number of procedural factors influencing method selection, such as response time of the analysis, its mode (discrete, intermittent or continuous) and its costs. For example, the needs of process control analysis are significantly different from those of a bench-scale procedure. A process control application might require a continuous measurement, but allow some reduction in sensitivity and selectivity. Simplicity of analysis and costs of maintenance and calibration would be a critical factor in any evaluation. A procedure for laboratory purposes, on the other hand, would likely be intermittent and might specify a higher degree of sensitivity and selectivity. Here, simplicity and costs of operation would not necessarily be limiting factors. This generalized comparison of two applications for an analytical method is merely presented to illustrate the importance of evaluating method adaptability along with detection sensitivity and selectivity.

AVAILABLE METHODS

In the following discussion, the various methods available for dissolved ozone analysis will be examined in detail. A summary of these analyses is presented in Table I for the purpose of comparing methods. In practice, two methods are employed most frequently and are con-

Table I. A Review of Selected Analytical Techniques for Ozone Analysis

Method	Principle	Advantages/Limitations
Iodometric Oxidation	Oxidation of I^- to I_2; detection of I_3^- by electrometric, titration or photometric methods	High detection sensitivity approaching 2 μg/l; most oxidants interfere; reaction stoichiometry questioned; ozone losses probable, due to sample collection and handling
Ultraviolet Spectrophotometry	Ozone molecule absorbs UV light at 254 nm; molar extinction coefficient approx. 2900 liter/mole-cm	High detection sensitivity to 20 μg/l with 50-cm cell path length; can be used as continuous process analyzer; potential interferences include organics and inorganics which absorb at 240- to 300-nm range, suspended particles and color.
Leuco Crystal Violet	Oxidation of leuco crystal violet to crystal violet; measure absorbance of crystal violet at 592 nm; molar absorptivity approx. 10^6 liter/mole-cm	Detection sensitivity approaching 1 μg/l; some oxidants interfere, notably manganese dioxide; major problem is test kit cell staining due to use of dye
Diethyl-p-phenylenediamine (DPD)	Oxidation of DPD; detection by titration or photometric methods	Detection sensitivity at sub-mg/l level, is available for use in kit form; suffers from interference from some oxidants, notably halogens and manganese; a major problem is reagent and product stability.
FACTS (Syringaldazine)	Oxidation of iodide to iodine; iodine oxidizes syringaldazine; photometric detection at 530 nm	Moderate sensitivity; requires addition of two reagents; subject to interference from other oxidizing agents

Syringaldazine—Glycine	Glycine is added to scavenge ozone; determination of oxidant is based on oxidation of iodide and syringaldazine; detection by titrimetric methods	Sensitivity at sub-mg/l level; requires separate determinations of total oxidant and (oxidant–ozone)
Indigo-blue	Bleaching of indigo dye on addition of ozone is measured spectrophotometrically at 600 nm	Good sensitivity at sub-mg/l level; peroxide does not interfere; manganese and chlorine do interfere, but can be corrected
Acid Chrome Violet K	Bleaching of ACVK dye measured spectrophotometrically at 550 nm	Moderate detection sensitivity
Amperometric Bare Electrodes	Reduction of ozone to oxygen directly in solution; either solution or electrode rotated to establish diffusion layer; current directly proportional to concentration.	Good sensitivity; applicable as continuous monitor; interferences from electrode fouling agents and other strong oxidants
Amperometric Steady-State Membrane Electrodes	Diffusion through films of ozone to be reduced to oxygen in a thin electrolyte film between electrode and membrane; current controlled by rate of membrane diffusion is directly proportional to concentration	Very good selectivity for molecular ozone in presence of other oxidants; membrane limits effects of fouling agents but can be fouled itself; good sensitivity; applicable for continuous monitoring; strong temperature dependence
Amperometric Pulsed Membrane Electrode	Diffusion through membrane limits interferences but does not control current; ozone in electrolyte film reduced to oxygen; current is directly proportional to concentration	Same advantages as above; membrane fouling and temperature dependence not as significant as with steady state electrodes

sidered to be standards against which other techniques can be compared. These two methods are the iodometric procedure and the direct ultraviolet (UV) absorbance method.

Iodometric Oxidation

The iodometric procedure originally was proposed in 1959 for gas-phase analysis and was adopted by the U.S. Environmental Protection Agency (EPA) as a reference technique for ozone determinations in air [16,17]. However, it has recently been abandoned in favor of the UV-absorbance method [1]. The iodometric procedure has been adapted [18] for analyzing for dissolved ozone after stripping it from aqueous solution into the gas phase.

The *Standard Methods* procedure [18] requires that the dissolved ozone be purged from solution and transferred into a trap containing a potassium iodide "scrubber" solution present in large excess. Iodide is oxidized to iodine while the ozone is reduced to oxygen. The coupled redox reaction is assumed to be as follows:

$$O_3 + 3I^- + H_2O \rightarrow O_2 + I_3^- + 2(OH)^-$$

In neutral solution, this reaction produces one mole of iodine for every mole of ozone present that is reduced. The quantity of liberated iodine then can be determined by several procedures and related stoichiometrically to the quantity of ozone originally present (in the absence of interferences).

Qualitatively, iodine or triiodide ion can be observed in the absence of other colored ions to levels of about 2×10^{-6} *M* (100 μg/l) [19]. More precise determinations are obtained by titration of the iodine with a reducing agent, such as sodium thiosulfate or phenylarsine oxide. The end-point of this reaction, or the point at which all of the liberated iodine is reduced to iodide, can be seen by the disappearance of the dark blue iodine-starch complex (formed when starch is added prior to the end-point) or by amperometric methods. The intensity of the iodine-starch color diminishes below detection limits at about 2×10^{-5} *M* (1.0 mg/l), whereas electrochemical techniques have sensitivities to 4×10^{-8} *M* (2 μg/l) [19]. The liberated iodine also may be determined colorimetrically, based on the ability of triiodide to absorb light at a wavelength of 352 nm [20]. The detection limit is reported to be 2×10^{-7} *M* (10 μg/l).

Often the purge step specified in the *Standard Methods* iodometric procedure will be omitted, with a corresponding loss in selectivity. Even with the purge procedure, ozone losses are likely because of the chemical instability of the molecule. Many times a known aliquot of the dissolved ozone solution is added directly to the potassium iodide reagent solution. Here the risk of interference may be enhanced because the purge procedure is abandoned. It is also possible, even with the purge step, that volatile compounds that interfere with the determination of ozone can be purged from solution along with the ozone, or that losses of ozone by decomposition in the purge assembly may negate efforts for improved accuracy.

A disagreement in the iodide-based procedures centers around the actual stoichiometry of the iodide oxidation reaction, particularly at low concentrations. Ingols [21] and Boyd et al. [22] have reported that additional iodine is produced at neutral pH, while others [23,24] confirm a stoichiometry of 1 mole of iodine per 1 mole ozone. More recent investigations [25–28] show that reduction of ozone, when present in low concentrations (<1 mg/l), results in the liberation of iodine in excess of the predicted stoichiometric amount. Flamm [25] reports that the buffer system (0.1 *M* Na_2HPO_4/0.1 *M* KH_2PO_4) employed with the potassium iodide reagent solution may be the source of a second product, tentatively identified as hydrogen peroxide, which, in turn, liberates additional iodine. He has suggested the use of a 0.1 *M* boric acid buffer solution to eliminate this inaccuracy in the iodometric method.

The iodometric method is commonly employed for the volumetric analysis of a large number of oxidants. Most oxidants have more positive reduction potential values than the standard redox potential of the iodine half-cell reaction which is $+0.53$ V [vs the Standard Calomel Electrode (SCE)] [19]. Similarly, iodine is used as an oxidizing agent for reductants of redox couples such as sulfide with more negative half-cell reaction potentials. Consequently, it is apparent that this technique is highly susceptible to inaccuracy due to either the liberation of additional iodine from reaction with oxidants other than ozone or to the reduction of liberated iodine in solution by reducing agents which react slowly with ozone.

To summarize, the iodometric procedure is based on the oxidation of iodide to iodine by ozone. It is highly susceptible to interference from other oxidizing and reducing agents, if present, due to the relative ease of oxidation of iodide and reduction of iodine. The boric acid buffer system is recommended, to avoid the potential stoichiometric error when measuring ozone concentrations of less than 1 mg/l by this procedure. Quantitative measurement of the liberated iodine can be accomplished based on visual examination or titration of the starch-iodine complex indicator solution, or absorbance by the liberated iodine as the triiodide ion, or with amperometric titration. It can be an extremely sensitive method, but suffers either from many interferences if measured directly in solution, or from losses if stripped through the gas phase.

Direct Ultraviolet Absorbance

In ozone-oxygen or ozone-air mixtures, the ozone molecule absorbs UV light over the 240- to 300-nm range [1,29]. The maximum absorption peak is at 254.7 nm, with a molar absorptivity of approximately 3000 liter/mol-cm. This procedure recently has been adopted by the EPA as the reference analytical method for ozone in the atmosphere [1]. Adler and Hill [30] first demonstrated that molecular ozone dissolved in water retains UV absorption characteristics similar to those of ozone in the gaseous phase. They observed a maximum absorption at a wavelength of 260 nm with a molar absorptivity of approximately 2900 liter/mol-cm at 20°C. Thus, the direct measurement of dissolved molecular ozone is possible once the absorption characteristics of the molecule in the particular solution under consideration have been defined.

Applying the Beer-Lambert principles with a molar absorptivity of only 2900 liter/mol-cm (at 20°C), it is possible to measure dissolved ozone at the 0.2 mg/l level. For example, using a 10-cm cell with an absorbance of 0.1, the concentration of ozone present is 3.45×10^{-6} M (165 μg/l). Beyond the problem of sensitivity, the most serious limitation to the use of the UV method in natural waters is the potential interference from dissolved organics and inorganics which absorb in the same wavelength region. Many oxidants used for water treatment purposes, such as chlorine or bromine compounds, and organic compounds such as aquatic humic materials, have been found to absorb UV light at wavelengths between 200 and 300 nm. UV absorbance measurements at 254 nm often are employed as a surrogate measurement for total organic carbon in natural waters. Problems also have been encountered in the direct measurement of ozone in highly turbid waters where light scattering effects influence the accuracy of the UV absorbance readings.

A variety of UV spectrophotometers are available having differing degrees of precision for ozone analysis. Several industrial process photometers have been designed based on the UV absorbance principle and applied successfully in water treatment applications. One design utilizes a dual light beam approach with additional compensation for optical and temperature effects (Figure 2) [31]. The reference beam reads a sample stream from which ozone or other oxidants present have been reduced chemically. The sample absorbance signal then is compared to the reference beam signal, with the difference being related to the amount of dissolved ozone present.

In summary, dissolved molecular ozone can be determined quantitatively based on the direct measurement of its UV absorbance at a wavelength of 250–260 nm. With the appropriate cell path length selection, most concentration ranges normally encountered in water treatment ap-

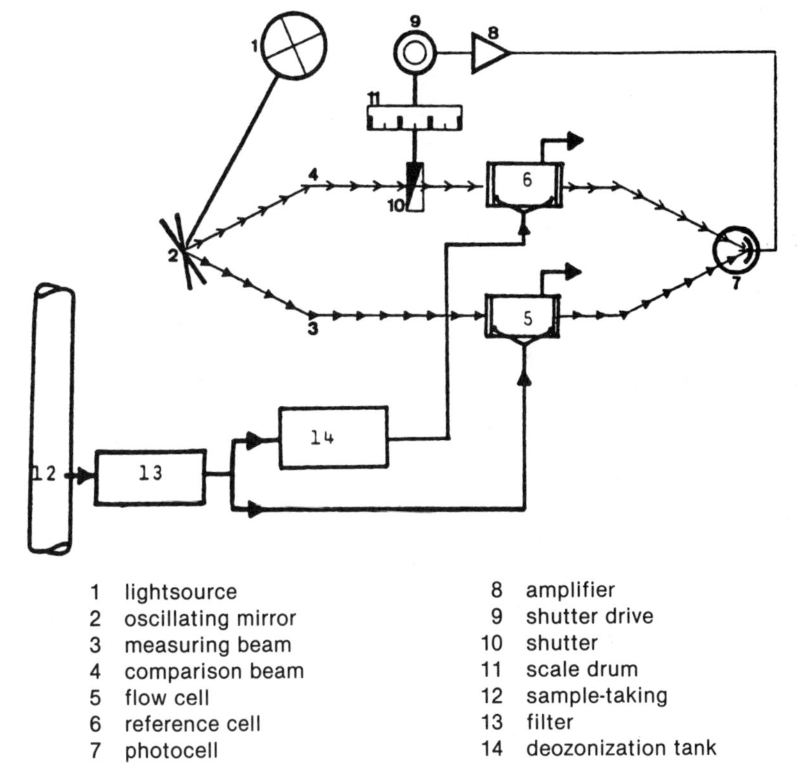

Figure 2. Differential photometric measurement of ultraviolet absorbance of ozone in solution at 254 nm [30].

plications can be measured with accuracy. A high degree of selectivity for molecular ozone is possible with this method.

The direct UV method is more selective for ozone, but is less sensitive than the iodometric method. Principal interferences are dissolved organic and inorganic materials that absorb UV light at or near the same wavelength region as does ozone. Particulates in suspension also affect measurement accuracy.

Other Colorimetric Techniques

The two methods we have discussed to this point are considered to be the standard techniques against which other analytical procedures must be compared. Both, however, suffer from a significant matrix of interferences. Yet there are a number of other methods employed to varying degrees which are procedurally simpler than the iodometric or direct UV absorbance methods. Most of these methods were developed for halogen analysis and have since been adapted to allow their use for the analysis of ozone. The major advantage of many of these methods is that they can be readily adapted for field testing. Sample color and turbidity are potential interferences in all methods to be described. The major disadvantage of these methods is lack of selectivity.

Leuco Crystal Violet

In the procedure, leuco crystal violet (LCV) is oxidized to crystal violet in the presence of an oxidant such as ozone [32]. The absorbance maximum for the bluish-purple oxidation product occurs at 592 nm with a molar absorptivity of 1×10^6 liter/mol-cm. At an absorbance of 0.10 using a 10-cm cell path length, the minimum concentration of ozone which can be detected is about 10^{-8} M (0.5 μg/l). The reagents and product have been found to be quite stable also. The extent of color development has been reported to be sensitive to reagent/sample mixing technique. It must be noted, however, that this technique is subject to most of the interferences of the iodometric procedure, since other oxidizing agents may also liberate crystal violet. Principal interferences are the halogens and the oxidized forms of manganese. Reducing agents are also possible interfering agents that may reduce crystal violet.

Diethyl-p-phenylenediamine

Ozone reacts directly with diethyl-p-phenylenediamine (DPD) indicator to produce a red product, which can be detected quantitatively by titrimetric methods with ferrous ammonium sulfate or by photometric analysis at 515 nm [33]. The method does not obey Beer's law at ozone levels above 2 mg/l and the indicator is unstable. Detection of ozone at the sub-mg/l level is possible with both the titrimetric procedure and photometric method, but this method is not as sensitive as the LCV or FACTS (see below) procedures. Oxidized manganese and free halogens again are the principal interferences. In the presence of free halogens, the use of an ozone scavenger, glycine, enables separate determination of residual ozone (total oxidant minus oxidant excluding ozone). The precision and accuracy of this procedure, however, are not well documented.

Free Available Chlorine Test with Syringaldazine (FACTS)

A saturated solution of syringaldazine (3,5-dimethoxy-4-hydroxybenzaldazine) in 2-propanol is employed as an indicator. Ozone is added to an iodide solution at pH 7.0. The liberated iodine then is added to a syringaldazine solution (pH 6.6) and the resulting color is measured spectrophotometrically at 530 nm. This method is an indirect analytical approach, in that the iodine actually oxidizes the syringaldazine indicator. The procedure obeys Beer's law over concentration ranges up to 4.5 mg/l O_3 with complete color development within 1 minute [34]. This method is slightly more sensitive than the DPD procedure. Again, however, interferences include all those compounds capable of oxidizing iodide to iodine and agents capable of reducing iodine.

Glycine also can be employed as an ozone scavenger, as noted in the DPD procedure. A measure of total oxidant is obtained using an alkaline iodide solution and the syringaldazine indicator solution [35]. A separate determination of total oxidant minus ozone is made with the added glycine scavenger. The difference between the two measurements represents the amount of ozone present in solution.

Indigo-Blue Dye

At a pH below 3.0, indigo di- or trisulfonate is bleached rapidly by ozone. The loss of color due to the ozonolysis of indigo then is determined photometrically and related to the concentration of ozone present. The maximum absorbance of the dye is at 600 nm with a molar absorptivity of 2.0×10^4 liter/mol-cm [36]. The precision of the measurement is affected by the initial concentration of indigo dye present, since it is a bleaching method. Detection at the 3-μg/l concentration level is possible, however. The method is subject to fewer interferences than most colorimetric and all iodometric procedures. At pH 2, chlorite, chlorate, hydrogen peroxide and manganese(II) do not interfere. Some other interferences can be corrected for by adding malonic acid.

Acid Chrome Violet K

Acid chrome violet K (ACVK) is readily bleached by ozone [37]. Using a 5-cm cell path length, the detection sensitivity approaches 25 μg/l as ozone. Chlorine and chlorate below 10 mg/l do not interfere. Perlauric acid does not interfere up to 100 mg/l and the ozonide of 2,3-hexene up to 40 mg/l does not interfere. Manganese does not interfere below 1 mg/l of potassium permanganate. The reagent, however, is not stable, especially when exposed to light.

Electrochemical Methods

The application of electrochemical methods for the analysis of ozone has been extensively examined over the past two decades. The success of these electrochemical systems for the measurement of oxygen and halogen compounds in water has promoted interest in development of a similar in situ system for ozone [38–40]. Galvanic, amperometric or steady-state voltammetric, and, more recently, pulsed voltammetric electrode systems have been employed successfully for this purpose. All of these techniques are based on the electrochemical reduction of ozone at an electroactive surface as follows [41]:

$$O_3 + H_2O + 2e^- \rightarrow O_2 + 2(OH^-)$$

The response of the electrochemical circuit current is directly proportional to the activity of molecular ozone in solution [39]. A complementary anodic electrode providing electrons is coupled with the active cathodic electrode where the above reaction takes place in an electrolytic medium. An additional feature often employed for greater measurement selectivity is the use of a gas-permeable membrane separating the electrochemical cell from the measured solution.

Bare Amperometric Electrodes

The first reported application of this procedure to ozone analysis was the use of a galvanic cell with a gold cathode–copper anode cell [41]. The response has been demonstrated to be linear with a reported detection sensitivity of 200 μg/l. A commercially available galvanic system employing gold and copper electrodes is illustrated in Figure 3. Other galvanic systems have been employed with sensitivities to 10 μg/l [42]. All bare amperometric electrode systems are subject to loss of sensitivity with use, due to the accumulation of surface impurities at the electrode surfaces. With uncovered electrode surfaces, fouling has been observed to be a significant problem, as was the case in earlier tests with oxygen electrodes [38,39]. Additionally, the response is influenced by numerous surface-active agents and also halogens and oxygen.

Membrane Amperometric Electrode Systems

A further step in the development of electrochemical methods for ozone analyses has been the application of a gas-permeable membrane for increased selectivity and prevention of electrode fouling [43]. The use of an external applied voltage to the electrochemical circuit has further increased selectivity for molecular ozone in the presence of other oxidants. Again, these modifications were patterned after similar developments with oxygen and halogen membrane electrodes.

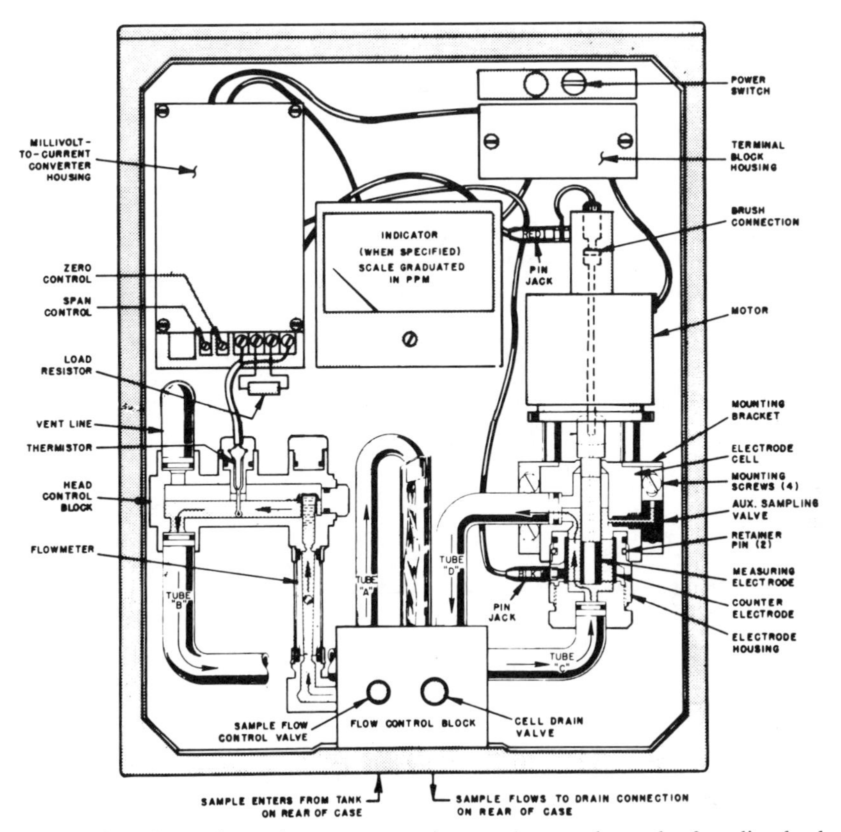

Figure 3. Bare-electrode amperometric membrane electrode for dissolved ozone [41].

More recently, various membrane materials, electrode materials, electrolyte media and applied voltage potentials have been examined in efforts to enhance both sensitivity and selectivity [44,45]. These efforts have produced an amperometric or steady-state membrane electrode with a demonstrated linear and reproducible response to dissolved molecular ozone in the presence of chlorine, chlorine dioxide, hydrogen peroxide and other oxidants. Current sensitivities have been reported on the order of 0.2–0.9 mA/mg/l, or detection limits near 60 μg/l. The electrode current output has a large temperature response of nearly 6%/°C. Stirring of the solution at the membrane electrode is required to maintain an adequate ozone concentration at the membrane surface. A commercially available amperometric membrane electrode design with these characteristics is illustrated in Figure 4. With this electrode design, a 95% response to ozone is observed in 20 sec.

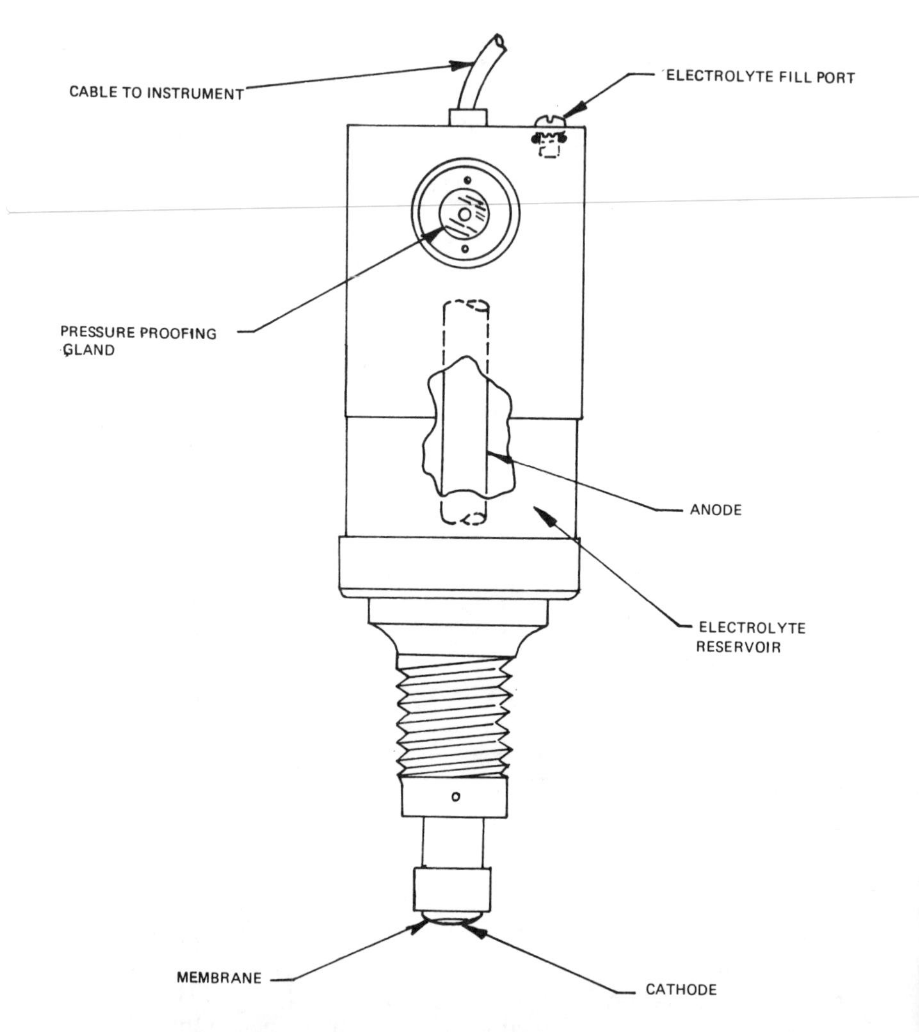

Figure 4. Dissolved ozone amperometric membrane electrode.

Recent work using pulse voltammetry techniques (as compared to steady-state) has virtually eliminated stirring and temperature effects [46]. Here, quasisteady-state conditions are created by numerous voltage pulses to the cathode. The current response to ozone reduction processes during the potential pulse is dependent primarily on the concentration of ozone and is nearly independent of membrane properties. With certain membranes, sensitivity has been enhanced by nearly 50-fold.

Selectivity of Membrane Amperometric Electrodes

The application of positive voltage potentials and the use of polymeric membranes that are selectively permeable to gases has enhanced the opportunity for selective measurement of molecular ozone. This is a very significant improvement over bare amperometric electrodes as well as colorimetric and titrimetric methods. With an applied voltage of +0.6 V (vs SCE) at the cathode, only the most powerful oxidizing agents can overcome the "resistance" of this anodic voltage and cause electron flow cathodically through the electrochemical circuit. The use of membranes that discourage transport of dissolved organic and inorganic species in favor of gaseous molecules is another major limiting factor for potential interferences. Studies [45] have demonstrated that the current response to some oxidants commonly found in water and wastewater was less than 2% of the response of the membrane amperometric electrode system to ozone. These oxidants included trichloramine, hypobromous acid, bromine, hypochlorous acid, hydrogen peroxide and chlorine dioxide. At the University of Michigan, a membrane amperometric electrode system was utilized in ozonated tap water, river water, and secondary treated sewage effluent and compared favorably to a referee procedure [46]. In fact, these researchers found the electrode response to be slightly higher in all instances than the referee method, suggesting that some ozone apparently was lost during sample collection and preparation. The results are promising and additional work in this area should continue with particular attention to selectivity and applications in ozonated waters.

Other Electrochemical Advances

In addition to their use as direct measures of ozone, amperometric methods are commonly used for endpoint detection in titrations. One recently proposed method uses a predetermined amount of phenylarsine oxide (PAO) added to a solution to reduce the dissolved ozone present [47]. The reaction mechanism is similar to that described earlier for PAO and iodine. The excess PAO then is measured quantitatively by pulse polarographic techniques. The difference between the amount of PAO added and the amount in solution after the ozone reacts with the excess PAO represents the amount of ozone initially present. Detection at the 2-to 3-μg/l level is reported with sample volumes as low as 5 ml. Interferences are minimized because of the selectivity of the PAO reaction. Free halogens thus would interfere, but not halamines or manganese.

WHICH METHOD TO USE?

As indicated earlier, there are four key considerations in selecton of a specific analytical method:

1. decide if a molecular ozone, residual, or if only total oxidant residual, is desired;
2. with the aqueous ozonation chemistry in mind, attempt to identify potential interferences that are present or will be formed during the ozonation step;
3. consider the ozonation process design and aqueous chemistry, and define the required analytical detection sensitivity; and
4. understand what information the analytical measurement is intended to provide and how it will be utilized.

The various analytical methods for ozone presently available have been reviewed, with considerable attention focusing on the two standard procedures, iodometric oxidation and direct UV absorbance. The promising electrochemical techniques also have been discussed.

Whenever possible, two independent procedures should be used: for example, a total oxidant method, such as an iodide-based amperometric titration using Flamm's buffer [25], the UV method [31], a membrane electrode [44] or a colorimetric procedure not using iodide [36]. This approach minimizes the possibility that interfering substances will go undetected in many applications. It may also allow for more accurate determination of ozone concentration levels, since many techniques suffer from varying degrees of ozone loss during sample collection and preparation.

When the iodometric procedure is employed, the effect of the purge step on the measurement should be examined by comparing it with direct addition of the sample to the potassium iodide. If possible, the purge step should be avoided when accuracy is not affected, so as to make the task of analysis simpler. Use of the boric acid buffer media suggested by Flamm [25] is recommended. It must be recognized that the iodometric method is very sensitive to interferences. For highest sensitivity, amperometric endpoint titration amplifying the current from a rotating electrode is recommended. Shechter's photometric method of analysis [20] is an acceptable alternative to the titration procedure in most cases.

The UV-absorbance method is highly selective and can be sensitive (with 50-cm cells). It is one technique which does not suffer interference from oxidizing agents, although it must be pointed out that many halogen species and organics absorb UV light in the 200- to 300-nm range. This method is unpredictable when turbidity levels increase above 10–30

NTU; thus it is most applicable in waters of high clarity. An advantage of the UV method is that process on-line analyzers are available which employ this technique.

The other colorimetric methods—DPD, LCV, FACTS, Indigo—are simple procedures for field applications and require little sophisticated or complicated equipment. DPD and FACTS procedures are subject to interferences from oxidizing agents since they are iodometric procedures. The LCV and Indigo Blue methods apparently are more selective, since these reagents react directly. The use of selective ozone reducing agents is a promising approach to improved selectivity, but the necessity for duplicate measurements is a drawback. Many of these colorimetric methods are particularly useful for field test applications because of the stability of the product formed by oxidation of the indicator chemical.

The use of membrane electrode systems is encouraged. These cells offer an advantage of in situ measurement with good sensitivity and potentially excellent selectivity. Their adoption as process control instruments could result in substantial reductions in ozonation costs as well as operational difficulties. Ozonation process performances also could be enhanced. However, the need for commercial development and application testing is apparent.

REFERENCES

1. Rehme, K. A., J. C. Puzak, M. E. Beard, G. F. Smith and R. J. Paur. "Evaluation of Ozone Calibration Procedures," U.S. EPA Report No. EPA-600/S4-80-050. NTIS Report No. PB-81-118,911 (1981).
2. Layton, R. F. "Analytical Methods for Ozone in Water and Wastewater Applications," in *Ozone in Water and Wastewater Treatment,* F. L. Evans, Ed. (Ann Arbor, MI: Ann Arbor Science Publishers, Inc., 1972), pp. 15–28.
3. Kinman, R. N. "Analysis of Ozone—Fundamental Principles," in *Proceedings of the First International Symposium on Ozone for Water and Wastewater Treatment,* R. G. Rice and M. E. Browning, Eds. (Vienna, VA: International Ozone Association, 1975), pp. 56–68.
4. Venosa, A. D., and E. J. Opatken. "Ozone Disinfection—State of the Art," paper presented at Pre-Conference Workshop on Wastewater Disinfection, 52nd Annual Water Pollution Control Federation Conference, Houston, TX, October 7, 1979.
5. Engelbrecht, R. S., et al. "Process Considerations in Design of Ozone Contactor for Disinfection," *J. Environ. Eng. Div., ASCE* 104(EE5): 835–847 (1978).
6. Hoigné, J., and H. Bader. "The Role of Hydroxyl Radical Reactions in Ozonation Processes in Aqueous Solutions," *Water Res.* 10:377–386 (1976).

7. Peleg, M. "The Chemistry of Ozone in the Treatment of Water," *Water Res.* 10:361–365 (1976).
8. Gurol, M. D. "Kinetic Behavior of Ozone in Aqueous Solution; Decomposition and Reaction with Phenol," PhD Dissertation, Dept. of Environmental Sciences and Engineering, Univ. North Carolina, Chapel Hill, NC (1980).
9. Roth, J., and D. E. Sullivan. "A Critical Review of Kinetics of Ozone Decomposition in Aqueous Solution," paper presented at Symposium on Advanced Ozone Technology, International Ozone Association, Toronto, Ontario, Canada, November 16–18, 1977.
10. Hoigné, J., and H. Bader. "Ozonation of Water: Selectivity and Rate of Oxidation of Solutes," *Ozone Sci. Eng.* 1(1):73–85 (1979).
11. Hoigné, J., and H. Bader. "Ozonation of Water: 'Oxidation-Competition Values' of Different Types of Waters Used in Switzerland," *Ozone Sci. Eng.* 1(4):357–372 (1979).
12. Niki, E., Y. Yamamoto, H. Shiokawa and Y. Kamiya. "Ozonation of Phenol in Water," paper presented at the Ozone Technology Symposium, International Ozone Association, Los Angeles, CA, May 1979.
13. Kosak-Channing, L., and G. R. Helz. "Ozone Reactivity With Seawater Components," *Ozone Sci. Eng.* 1(1):39–48 (1979).
14. Crecelius, E. A. "The Production of Bromine and Bromate in Seawater by Ozonation," presented at the Symposium on Advanced Ozone Technology, International Ozone Association, Toronto, Ontario, Canada, November 16–18, 1977.
15. Richard, Y., and L. Brener. "Mesure et Régulation de l'Ozone en Présence de Chlore," in *Applications de l'Ozone au Traitement des Eaux* (Paris, France: European Committee, International Ozone Assoc., 1979), pp. 437–468.
16. Saltzman, B. E., and N. Gilbert. "Iodometric Microdetermination of Organic Oxidants and Ozone," *Anal. Chem.* 31(11):1914–1920 (1959).
17. *Federal Register* 39(223):22392 (1971).
18. *Standard Methods for the Examination of Water and Wastewater,* 13th Ed. (Washington, DC: Am. Public Health Association, 1971), pp. 107–143, 271–275.
19. Kolthoff, I. M., and R. Belcher. *Volumetric Analysis, Vol. 3* (New York: Interscience Publishers, Inc., 1957).
20. Shechter, H. "Spectrophotometric Method for Determination of Ozone in Aqueous Solution," *Water Res.* 7:729–739 (1973).
21. Ingols, R. S. "Comments on Analysis of Ozone in Water and Air," in *Ozone: Analytical Aspects and Odor Control,* R. G. Rice and M. E. Browning, Eds. (Vienna, VA: International Ozone Association, 1976), pp. 10–11.
22. Boyd, A. W., C. Willis and R. Cyr. "New Determination of Stoichiometry of the Iodometric Method for Ozone Analysis at pH 7.0," *Anal. Chem.* 42: 670–672 (1970).
23. Hodgeson, J. A., et al. "Stoichiometry in the Neutral Iodometric Procedure for Ozone by Gas Phase Titration with Nitric Oxide," *Anal. Chem.* 43(8):1123–1126 (1971).
24. Hofmann, P., and P. Stern, "Spectrophotometric Determination of Ozone in the Presence of Chlorine and Its Compounds," *Anal. Chim. Acta* 25(1):149–155 (1969); Chem. Abstr. 70:90663r (1969).

25. Flamm, D. L. "Analysis of Ozone at Low Concentrations with Boric Acid Buffered KI," *Environ. Sci. Technol.* 11(10):978–983 (1977).
26. Demore, W. B., and M. Patapoff. "Comparison of Ozone Determinations by UV Photometry and Gas Phase Titrations," *Environ. Sci. Technol.* 10(9):897 (1976).
27. Parry, E. O., and D. H. Hern. "Stoichiometry of Ozone-Iodide Reaction: Significance of Iodate Formation," *Environ. Sci. Technol.* 7(1):65–66 (1973).
28. Pitts, J. N., et al. "Long-Path Infrared Spectroscopic Investigation at Ambient Concentrations of the 2% NBKI Method for Determination of Ozone," *Environ. Sci. Technol.* 10(8):787 (1976).
29. Inn, E. C. Y., and Y. Tanaka. "Ozone Absorption Coefficients in the Visible and Ultraviolet Regions," in *Ozone Chemistry and Technology,* Advances in Chemistry Series, No. 21 (Washington, DC: American Chemical Society, 1959), pp. 263–268.
30. Adler, M. G., and G. R. Hill. "The Kinetics and Mechanism of Hydroxide Ion Catalyzed Ozone Decomposition in Aqueous Solution," *J. Am. Chem. Soc.* 72:1884–1887 (1950).
31. Sigrist, W. "Monitoring of Ozone in Water," Gubelin Industries, Mt. Kosco, NY (1978).
32. Layton, R. F., and R. N. Kinman. "A New Analytical Method for Ozone," in *Proceedings of the National Specialty Conference on Disinfection* (New York: American Society of Civil Engineers, 1970), pp. 285–305.
33. Palin, A. T. "Analytical Control of Water Disinfection With Special Reference to Differential DPD Methods for Chlorine, Chlorine Dioxide, Bromine, Iodine and Ozone," *J. Inst. Water Eng.* 28(3):129–154 (1974).
34. Liebermann, J., Jr., N. M. Roscher, E. P. Meier and W. J. Cooper. "Development of the FACTS Procedure for Combined Forms of Chlorine and Ozone in Aqueous Solutions," *Environ. Sci. Technol.* 14(11):1395–1400 (1980).
35. Richard, Y. "Chlorine-Ozone Interferences in Water Treatment," in *Proceedings of the Second International Symposium on Ozone Technology,* R. G. Rice, P. Pichet and M.-A. Vincent, Eds. (Vienna, VA: International Ozone Association, 1976), pp. 169–180.
36. Bader, H., and J. Hoigné. "Determination of Ozone in Water by the Indigo Method," *Water Res.* 15(4):449–456 (1981).
37. Masschelein, W. J., and G. Fransolet, "Spectrophotometric Determination of Residual Ozone in Water With ACVK," *J. Am. Water Works Assoc.* 69:461–462 (1977).
38. Clark, L. C., R. G. Weld and Z. Taylor. "Continuous Recording of Blood Oxygen Tension," *J. Appl. Physiol.* 6:189–193 (1953).
39. Mancy, K. H., D. A. Okun and C. N. Reilley. "A Galvanic Cell Oxygen Analyzer," *J. Electroanal. Chem.* 4:65–92 (1962).
40. Johnson, J. D., J. W. Edwards and F. K. Keeslar. "Real Free Chlorine Residual Probe: HOCl Amperometric Membrane Electrode," *J. Am. Water Works Assoc.* 70(6):341–348 (1978).
41. Johnson, D. C., D. T. Napp and S. Bruckstein. "Electrochemical Reduction of Ozone in Acidic Media," *Anal. Chem.* 40(3):482–488 (1968).
42. Dailey, L., and J. Morrow. "On Stream Analysis of Ozone Residual," in *Proceedings of the First International Symposium on Ozone for Water and*

Wastewater Treatment, R. G. Rice and M. E. Browning, Eds. (Vienna, VA: International Ozone Association, 1975), pp. 69–83.
43. Dunn, J. F., and J. D. Johnson. "Ozone Amperometric Membrane Electrode," in *Proceedings of the Second International Symposium on Ozone Technology,* R. G. Rice, P. Pichet and M.-A. Vincent, Eds. (Vienna, VA: International Ozone Association, 1976), pp. 132–142.
44. Stanley, J. H., and J. D. Johnson. "Amperometric Membrane Electrode for Measurement of Ozone in Water," *Anal. Chem.* 51(13):2144–2147 (1979).
45. Smart, R. B., R. Dormond-Herrera and K. H. Mancy. "In Situ Voltammetric Membrane Ozone Electrode," *Anal. Chem.* 51(14):2315–2319 (1979).
46. Smart, R. B., J. H. Lowry and K. H. Mancy. "Analysis for Ozone and Residual Chlorine by Differential Pulse Polarography of Phenylarsine Oxide," *Environ. Sci. Technol.* 13(1):89–92 (1979).

CHAPTER 9

INTERFERENCES BETWEEN OZONE AND CHLORINE

Y. Richard and L. Brener
Société Degrémont
Suresnes, France

Ozone is the most powerful oxidizing and viricidal agent used in drinking water treatment. Its effects (elimination of organics, tastes, odors and color) have been described widely.

In numerous cases, however, it has been observed that after post-ozonation treatment, nonpathogenic organisms proliferated in the water distribution network, whereas they were absent from the treated water at the outlet of the plant [1–7]. To avoid this bacterial growth, (caused by rapid decomposition of ozone in water) a more stable oxidizing agent, such as chlorine or chlorine dioxide, is added continuously.

The dose of chlorine or chlorine dioxide that must be added in the presence of ozone for maintaining water disinfection in the distribution network is affected by:

1. the level of each specific oxidizing agent present;
2. specific reactions of ozone with chlorine or chlorine dioxide; and
3. the effects of the two combined oxidants.

INTERFERENCE IN MEASUREMENT

Monitoring of the final oxidation treatment requires measurement of the residual oxidant. It is necessary to verify that the amount of residual oxidant corresponds to the minimal dose to be applied to the water to be treated during the minimal period of time. To know the effect of each oxidizing agent in the treatment process, it is necessary to measure these oxidants separately.

Colorimetric and Spectrophotometric Methods

Ward and Larder [8] measured the ozone content of a water in the presence of chlorine by means of 1,5-*bis*-(4-methylphenylamino-2-sodium sulfate)-9,10-anthraquinone. This method is not suitable for analyzing mixtures of ozone and chlorine dioxide.

Palin [9] measured ozone, chlorine and chlorine dioxide contents separately and spectrophotometrically by colorimetric reactions with N,N-diethyl-*p*-phenylenediamine (DPD).

Richard [4] titrated the amount of ozone in the presence of chlorine by means of syringaldazine. The total oxidant (O_3 and Cl_2) is measured in the presence of an alkaline iodide. The total chlorine content is obtained after destruction of the ozone by addition of glycine. The amount of ozone is obtained by subtracting the two results.

Electrometric Methods

Coin et al. [10] proved that to disinfect water—to destroy all pathogenic bacteria and viral particles—it is necessary to maintain a 0.4-mg/l ozone residual in the water for 4 min.

Introduction of ozone into an oxidation tower in a drinking water treatment plant is controlled by measurement of residual ozone. Management and automation are effected by continuous measurement of the residual ozone using galvanic cells.

Richard and Brener [11] proved the interaction of ozone and chlorine in an ozonation tower by using a continuous differential measurement of two galvanic cells fitted with a gold-copper metallic couple. They noticed a reduction of the chlorine content in the tower (Figure 1). Monitoring of oxidant content with a simple galvanic cell fitted with a gold-copper couple is not selective for ozone. The instrument reads greater than the actual ozone content, resulting, in this case, in insufficient ozone treatment and, hence, incomplete disinfection of water (Figure 2).

Masschelein [12] described a nickel-silver galvanic couple to continuously and selectively measure ozone at the outlet of the ozonation towers in the presence of chlorine. However, this result was obtained when chlorine preoxidation of drinking water was employed.

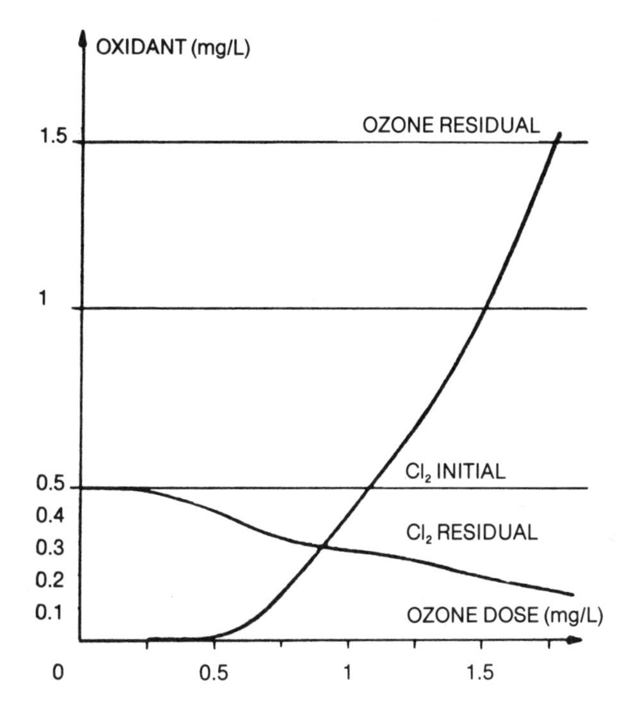

Figure 1. Effect of ozone on chlorine consumption in an ozonation tower.

REACTION OF OZONE WITH CHLORINE OR CHLORINE DIOXIDE

Action of Ozone on Chlorine

Kölle [13] showed that ozonation of a 20-mg/l chlorine (hypochlorite) solution lowers the pH and causes formation of chlorine derivatives of higher oxidation number. At pH values less than 7, ozone oxidation of chlorine leads to the formation of chlorate and perchlorate ions (Figure 3). At very low chlorine and ozone concentrations, it has been noted that the shorter the contact time between chlorine and ozone, the lower the partial deactivation of the two oxidants.

However, Buydens and Fransolet [14] did not notice any interaction of ozone and chlorine. The dose of ozone introduced was not modified, at least during the time necessary for dissolution of ozone and within the limits of the concentrations tested. They noted, however, that the resid-

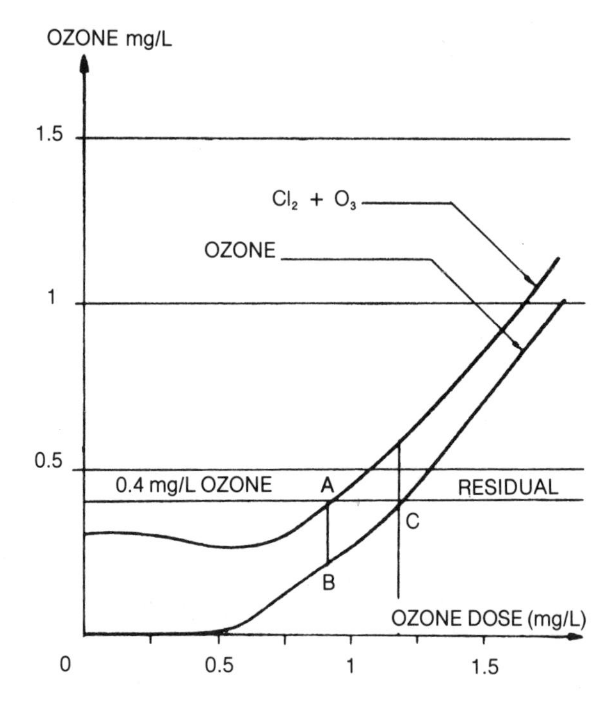

Figure 2. Influence of ozone measurement during water disinfection.

ual chlorine concentration decreased or disappeared. This was attributed to the action of chlorine on the organic matter fractionated by the injected ozone. In other words, ozonation of the organic matter made it reactive to chlorine.

Action of Ozone on Chlorine Dioxide

Kölle [13] proved that the reaction kinetics of ozone with chlorine dioxide are very fast. The resulting products of this oxidation reaction are chlorate and perchlorate.

Buydens and Fransolet [14] noted that the introduction of ClO_2 to waters containing residual ozone caused a drop in the ozone content, resulting from:

$$2ClO_2 + O_3 \rightarrow 2ClO_3^- + O_2$$

In the course of the same study they observed that 0.6 and 0.3 mg/l of

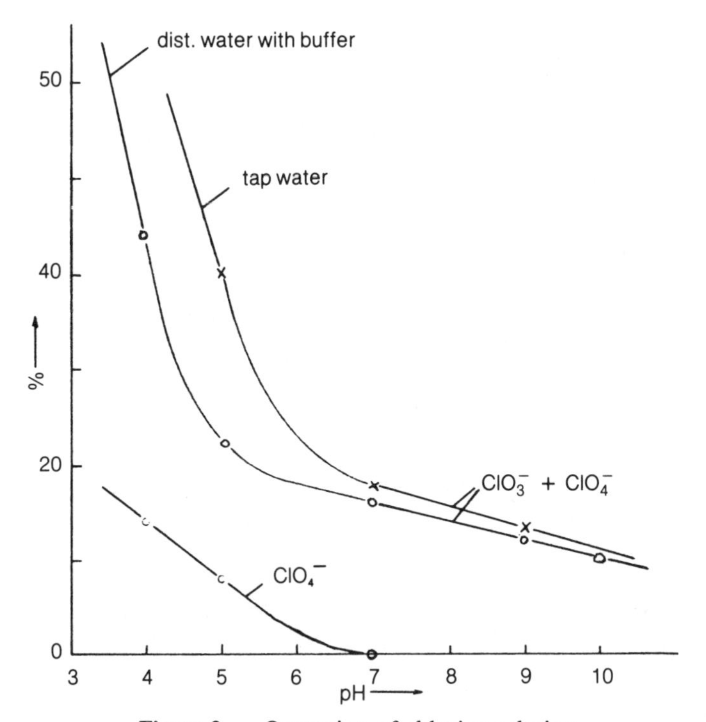

Figure 3. Ozonation of chlorine solution.

ClO_2^- reduced the ozone residual levels by 0.34 and 0.21 mg/l, respectively, an average ratio of 1.44, which they expressed as:

$$ClO_2^- + O_3 \rightarrow ClO_3^- + O_2$$

A stoichiometric ratio of 1.40 results from this equation. Therefore, the oxidation of ClO_2^- requires twice as much ozone as ClO_2.

Masschelein [15] noted that in an aqueous solution of pH 5–6, ozonation of chlorite ions leads to the formation of chlorine dioxide:

$$2NaClO_2 + O_3 + H_2O \rightarrow 2NaOH + ClO_2 + O_2$$

Richard and Brener [16] showed that ozone reacts rapidly with ClO_2 and ClO_2^-. Figure 4 shows that ClO_2^- is first oxidized to ClO_2, which is further oxidized by ozone.

It can be concluded, therefore, that when a postozonated drinking water is to undergo a final disinfection with chlorine dioxide, the injection point of this chemical must be sufficiently distant from the ozone

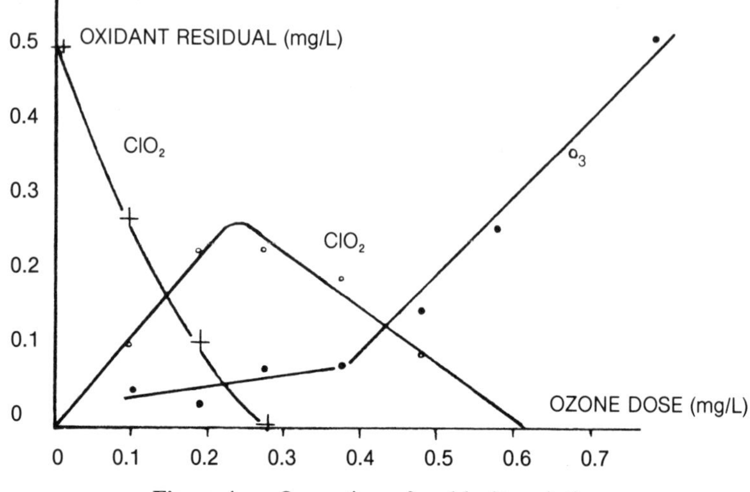

Figure 4. Ozonation of a chlorite solution.

treatment to allow the ozone to disappear with time, to avoid any interference of the chlorine dioxide with the ozone.

INTERFERENCE WITH OZONE

Richard [4] noted that ozonation interferes with absorbed chlorine levels. Ozone is a powerful oxidizing agent that breaks up high-molecular-weight molecules into smaller molecules, thus creating a new chlorine demand in the water (Figure 5).

Somiya and Goda [17] proved that ozonation of nitrogenous substances, such as amino acids, proteins and ethylamine, is accompanied by an ammonification reaction, and, if the pH is favorable, by nitrification.

It can be expected that ozonation in this case will be accompanied by a new chlorine demand. Chlorine reacts with ammonia to form chloramines, the disinfection capability of which is weaker.

Stettler [5] noted that in many Swiss plants the addition of chlorine to an ozonated water led to a marked proliferation of saprophytic microorganisms in the distribution networks. Ozone treatment was totally stopped in one or two plants in favor of chlorine, and the microorganisms totally disappeared with a low dose of chlorine. To avoid these interferences of chlorine and ozone he suggests:

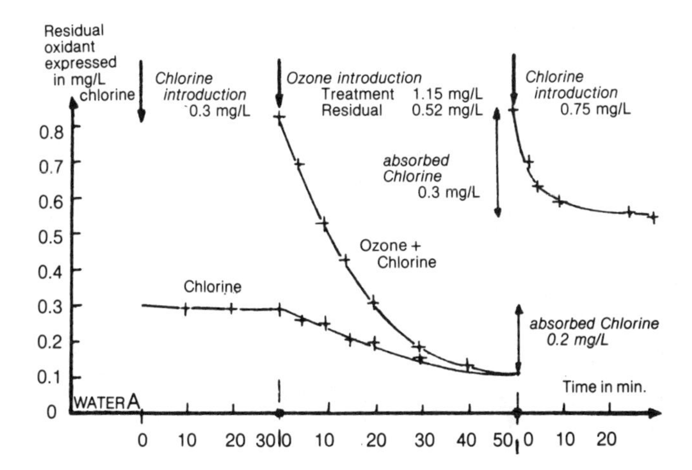

Figure 5. Effect of ozone treatment on chlorine absorption.

1. Residual ozone should be destroyed by granular activated carbon before chlorine and chlorine dioxide are added.
2. If there is no filtration through activated carbon, ozonation should be adjusted so that the ozone residual is reduced to 0.05 mg/l before chlorine or chlorine dioxide is added.

CONCLUSIONS

Once ozone has been used at the end of the drinking water treatment line, it is generally necessary to use chlorine or chlorine dioxide as a safeguard against bacterial growth in the distribution network. Use of ozone after clarification often leads to an increase in chlorine demand in the water. The use of granular activated carbon partially makes up for this inconvenience and results in a water with a low chlorine demand.

It is important to check with care each residual oxidant that is used for water treatment to be certain of its action. This is particularly important when ozone is used for its viricidal properties. A chlorine dioxide preoxidation leads to a significant increase in the ozone treatment rate.

REFERENCES

1. Schalekamp. M. "Untersuchungen zur Abklärung des Phänomens der Wiederkeimung im Zusammhang mit Ozonung," *Gas Wasser Abwasser* (8) (1969).

2. Vaillant, C. J. "L'ozonation et la filtration lente dans le problème de la prolifération microbienne dans le réseau de distribution," *Gas, Eaux, Eaux Usées* 50(3):67–70 (1970).
3. Dietlicher, K. "Wiederkeimung ozonisierter Schnellfiltrate im Rohrhetz," *Wasserversorgung Zürich* (October 1973).
4. Richard, Y. "Chlorine-Ozone Interferences in Water Treatment," in *Proceedings of the Second International Symposium on Ozone Technology,* R. G. Rice, P. Pichet and M.-A. Vincent, Eds. (Vienna, VA: International Ozone Association, 1976), pp. 169–180.
5. Stettler, R. "Utilisation de l'ozone dans le traitement des eaux de boisson," *Gas, Eaux, Eaux Usées* 57(1):81–95 (1977).
6. Lamblin, H. "Protection d'un réseau d'eau ozonée vis-à-vis des germes," paper presented at the International Ozone Association, Third International Congress, Paris, 1977.
7. Köhler, H. G., and O. Schatz. "Dokumentation ueber die Versuche einer Trinkwasserentkeimung durch Ozone," *Gas. Wasser Abwasser* 116(9): 417–423 (1975).
8. Ward, S. B., and D. W. Larder. "The Determination of Ozone in the Presence of Chlorine," paper presented at the Symposium of the Society for Water Treatment and Examination, London, March 29, 1973.
9. Palin, A. T. "Analytical Control of Water Disinfection with Special Reference to Differential DPD Methods for Chlorine, Chlorine Dioxide, Bromine, Iodine and Ozone," in *Proceedings of the Society for Water Treatment and Examination* (1975), pp. 139–154.
10. Coin, L., C. Hannoun, and C. Gomella. "Inactivation of Poliomyelitis Virus by Ozone in the Presence of Water," *Presse Médicale* 72(37):2153–2156 (1964); Coin, L., C. Gomella, C. Hannoun and J. C. Trimoreau. "L'Inactivation par l'Ozone du virus de la poliomyélite présent dans le Eaux," *Presse Médicale* 75(38):1883–1884 (1967).
11. Richard, Y., and L. Brener. "Mesure et régulation de l'ozone en présence de chlore," in *Applications de l'Ozone aux Traitements des Eaux* (Paris: Comité Européen, International Ozone Association, 1979), pp. 437–468.
12. Masschelein, W. J. "Mesure de l'ozone résiduel dans l'eau—Spécificité-Automatisation," paper presented at the International Ozone Association Third International Congress, Paris, 1977.
13. Kölle, W. "Problem der gemeinsamen Anwendung verschiedener Oxydationsmittel bei der Wasseraufbereitung," *Vom Wasser* (1968), pp. 367–381.
14. Buydens, R., and G. Fransolet. "L'action de l'ozone sur le chlore, le dioxyde de chlore et le chlorite contenus dans les eaux traitées," *Tribune Cébédeau* 326(1):4–6 (1971).
15. Masschelein, W. J. *Chlorine Dioxide: Chemistry and Environmental Impact of Oxychlorine Compounds* (Ann Arbor, MI: Ann Arbor Science Publishers, Inc., 1979).
16. Richard, Y., and L. Brener. "Ozonation des solutions de chlorite de sodium et de dioxyde de chlore," unpublished results.
17. Somiya, I., and J. Goda. "Ozonation of Nitrogeneous Compounds in Water," paper presented at the International Ozone Association Symposium on Advanced Ozone Technology, Toronto, Ontario, 1977.

CHAPTER 10

ANALYTICAL INSTRUMENTATION FOR CONTROL OF OZONATION SYSTEMS

M. Pare

Société Degrémont
Rueil-Malmaison, France

Ozone occurs naturally in the atmosphere; the layer it forms in the upper atmosphere provides a barrier to short-wavelength ultraviolet (UV) radiation, thus protecting the earth. Ozone is also an atmospheric pollutant: in sunlight, UV rays react with nitric oxides in the presence of hydrocarbons to form ozone.

Ozone is used increasingly in industry, particularly in water treatment, for its high oxidizing and bactericidal powers. Commercial production of ozone is obtained in reactors, known as ozonators, using the corona discharge process whereby air or oxygen is passed through a strong electric field. Ozone utilization consists of placing the ozonated gas into contact with the product to be treated. In the case of water treatment, for instance, a specific treatment parameter is the amount of residual ozone measured in the liquid after a certain contact time. Any leaks occurring during ozone production or application processes may be hazardous to operating personnel and surrounding equipment. These various areas of ozone occurrence or application require that ozone be measured continuously, automatically and accurately in gases and liquids over wide concentration ranges.

Three types of instruments are available to measure (1) atmospheric ozone or ozone present in the ambient air; (2) commercially produced ozone; and (3) ozone in liquids, particularly in water. For these three fields, the orders of magnitude of concentrations, expressed in their usual units, are as follows:

1. ozone in the atmosphere or the ambient air: 1–20 ppm;
2. ozone produced commercially: from air—1–2% by weight, from oxygen—2–10% by weight; and
3. residual ozone in water 0.1–2 g/m^3 (mg/l) of water.

As a reminder, under standard conditions:

- 1 ppm = 2.14×10^{-6} kg/m^3 = 1.66×10^{-4}% by weight
- 1% by weight in air = 12.9 g/m^3 $\simeq$ 6000 ppm
- 1% by weight in oxygen = 14.3 g/m^3 $\simeq$ 6700 ppm

MEASUREMENT METHODS

Depending on the fields of ozone generation or application, different continuous measurement methods are used. Atmospheric ozone or ozone in ambient air is measured by:

1. absorption of UV radiation,
2. ethylene or chemiluminescence, or
3. amperometry.

Commercially produced ozone is measured by:

1. absorption of UV radiation or
2. calorimetry.

Residual ozone in water is measured by

1. absorption of UV radiation,
2. amperometry, or
3. potentiometry.

OZONE MEASUREMENT BY ABSORPTION OF RADIATION

Principle

Absorption of radiation by a gas or liquid is governed by the Lambert-Beer law:

$$I = I_0 e^{-KlC}$$

where I_o = intensity of the incident light beam with no sample present
I = intensity of the light beam after passing through the sample (gas or liquid)
l = length of the light path through the sample
C = concentration of absorbing material in the sample
K = specific absorptivity coefficient of the material for the wavelength of the light

This type of measurement implies accurate knowledge of K for the material at a given wavelength.

Ozone Measurement

Ozone absorbs light in the short-ultraviolet wavelength region (200–300 nm, the Hartley band, with maximum absorption at 253.7 nm (Figure 1). For the 253.7-nm wavelength, the generally accepted value for the absorptivity coefficient ranges from 303.9 to 313.2 cm^{-1}, base e, at 273°K and 760 torr [2]. Researchers have verified the value of 3024 cm^{-1}-mol^{-1}-l [3].

Available Instruments

Ozone measurement by absorption of UV radiation is used by several manufacturers, including Dasibi, Mast, Sigrist, and Bran and Lübbe.

Dasibi and Mast

Dasibi (Glendale, CA) and Mast Development Co. (Davenport, IA) sell instruments that work on the same principle; the difference lies in the fact that Dasibi units are more sophisticated. Figure 2 shows a schematic diagram of a Dasibi analyzer designed to measure ozone in ambient air.

On entering the instrument, the sample is diverted by a valve into a catalytic converter that selectively decomposes any ozone present to oxygen. This treated sample, now containing no ozone, then passes through an absorption chamber, where a detector measures the amount of UV light transmitted through it. This measurement is made digitally, and its value is stored in the instrument. This value includes those of all the other gases and particulates present in the sample, and serves as the reference measurement.

On completion of this reference measurement, the valve is then activated, the absorption chamber is flushed with the sample containing

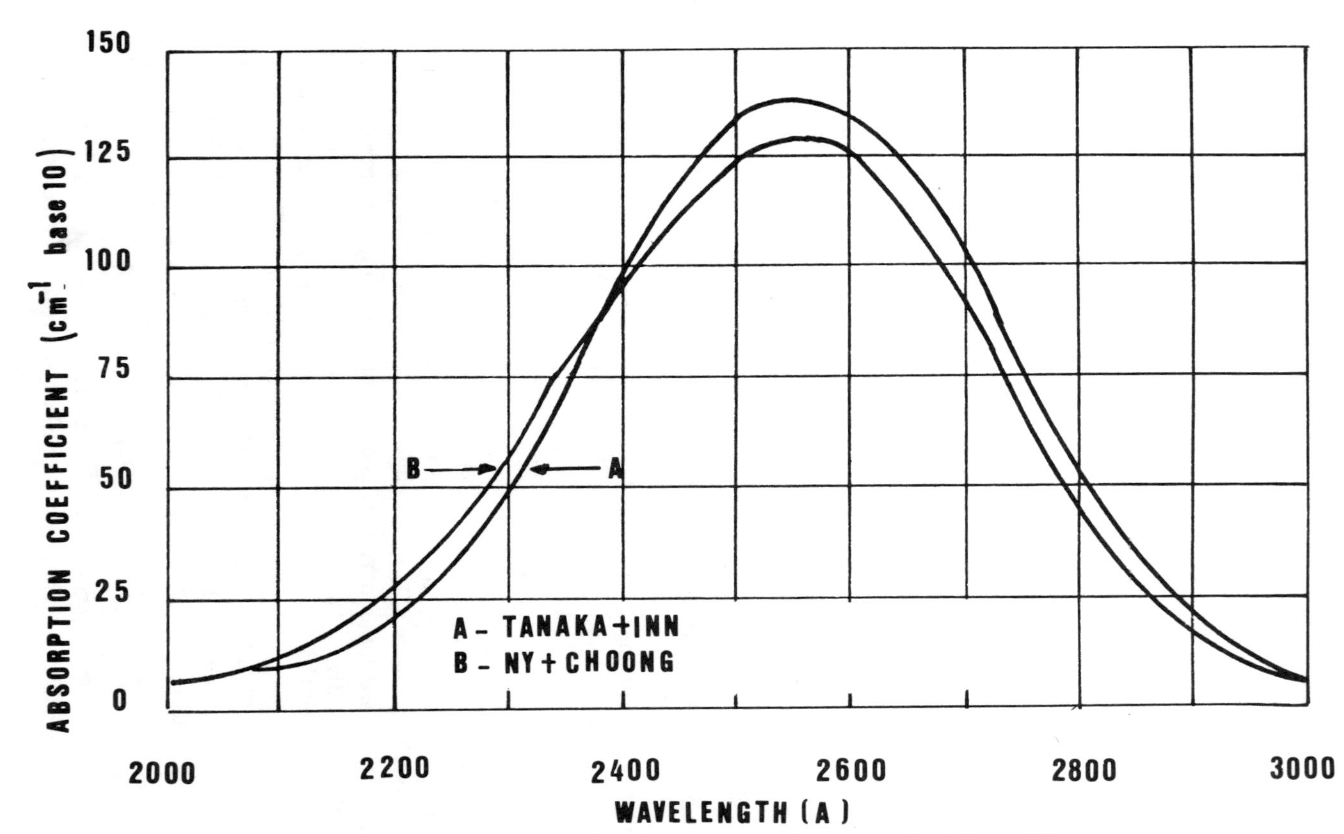

Figure 1. Absorption coefficient of ozone in the region 200–300 nm [1].

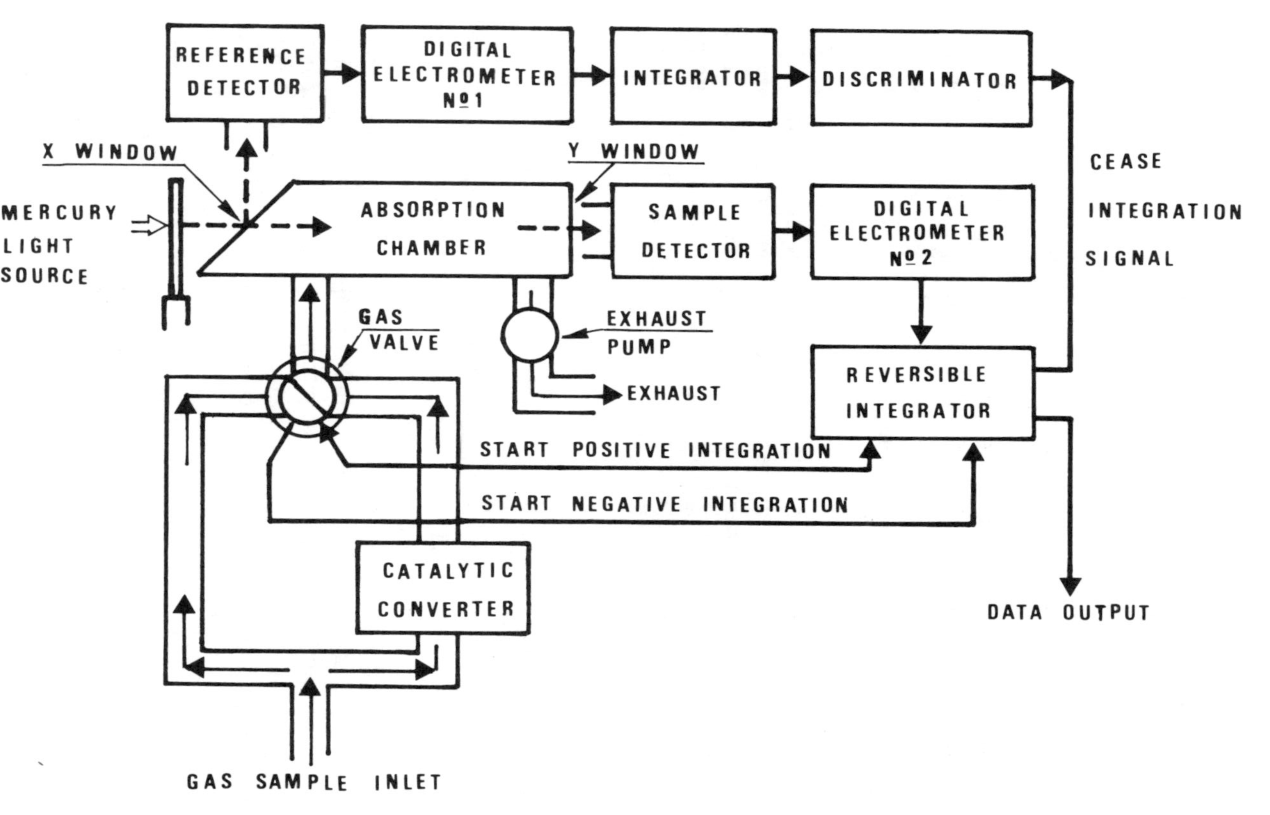

Figure 2. Schematic diagram of Dasibi ozone analyzer.

ozone, which now is ready to be measured. A digital counter now measures the reduced amount of UV light the detector sees with the ozone in the sample. The value of this measurement is subtracted from that of the stored reference measurement, and the difference is displayed as parts per million of ozone, or grams per cubic meter.

Sigrist

The Sigrist (Zürich, Switzerland) instrument uses a double optical beam, as shown in Figure 3. A light source directs a beam to the swinging mirror, which alternately reflects back a measurement beam and a reference beam at an approximate frequency of 600 Hz. These beams pass through the sample cell and the reference cell, respectively, and arrive at the photoelectric cell. The current emitted by this cell is amplified by the unit to control a servomotor. The servomotor actuates an optical iris, thus modifying the intensity of the reference beam until both beams have the same light intensity when they reach the photoelectric cell. In this state of balance, the position of the iris corresponds to the ozone concentration in the sample under examination. Thus unit is suitable for measurement of ozone concentrations both in air and water.

Bran and Lübbe

The principle of the Bran and Lübbe (Hamburg, Federal Republic of Germany) instrument is shown in Figure 4. The radiation of a mercury lamp is focused by a condenser to form a parallel beam that passes through the measuring cell on the light receiver. A portion of the radiation is deflected by the light divider and is projected to a second light receiver for reference measuring. The light intensity is adjusted to identical levels by a variable iris. The light receivers are in a bridge circuit. Light absorption in the measuring cuvette causes unbalance of the bridge, and a servomotor provides for rebalancing. This correction movement corresponds to the light absorption. This instrument incorporates automatic zero compensation. When ozone is measured in air, an inert gas is introduced into the measuring cell through the solenoid valve. When ozone is measured in water, a standard solution is introduced into the measuring cell.

All of these instruments are guaranteed to provide an absolute measurement of ozone concentration, and theoretically require no calibration.

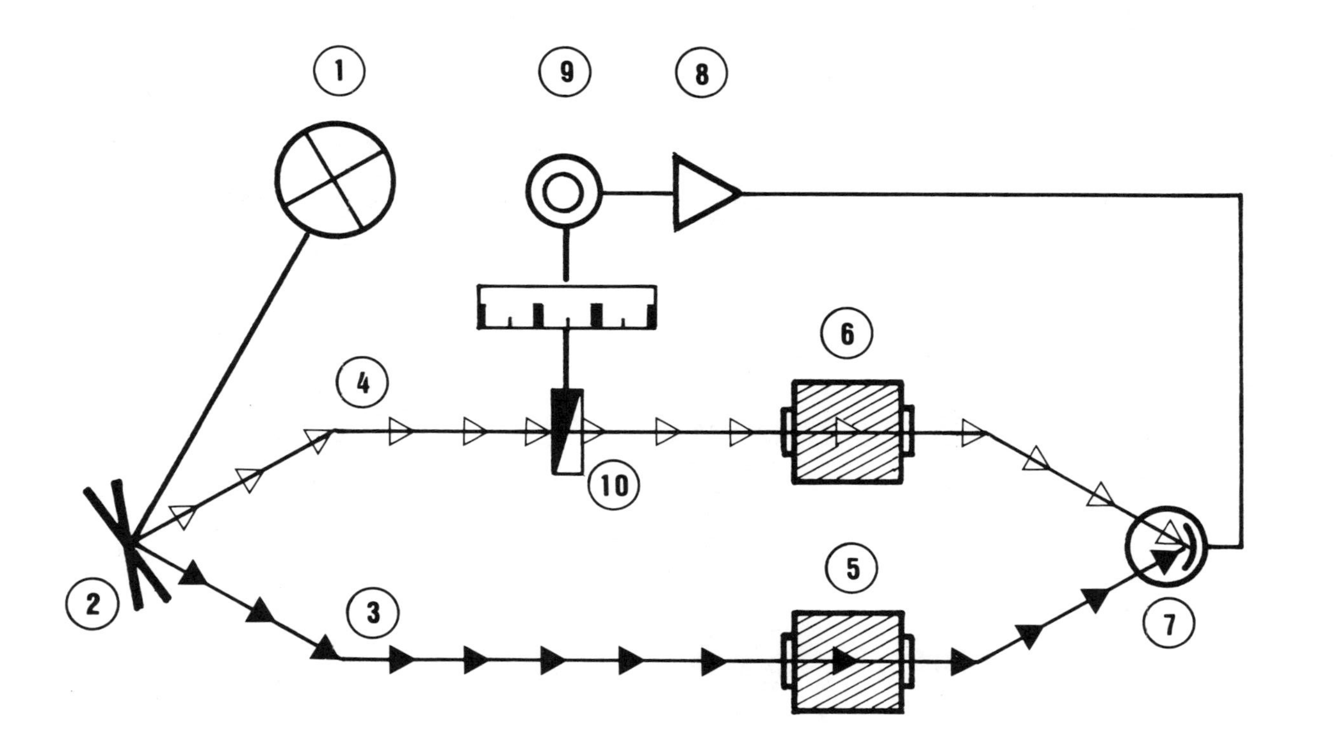

Figure 3. Schematic diagram of the Sigrist double-beam ozone analyzer. 1. Light source; 2. swinging mirror; 3. measurement beam; 4. reference beam; 5. sample cell; 6. reference cell; 7. photoelectric cell; 8. amplifier unit; 9. servomotor; 10. optical iris.

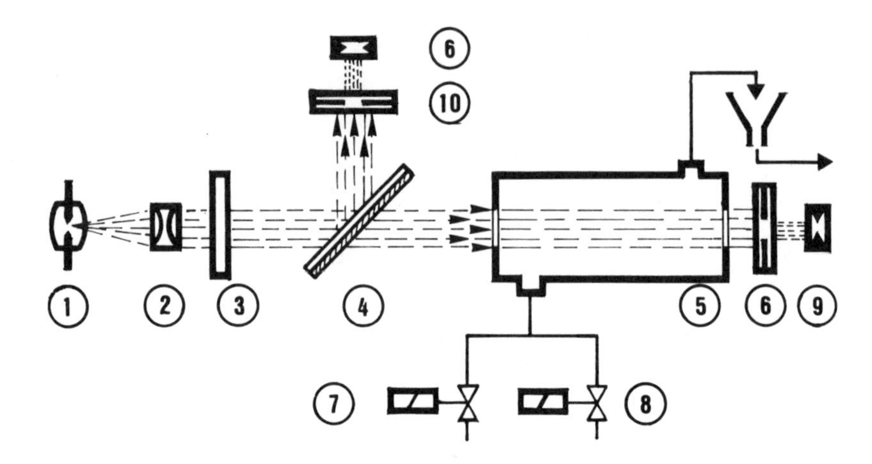

Figure 4. Schematic diagram of the Bran and Lübbe ozone analyzer. (1) mercury lamp; (2) condenser; (3) parallel beam; (4) light divider; (5) measuring cell; (6) second light receiver; (7) solenoid valve; (8) inert gas; (9) light receiver; (10) variable iris.

OZONE MEASUREMENT BY ETHYLENE CHEMILUMINESCENCE

Principle

Measurement of ozone by ethylene chemiluminescence uses the reaction of ozone with ethylene in the gas phase, which is accompanied by the emission of light. Ozone reacts with ethylene to give formaldehyde and oxygen. The formaldehyde molecules thus formed are initially in an excited state, and their deactivation is accompanied by the emission of photons. The intensity of light emission, which is in the region of 10 nA/ppm of ozone, is proportional to the concentration of ozone in the sample and can be measured using a photomultiplier tube.

This method is specific to ozone, and is suitable for ozone measurement in the ambient air, because any other compounds present in the sample have no effect on this measurement.

Available Instruments

The manufacturers of instruments applying this method include:

- Analytical Instrument Development (Avondale, PA)
- Beckman (Fullerton, CA)

- Meloy Laboratories (Springfield, VA)
- Monitor Labs (San Diego, CA)
- REM (Santa Monica, CA)

Beckman

As an example, the unit marketed by Beckman is shown in Figure 5. Ethylene is introduced into the reaction chamber at an approximate rate of 20 cm^3/min, through a solenoid valve, which automatically shuts off gas delivery in the event of a power failure. The air sample is delivered to the reaction chamber at a constant rate by an internal pump and a regulator. Zero setting is obtained by passing chemically purified, completely ozone-free ambient air through the reaction chamber. To prevent the explosion risks inherent in the use of ethylene, Beckman recommends the use of a mixture of 10% ethylene and 90% CO_2.

Because this method does not provide an absolute measurement of ozone, all marketed instruments must be calibrated against an accepted reference method, e.g., the iodometric technique.

OZONE MEASUREMENT IN AIR BY AMPEROMETRY

An analyzer based on this principle is sold by the Mast Development Company (Figure 6). A chemical solution containing the proper amounts of sensing reagent is metered into the sensor through the solution supply tube. The solution flows in a fine film down the electrode support, on which are wound many turns of a fine-wire cathode and a single turn of a wire anode. An air sample is pumped through the sensor, where it comes into contact with the solution contained on the electrode support, and exits through a precision vacuum pump.

Sensing of ozone in the air sample is accomplished by oxidation-reduction of potassium iodide in the sensing solution. This reaction takes place on the cathode portion of the electrode support. In this region, any ozone in the air sample reacts with the sensing solution. When no voltage is applied between the electrodes, a thin layer of hydrogen gas is produced at the cathode by a polarization current. When a voltage is applied to the electrodes, the hydrogen layer builds to its maximum and the polarization current ceases to flow. When free iodine (I_2) is produced by the reaction with ozone, it immediately reacts with the H_2, and reduces it. The removal of hydrogen from the cathode causes a repolarization current to flow in the external circuit, reestablishing equilibrium. The current is directly proportional to the mass of ozone entering the sensor per unit time.

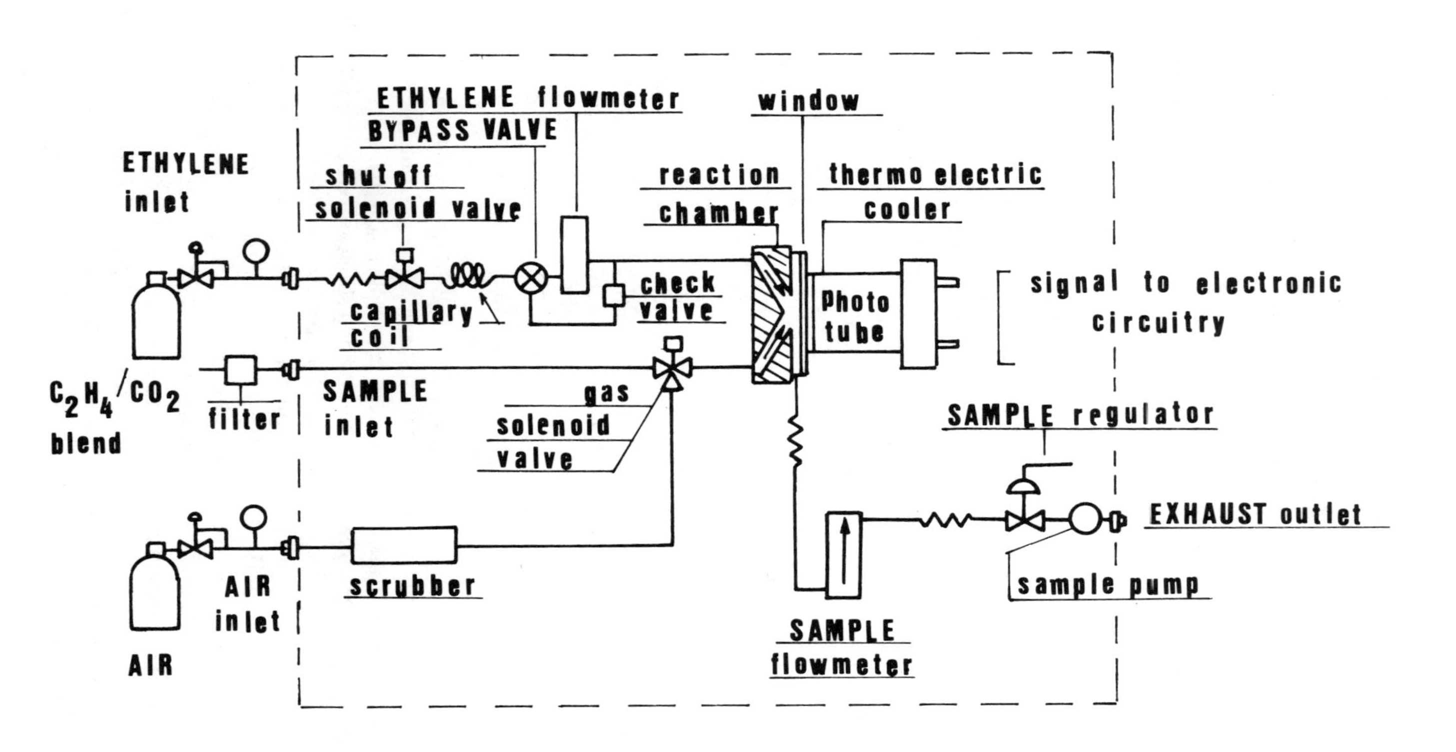

Figure 5. Schematic diagram of the Beckman analyzer (ozone in air).

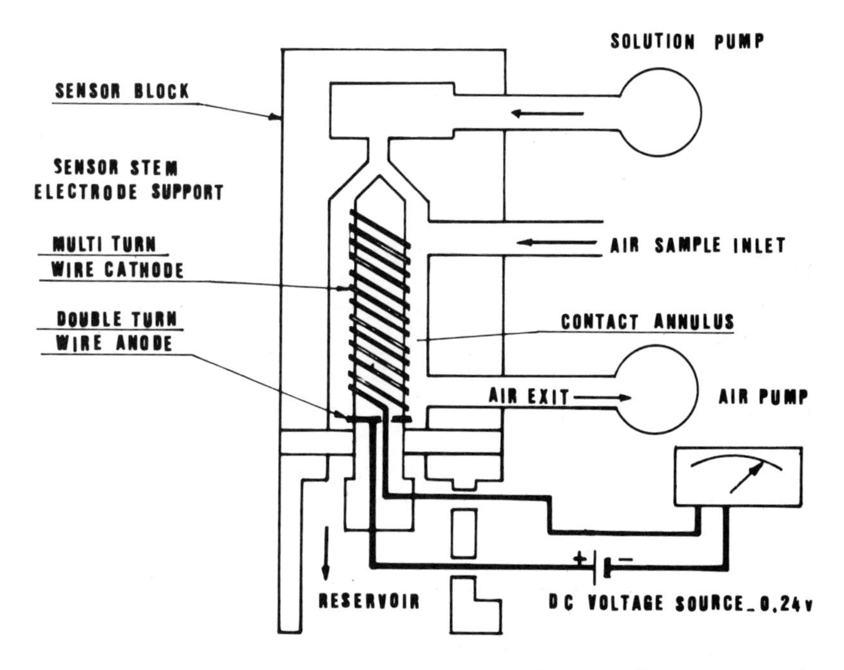

Figure 6. Schematic diagram of the Mast Development Co. amperometric ozone analyzer.

OZONE MEASUREMENT IN WATER BY AMPEROMETRY

Principle

A sample cell, continuously fed with the water to be analyzed, contains a nonconsumable but polarizable gold or platinum cathode, and a copper, cadmium or silver anode.

When there is no oxidant present, the galvanic cell thus formed is polarized by a layer of hydrogen; a very low current flows. When the water to be analyzed contains an oxidant, the depolarization of the cell, and hence the intensity of the current it delivers, are proportional to the concentration of oxidant reduced at the cathode.

Available Instruments

The analyzers applying this method include the units marketed by Fisher and Porter, Wallace and Tiernan, and Degrémont.

Fisher and Porter

In the Fisher and Porter (Horsham, PA) analyzer (Figure 7) the measuring electrode is made of gold, and the counterelectrode is made of copper. Both electrodes are concentric. The sample to be measured flows in the annulus between them. The gold electrode, driven by a motor, rotates at a high speed (1550 rpm). The annulus contains plastic pellets, which are continuously stirred by the water flow, to clean the electrodes. The sample flowrate is not a critical parameter, since the relative speed of the water with respect to the measuring electrode is ensured by the rotation of this electrode.

Wallace and Tiernan

In the Wallace and Tiernan (Günzburg, Federal Republic of Germany) analyzer (Figure 8), the measuring electrode is a platinum wire arranged

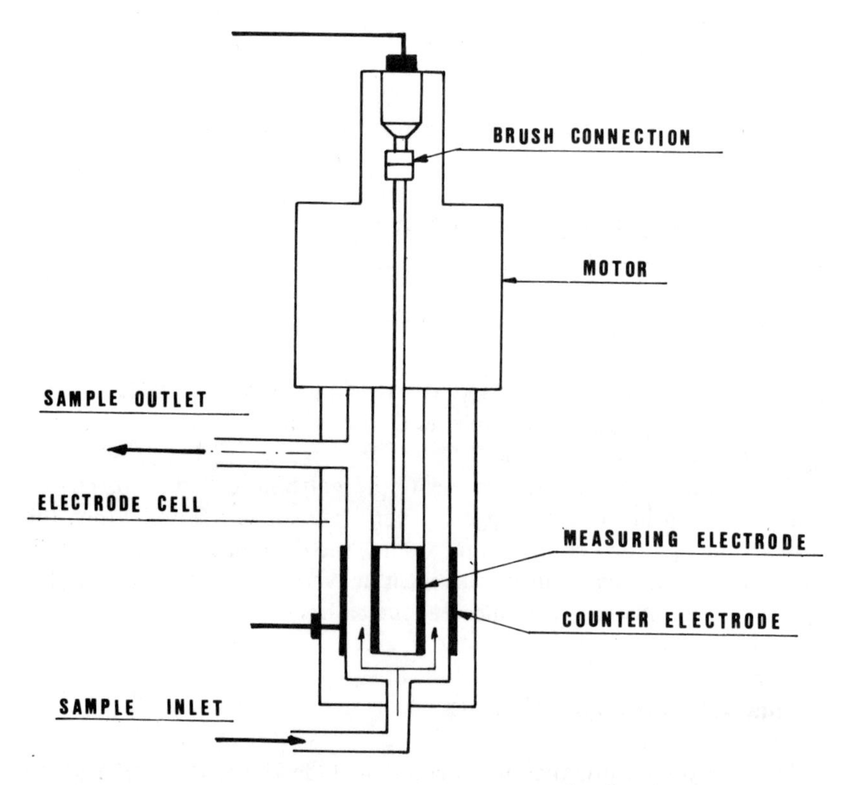

Figure 7. Fisher and Porter amperometric analyzer for ozone in water.

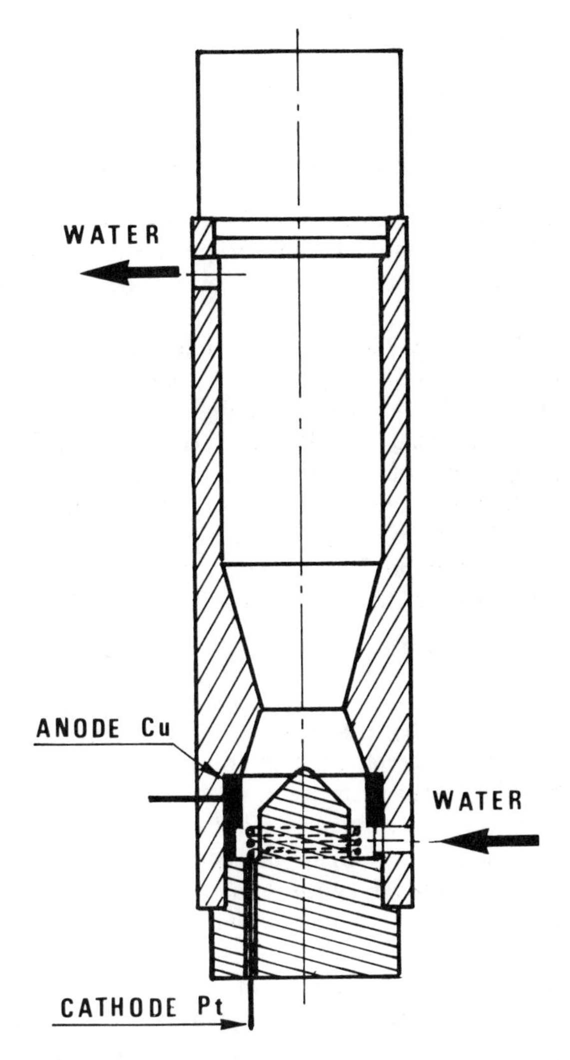

Figure 8. Wallace and Tiernan amperometric analyzer.

concentrically inside a ring-shaped copper electrode. The space between the electrodes contains a certain amount of calibrated quartz grains. The flow of water to be analyzed enters the cell tangentially, and causes the quartz grains to circulate and clean the electrodes.

These two analyzers give satisfactory results provided there is no other oxidant than ozone in the sample. In the presence of chlorine, for instance, ozone concentration measurement is erroneous. Both ozone and chlorine will be determined, and the value of "total oxidant" measured will be the sum of the concentrations of ozone and chlorine, as well as other oxidants which may be present in solution.

Ozone Measurement in the Presence of Chlorine

The problem of ozone measurement in the presence of chlorine led Masschelein [4] and Richard and Brener [5] to develop an electrode couple which, under certain conditions, proves to be specific for ozone measurement.

Masschelein designed a nickel-silver couple (actually nickel–silver–silver chloride) wherein no hydrogen forms at the cathode; thus, the cell cannot be depolarized by the chlorine present.

After experiments, Richard and Brener [5] concluded that ozone measurement was dependent on the nature of the chlorine in the sample, according to whether it is present in the form of free chlorine or chloramines. When calibration is made in the chloramine phase, the calibration curve is displaced to the right by a constant value in the presence of free chlorine, regardless of the amount of the latter.

Degrémont

These conclusions led Degrémont to introduce a nickel-silver cell unit fitted with a continuous ammonium chloride feed device (Figure 9) to convert the chlorine present in the water to chloramines. The measuring cell consists of two planar, circular, parallel electrodes. The flow of sample enters the interelectrode space tangentially and initiates the rotation of a bed of glass beads designed to favor oxidant diffusion toward the cathode and to ensure continuous cleaning of the electrodes. To feed the cell at a constant rate, the sample is introduced under a constant head, providing a flowrate of the order of 60 liter/hr.

Ammonium chloride is fed at a constant flowrate by a compressor bubbler unit designed to apply a constant air pressure (corresponding to the immersion depth of the tube in the bubbler) to the ammonium chloride storage tank. A capillary tube on the delivery pipe from the compressor permits the output of the latter to be limited.

The storage tank is fitted out as a Mariotte bottle. Two air inlet and injection tubes are immersed in the ammonium chloride solution. The injection tube, fitted with a capillary tube, plunges deeper in the liquid than does the air inlet tube. This arrangement allows the capillary tube to be placed under a constant head regardless of the level of the chemical solution in the tank, and thus permits ammonium chloride to be fed regularly at a low flowrate.

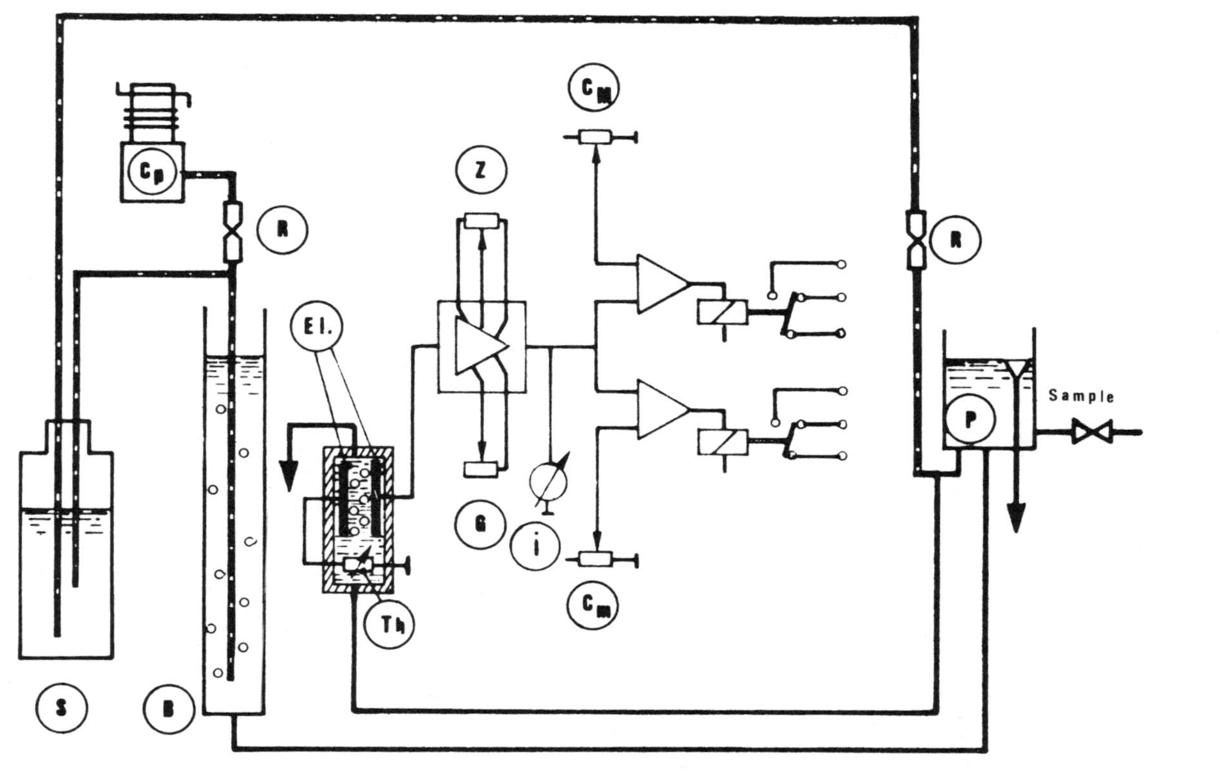

Figure 9. Degrémont nickel-silver amperometric analyzer of ozone in water. P = constant head; C_p = compressor; B = bubbler; S = ammonium chloride storage tank; R = capillary tube; Th = thermistor; i = galvanometer; Z = zero setting; G = potentiometer; C_m = minimum setting potentiometer; C_M = maximum setting potentiometer.

The signal picked up at the electrodes of the cell is corrected by a thermistor as a function of sample temperature, and next is passed into an electronic box where it is amplified to be displayed on a galvanometer as mg O_3/l. Zero setting nullifies the residual current, and a second potentiometer adjusts the scale range. The electronic box also sends a signal to an external recorder or an ozone production control device. Additionally, two comparators permit the sensing of minimum and maximum set points adjustable from 0 to 100% of the scale, by means of two potentiometers. Whenever the set point is overrun, a changeover switch initiates the control of an external element.

OZONE MEASUREMENT BY CALORIMETRY

This method is based on the enthalpic decomposition of ozone in the presence of a catalyst. Knowing the enthalpy of destruction of ozone, i.e., $\Delta H = 144.41$ kJ/mol, ozone concentration in a gaseous mixture can be determined by measuring the differential temperature of the gas between the inlet and outlet after passing over a catalyst.

Ozone concentration is given by the following formula:

$$[O_3] = \Delta T \left(\frac{MC}{\Delta H} \right)$$

where ΔT = differential temperature measured
M = molecular weight of ozone
C = specific heat of the gas (air or oxygen)

Messer Griesheim (Düsseldorf, Federal Republic of Germany) manufactures a unit based on this principle (Figure 10). In this analyzer, a constant flow of gas (air or oxygen) containing ozone is directed to a heat-insulated measurement cell containing a proprietary catalyst. This cell is fitted with thermoelements that measure the gas temperatures at inlet and outlet. Any nitrogen oxides in the carrier gas, liable to poison the catalyst, are retained on an adsorbent, which does not destroy ozone. This adsorbent has a service life of 4–5 weeks.

OZONE MEASUREMENT BY POTENTIOMETRY

This method also is reserved for the measurement of ozone in water. The Thiedig Company (Berlin, Federal Republic of Germany) manufactures a unit using this principle, known as the Digox "OZ-2" analyzer.

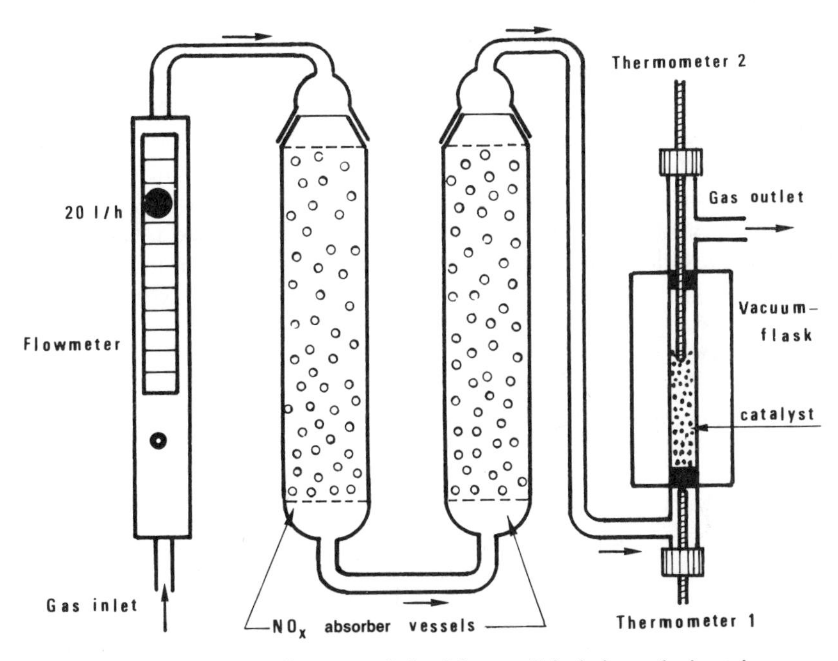

Figure 10. Schematic diagram of the Messer Griesheim calorimetric ozone analyzer.

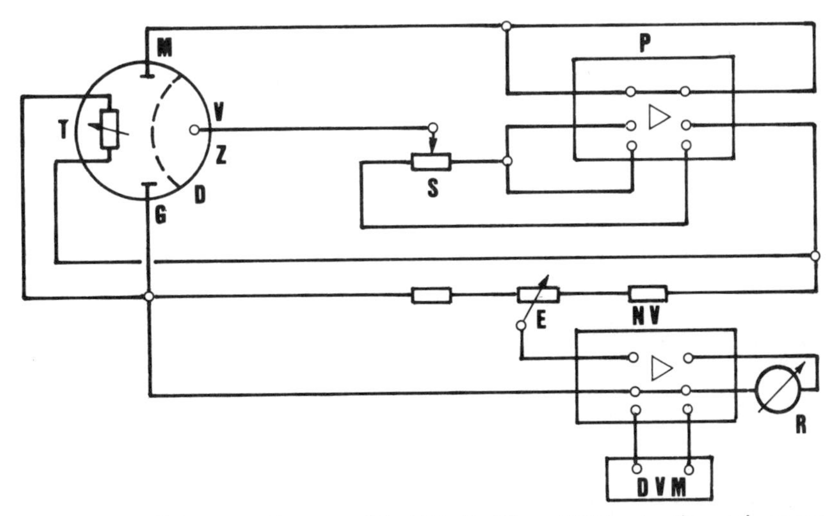

Figure 11. Schematic diagram of the Thiedig Digox OZ-2 potentiometric ozone analyzer. Z = measuring cell; M = measuring electrode (gold); G = counter-electrode (18/12 Mo stainless steel); V = reference electrode (phthalamide); T = thermistor; P = potentiostat; S = range selector switch; NV = amplifier; DVM = digital voltmeter; D = diaphragm; E = calibration potentiometer.

In this unit (Figure 11), the measuring cell includes a gold electrode and a stainless steel counterelectrode, between which flows the sample to be analyzed. A reference electrode is in contact with the measuring cell through a KCl bridge and a diaphragm. The measuring electrode is maintained at a constant potential with respect to the reference electrode by a potentiostat. The cell thus delivers a current whose intensity is only dependent on the amount of ozone in the water. Other oxidants, such as chlorine or chloramines, do not interfere with the measurement.

SUMMARY

Table I compares the various analyzers described in this chapter.

Table I. Characteristics of Various Ozone Analyzers

Measurement Principle	Manufacturer	Type of Unit	Phase of Measurement	Max. Measuring Range
UV radiation	Dasibi	1003 AH	Atm.	20 ppm
		1008 AH	Atm.	20 ppm
		1003 HC	Gas	10%
		1008 HC	Gas	10%
	Mast	727-2	Atm.	10 ppm
	Sigrist	UP 62 B6	Gas	≃13%
		UP 62 B6	Water	0.5 g/m3
	Bran and Lübbe	Uvameter	Gas	≃2%
		Uvameter	Water	0.5 g/m^3
Chemiluminescence of ethylene	AID[a]	560	Atm.	10 ppm
	Beckman	950	Atm.	2.5 ppm
	Meloy	OA-300	Atm.	10 ppm
	Monitor Labs	8410 A	Atm.	10 ppm
	REM	612	Atm.	2 ppm
Amperometry	Mast	725-3 CS	Atm.	1 ppm
	Fisher and Porter	17 L 2000	Water	1 g/m^3
	Wallace and Tiernan	Depolox 3	Water	2 g/m^3
	Degrémont	Amperazur	Water	1 g/m^3
Calorimetry	Messer Griesheim		Gas	7%
Potentiometry	Thiedig	Digox OZ-2	Water	1 g/m^3

[a] Analytical Instrument Development

REFERENCES

1. Inn, E. C. Y., and Y. Tanaka. "Absorption Coefficient of Ozone in the Ultraviolet and Visible Regions," *J. Optical Soc. Am.* 44(10):870–873 (1953).
2. Gourdon, F. "Mesure de l'Ozone et Etalonnage d'Analyseurs par Absorption U.V.," *Poll. Atmos.* 81:15–23 (1979).
3. Leitzke, O. "Instrumentelle Ozonanalytik in der Wasser und in der Gasphase," in *Internationales Symposium Ozon und Wasser* (Berlin: Colloquium Verlag Otto H. Hess, 1977), pp. 164–177.
4. Masschelein, W., G. Fransolet and R. Goossens. "Mesure de l'Ozone Residuel dans l'Eau," paper presented at the Third International Congress on Ozone, International Ozone Association, Paris, France, 1977.
5. Richard, Y., and L. Brener. "Mesure et Regulation de l'Ozone en Présence de Chlore," Journées Européennes sur les Applications de l'Ozone au Traitment des Eaux (Paris, France: Comité Europeen de l'Association Internationale de l'Ozone, 1979), pp. 437–468.

SECTION 5

ENGINEERING ASPECTS

CHAPTER 11

DESIGN ENGINEERING ASPECTS OF OZONATION SYSTEMS

C. Michael Robson

Purdue University
West Lafayette, Indiana and
City of Indianapolis
Indianapolis, Indiana

The technology of individual components of ozonation systems (gas preparation, ozone generation and gas diffusion) is discussed in other chapters of this book. However, none of these chapters includes a consideration of the details required to provide an operable integrated ozonation system.

The author has had an opportunity to evaluate large, operating European and Canadian drinking water ozonation facilities, and is now associated with the design and construction of two major U.S. wastewater ozonation facilities. He has reviewed the design and startup experience of many U.S. facilities, and seeks in this chapter to address details of concern to the designers and operators responsible for ozonation systems, including those treating drinking water, domestic wastewater and industrial liquids.

FEASIBILITY STUDIES

Initial Considerations

There must be at the outset of any project some initial consideration that ozonation is an appropriate application. This consideration may be based on a literature review or on direct knowledge of other successful applications. The potential multiple benefits of the application of ozone, such as oxidation, oxygenation and disinfection, in drinking water treatment must be considered. In any case, ozone should not be considered as merely an expensive substitute for chlorination as a terminal disinfectant.

The sparsity of U.S. operational facilities and authoritative literature increases the difficulty of identifying ozonation as an appropriate process application. However, it is expected that documents such as this book and periodicals such as *Ozone Science & Engineering* (the journal of the International Ozone Association) will increase the awareness of the many process applications of ozone. Selection of ozonation for further study for specific applications will be based on considerations, including:

1. liquid stream quality, such as strength, color and turbidity;
2. quantity of liquid to be treated, as well as flow variations; and
3. applications of ozone, such as disinfection, taste and odor removal, color removal, oxidation, and oxygenation.

Bench-Scale and Pilot-Plant Testing

There is almost universal agreement that some level of testing is required before further consideration of ozonation as a treatment process. Testing can range from a simple bench-scale procedure to a complex pilot-plant testing program.

Nebel [1] summarizes the questions that must be addressed in the pilot program as follows:

1. quantity of ozone required;
2. point of ozone application: (1) pretreatment, (2) posttreatment or (3) multiple injection points;
3. best method of ozone application;
4. optimum ozone application concentration;
5. optimum contact time;
6. applicable ozonation control systems; and
7. possible requirement for ozonated liquid retention basins.

An initial consideration is the ozone demand of the liquid at the proposed point of application. Légeron [2] suggests the use of an ozone-demand flask, into which known quantities of ozone and water are introduced and manually shaken. After a predetermined shaking period, residual ozone and other significant parameters are measured. The test is repeated to provide data for test results with differing ozone dosages.

Bench-scale static testing involves a higher degree of testing sophistication. This type of test generally will provide answers to the following questions:

1. Is the ozonation process technically feasible?
2. Is the ozonation process economically feasible?

The static test is conducted in a container filled with a sample of the liquid stream into which ozonated gas is diffused. The ozone concentrations of influent and effluent gas streams are determined. Also, samples of the liquid are analyzed for selected parameters at predetermined periods. The quantity of ozone consumed, in achieving a desired degree of treatment, can be determined using this data. Figure 1 illustrates a static test equipment schematic. Figure 2 illustrates the type of information available through this level of testing.

A further level of testing, the dynamic pilot plant, is used to determine criteria for the proposed full-scale facility. As such, it perhaps could be addressed under Design, but much of the information is also valuable in the initial feasibility study considerations. Figure 3 illustrates a dynamic pilot-plant schematic.

Dynamic pilot-plant testing can be used to determine the optimum ozone contact time and other design parameters. The relationship of a particular dynamic pilot-plant contactor to a prototype unit is not totally clear, but a diffuser column or series of diffuser columns appears to be a common type of pilot-plant contactor currently used. These types of facilities are available by loan or by lease from ozonation equipment suppliers. However, the value of operational experience and expanded data to be derived from a pilot unit purchased and operated for a prolonged period of time should be compared to the cost of a relatively short test period with a leased unit.

Operational Facility Visits

It is important for design and operational personnel to make inspection visits to operational facilities incorporating equipment or processes similar to those under consideration for the facility in question. This is the ultimate test of the equipment or process vendor's optimism, and is particularly important in the field of ozonation. There is little published information describing design and operational experience of North American ozonation facilities, and site inspection trips appear to be the only current way to obtain authoritative information on operational

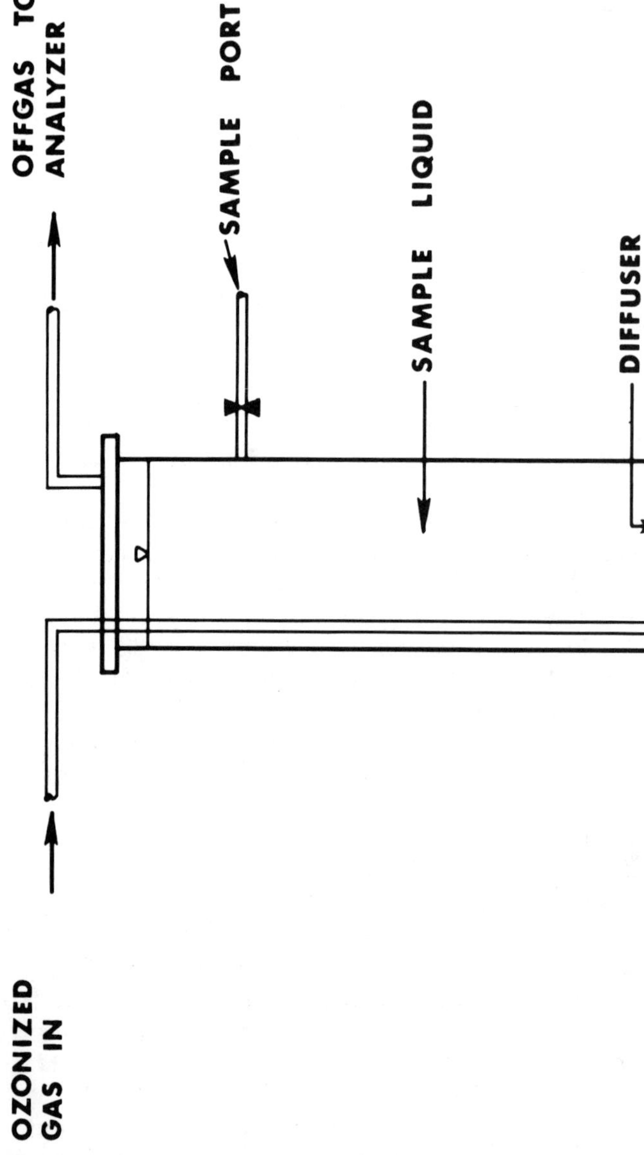

Figure 1. Static ozonation test equipment arrangement.

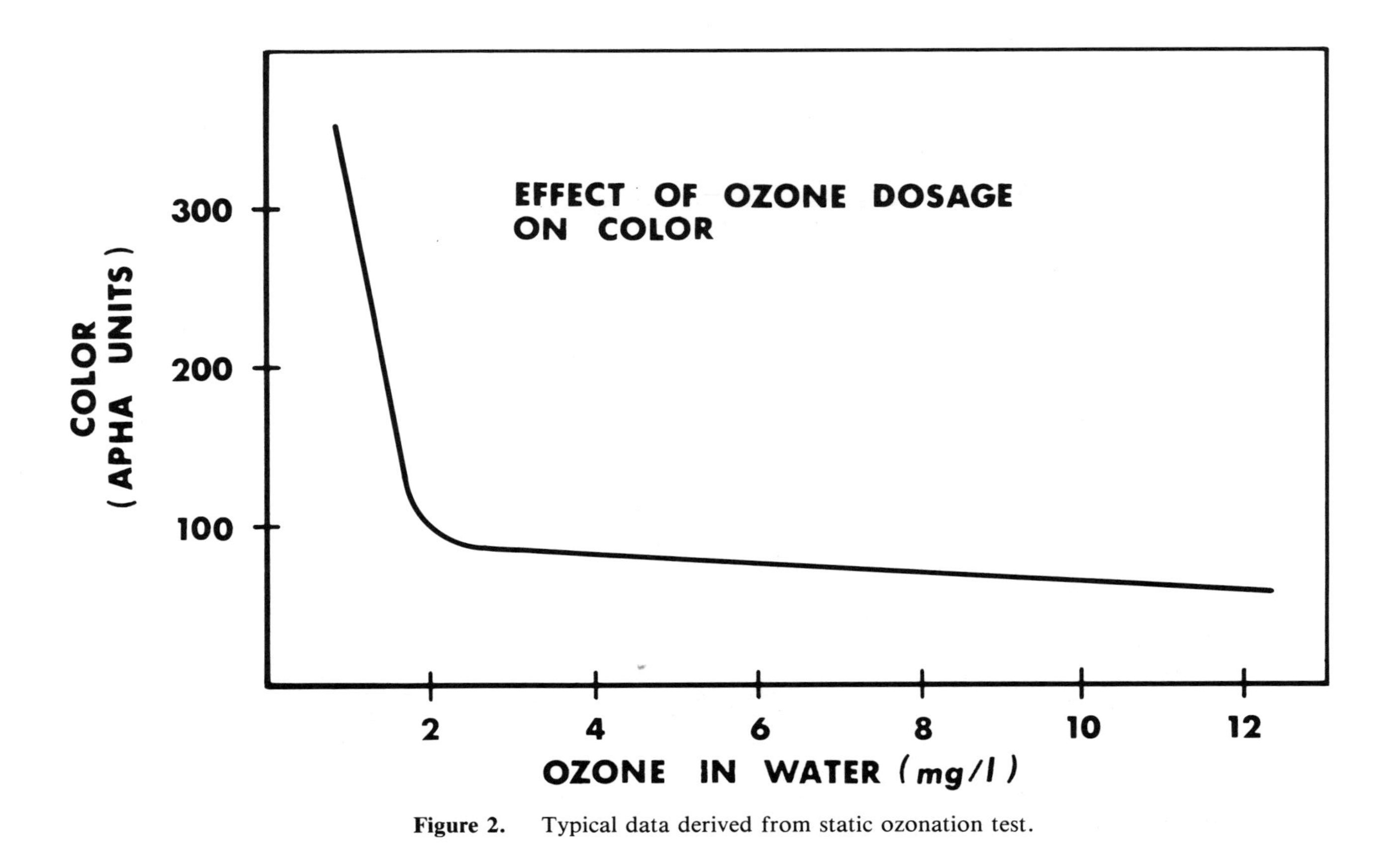

Figure 2. Typical data derived from static ozonation test.

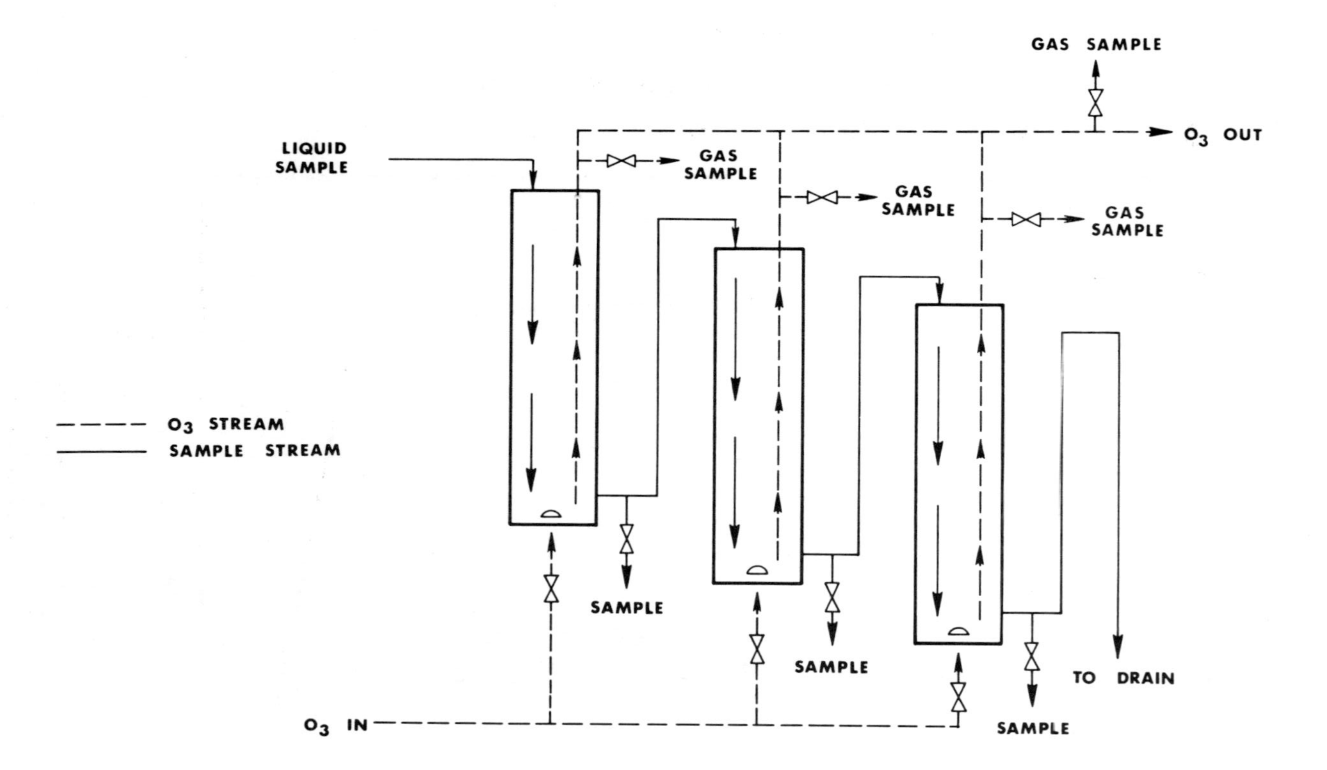

Figure 3. Ozonation pilot-plant schematic.

facilities. This is necessary to avoid repetition of previous design errors. It is also important that operational personnel discuss operation and maintenance problems with their peers who are involved with the day-to-day operation of an ozonation facility.

A list of the operational ozonation facilities treating drinking water, domestic wastewater and industrial liquids in North America is comparatively short, but growing. The installations range from small (334 kg O_3/day, Upper Thompson Sanitation District, Estes Park, CO, to the very large, (3300 kg O_3/day, Charles J. Des Baillets Water Treatment Plant, Montreal, Quebec, Canada). The operational facilities represent a wide range of types of ozone generators and contactors.

Preliminary Design

Preparation of a preliminary design is required to determine site requirements, unit dimensions, control strategies and other considerations necessary to prepare a feasibility study report. Questions that should be addressed in the feasibility study include the following:

1. applicability of ozone to the liquid or gas treatment system at one or more points in the system;
2. ozone dosage(s) required to perform the proposed ozonation task(s);
3. determination of whether the proposed ozonation application(s) is/are mass-transfer–or reaction-rate–limited and, therefore, some idea of contactor types and size;
4. the ozone generation feed gas to be used such as: (1) high-purity oxygen, (2) recycled high-purity oxygen, or (3) air; and
5. an estimate of the ozonation system cost.

Cost Estimating

Total cost is a major factor in deciding to incorporate a new process in a treatment system. Total cost includes original capital cost amortization as well as operational and maintenance expenses. Very little authoritative information can be offered at this time, although it is addressed in other chapters of this book. Development of cost estimating data from recently bid ozonation projects is difficult, as many of the projects are significantly different in terms of feed gas, ozone generators and contactors. Another complicating factor is the entry of a number of ozone generator suppliers into the U.S. municipal marketplace who may bid an unrealistically low price to achieve an initial installation of their equipment.

A cost estimate for an ozonation facility should be prepared with as much detail as possible, in view of the site-specific nature of the application. It is recommended that the cost estimate address the factors shown in Table I.

Table I. Factors to Be Addressed in an Ozonation Facility Cost Estimate

	Estimated Cost		
Subunit Description	**Capital**	**Operation**	**Maintenance**
Gas Preparation	a	b	c
Ozone Generation and Power Supply	a	b	d
Ozone Dissolution	e	f	
Offgas Destruction		b	g
Controls and Instrumentation	h		
Interconnecting Piping	i		
Process Housing, Including Environmental Considerations	j		

[a] Vendor-supplied data will probably group items 1 and 2.
[b] Will include electrical power.
[c] Will include labor and gas filtration media replacement.
[d] Will include labor costs and dielectric replacement.
[e] Capital cost is a function of type and size. An appropriate basis for a first estimate would be a two-stage, porous diffuser unit with the appropriate retention time.
[f] Will be a relatively minimum cost unless a turbine-type contactor is used, or if addition of the ozonation process requires additional pumping head.
[g] Would include labor and possibly media replacement, depending on ozone destruction system selected.
[h] Cost will depend on degree of system sophistication control desired.
[i] Costs should take into account special materials for handling "dry" and "wet" ozonated gas.
[j] Space required should be estimated conservatively, depending on the wide range of equipment available. Environmental control of equipment noise and ozonated gas must be included.

DESIGN

The formal design phase begins with the owner's acceptance of the feasibility study and the decision to incorporate ozonation into the gas or liquid treatment system. Design of the ozonation facility currently must be performed without the formalized guidance of design standards or other rigorous guides.

Process and Instrumentation Diagram

Preparation of a process and instrumentation (P&I) diagram is an essential first step in the design. Although it is subject to revision throughout the project, the initial P&I diagram presents, on one or more drawings, the initial conception of the ozonation facility. The following items would be addressed in the P&I diagram:

1. design characteristics of liquid or gas to be treated, including flow;
2. design ozone dosages;
3. design feed gas quality and quantity;
4. number of gas ozonating trains;
5. number and size of contactors;
6. types and number of primary monitoring elements, such as flow, pressure, temperature, ozone levels and dew point;
7. system controls, including interlocks and automatic shutdown capability; and
8. ozone generator cooling systems.

The P&I diagram serves as a strong basis for the facility design and as a key document for process and instrumentation design review by those experienced in the field. Although the designer cannot abdicate final responsibility for the design of the facility, it is currently necessary to take advantage of the experience and insight of ozonation system vendors, and others in the field, to avoid needless repetition of previous mistakes manifested in other operational ozonation facilities. This is particularly important in areas of the world where ozonation is not an accepted and well understood technology.

Equipment Selection

Carlins and Clark [3] describe the wide range of size and type of ozone generation systems available. Although it is recognized in North America that publicly owned facilities normally require that more than one supplier be allowed to bid, the bid documents must clearly establish the standard of quality to be provided. As the market for ozonation equipment continues to expand, even more suppliers may enter the field. The facility owner must be assured of receiving operable equipment of predictable longevity. It may be noted, in passing, that the Otto-type ozone generators installed in the Bon Voyage Water Treatment Plant of Nice, France, operated from 1906 to 1970, at which time the Bon Voyage Plant was superseded by the Super Rimiez facility. Although it is recommended that the supply of at least the gas preparation, ozone generation and electrical power subunits be the responsibility of a single vendor, minimum standards of quality of the individual subunit components must be established. The subunits to be addressed should include the following:

1. gas preparation;
2. electrical power supply;
3. ozone generation;
4. dissolution, including off-gas destruction;
5. controls and instrumentation; and
6. piping and valving.

The need to establish ozonation system quality standards is clearly illustrated in deficiencies manifested in recent U.S. installations constructed as part of the aggressively bid U.S. Environmental Protection Agency funded wastewater treatment plant construction program.

Gas Preparation

Gas preparation subunits of some type are part of every ozonation system. For once-through, high-purity oxygen systems, the gas preparation subunit may consist of little more than a gas pressurization device. However, more complex systems are required for a system using recycled oxygen or ambient air as the feed gas. Figures 4, 5 and 6 illustrate three types of gas preparation systems used for preparing recycled oxygen or air to be the feed gas for an ozone generator.

Gas preparation subunits are not unique to the ozonation process. Similar subunits are required for pneumatic control systems and other air-handling applications. However, a reliable supply of a gas with a dew point of −60°C and free from specific contaminants such as hydrocarbons is essential for high levels of ozone production. The gas preparation system warrants careful selection to ensure that it is appropriate for the ambient conditions of the raw feed gas. Air preparation systems are well addressed in the literature [4,5], and will not be covered further in this chapter. Provision of a conservatively designed gas preparation subunit is inexpensive insurance against the problems that have resulted from marginal or deficient gas preparation subunits.

Electrical Power Supply

Ozone generation is a function of applied voltage and frequency as well as gas quality, temperature and pressure. Electrical distribution

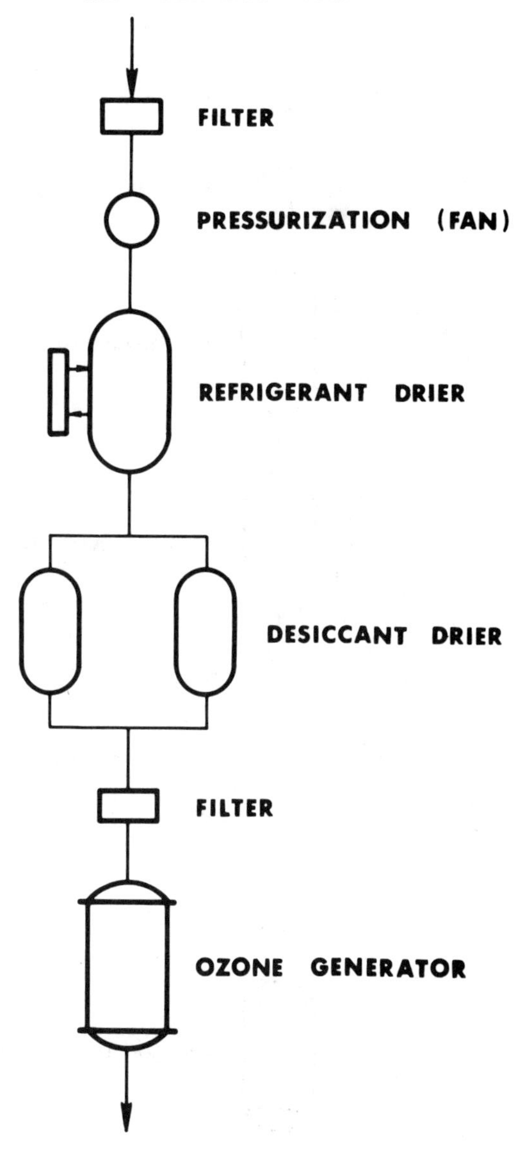

Figure 4. Process diagram of low-pressure gas preparation system.

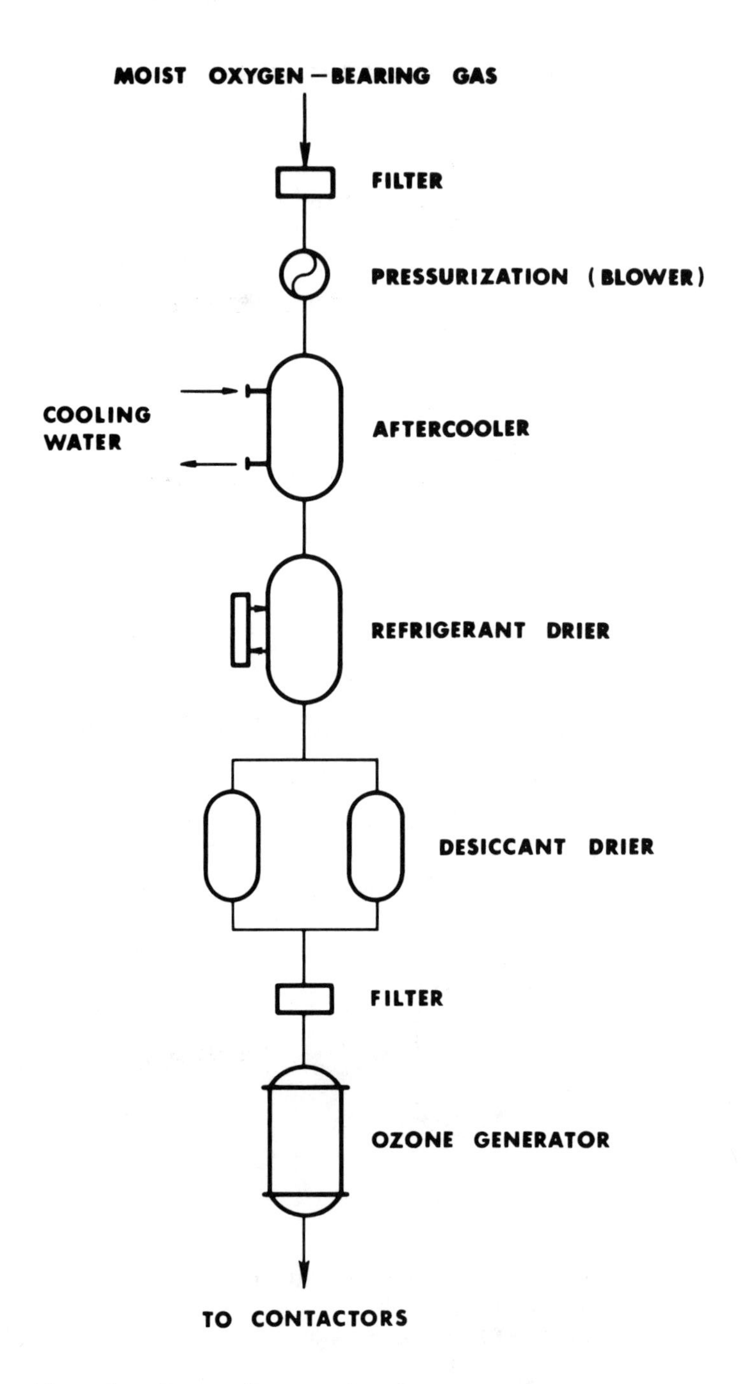

Figure 5. Process diagram of medium-pressure gas preparation system.

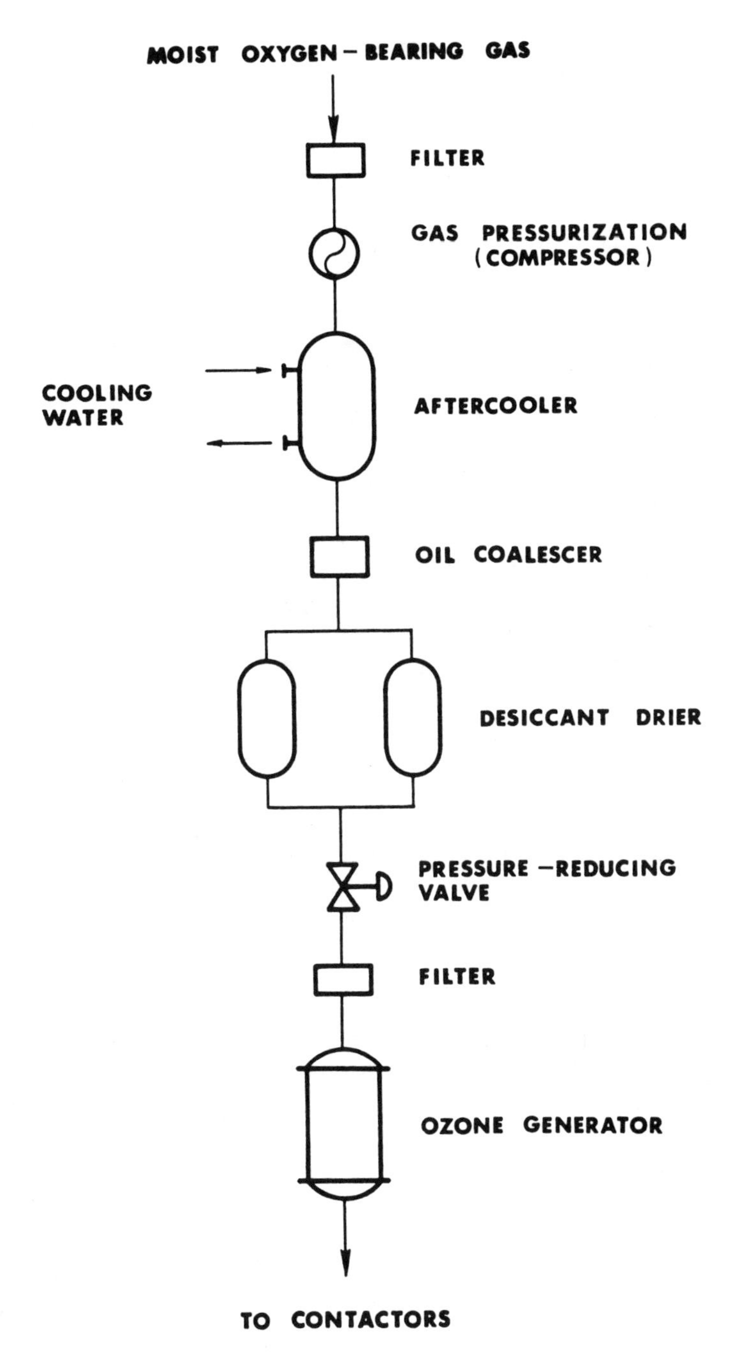

Figure 6. Process diagram of high-pressure gas preparation system.

system voltage and frequency are modified within the ozonation system electrical power subunit to provide the particular needs of the ozonation system. Electrical power supply to an ozone generation system remains a highly proprietary portion of the ozonation equipment supplier's package. The supplier's method of controlling the high voltages, and in many cases high frequencies, is part of the system's electrical power supply subsystem. The voltages and frequencies used to generate ozone commercially can be classified as follows:

Frequency Category	Frequency (Hz)	Voltage (V)
Low (or Line)	50 or 60	25,000
Medium	600	20,000
High	2,000	10,000

While the low-frequency (high-voltage) system has been the most common contemporary ozone installation, recent advances in electronic circuitry have led to higher frequencies often being used for new installations.

Reliability of operation is a major factor in an ozone generator power supply unit. The specific requirements of generating ozone leave the selection of subunit equipment almost exclusively as the responsibility of the generator supplier. However, some level of responsibility must be borne by the designer.

Ozone Generation

Larocque [6] defines ozonation systems as follows:

- small: 0–200 lb (0–440 kg) of ozone per day;
- medium: 201–1000 lb (442–2200 kg) of ozone per day; and
- large: >1000 lb (2200 kg) per day.

Ozone generators may be provided as completely preengineered and prefabricated modular systems or as custom-designed systems. The modular system, consisting of a frame-mounted ozone generator, air preparation subunit, electrical power supply subunit and control package, generally is applied for small- and medium-sized installations. Custom-designed units with each unit supplied separately, but under a single supplier responsibility, are used for larger-medium and large installations. Larocque [6] discusses the advantages of each type of unit, but they can be summarized as follows:

- modular system: standardized, shop-tested units, each unit to be operated as a single train; and

- custom design: most flexible installation which requires more design expertise and knowledge and a clear definition of the supplier's responsibility.

Questions to be answered in selecting an ozone generator include the following:

1. What is the design capacity of the installation?
2. What is to be the feed gas: air, high-purity oxygen or recycled high-purity oxygen?
3. What are the projected variations of ozone demand? Are the units likely to be operated for prolonged periods at their rated capacity?
4. What level of control is desired?
5. What level of maintenance is expected?

Ozone Dissolution

The wide range of ozone dissolution devices or contactors is illustrated by Masschelein [7]. Mignot [8] provides further information. There are several types of ozone contactors that can be employed for a given application; however, a conservative approach, using contactors of proven process application and established ozone transfer efficiency, should be employed in selecting a contactor. For example: ceramic diffuser, multistage contactors frequently have been applied for ozonation of highly purified water and wastewater. On the other hand, turbine contactors are frequently used for raw water ozonation.

Questions to be answered when selecting an ozone contactor include the following:

1. Is the ozonation process mass transfer- or reaction rate-limited? For example: disinfection is mass transfer-limited while chemical oxidation of chlorinated organics many times is reaction rate-controlled.
2. What hydraulic head is available?
3. What is the gas pressure available throughout the ozone generation system?
4. What level of ozone utilization is required?
5. What is the ozone uptake rate of the liquid in question?

Selection of an ozone dissolution subunit will be based not only on the answers to the preceding questions, but also on other basic considerations. For example: site constraints might require a very deep diffuser-type contactor, or bidding constraints might preclude the use of a proprietary type of turbine contactor.

Contactor Offgas Treatment

Ozone contactor offgas treatment normally is included as part of the contactor designer's task but is discussed separately in this presentation, for clarity. Some level of treatment is required to minimize personnel health hazards and possible deleterious effects within the plant structures or in the surrounding area.

An acceptable ambient ozone level of less than 0.1 ppm by volume normally requires consideration of offgas treatment, since a 90–95% ozone transfer contactor efficiency would still result in a contactor offgas concentration of 2000 ppm of ozone on a weight basis. Contactor offgas treatment methods include the following:

1. dilution by prevailing winds or by supplementary air;
2. wet granular activated carbon (dry GAC treatment is not recommended);
3. thermal destruction;
4. catalytic thermal destruction; or
5. recycling to other parts of the process.

In addition to Masschelein [7], Tritschler [9] and Schalekamp [10] address the various methods of contactor offgas treatment.

The ozone destructor type selected must reduce ozone concentrations to acceptable limits and satisfy the following criteria:

- simple to operate,
- low capital cost,
- low operational cost,
- energy efficient,
- reliable,
- long service life, and
- compactness

Ozonation Control

Ozonation control systems vary considerably in operational ozonation facilities. French drinking water treatment plants, using ozonation primarily for disinfection and viral inactivation, incorporate a closed-loop control system, in which the residual dissolved ozone level in the ozonated water is used to control the amount of ozone supplied to maintain that ozone residual. The key to successful operation of such a system is an accurate and reliable dissolved ozone analyzer. On the other hand, many German drinking water treatment plant ozonation systems, which are used primarily for iron, manganese and/or organics oxidation, are manually controlled by means of a periodic manual "sniff" test of offgas from the holding tanks after ozonation or by visual observation of the permanganate pink color which identifies oxidized manganese in the ozonated water. Even in these plants, there is a trend to higher degrees of control system automation. U.S. wastewater treatment ozonation facilities currently use flow-paced, ozone dosage control because of the current lack of a recognized parameter that can be measured downstream of the process to enable closed-loop control.

Continuous residual dissolved ozone monitoring equipment may be applied successfully to drinking water that has already received a high degree of treatment. This has been observed in the most common French application of ozone as the primary means of disinfection near the end of the treatment system. However, a more cautious approach must be taken with the application of continuous, dissolved ozone monitoring equipment for water that has only received chemical clarification, or for treated wastewater, because the ozone demand has not yet been completely satisfied and therefore the measured residual dissolved ozone level is not stable. Continuous monitoring of the ozone concentration in the gas phase appears to be reliable due to greater experience and the availability of more well proven equipment, which has been developed for air pollution monitoring.

There are various methods for controlling ozone production rates, which are required for several reasons. First, since excess ozone cannot be stored, power is wasted in generating needless excess ozone. Second, excess ozone in the contactor offgases must be destroyed before being discharged to the environment for safety and health reasons. This also requires energy. On the other hand, when water quality decreases, additional ozone is required immediately, since underozonation of certain organic contaminants present in the liquid can produce partially oxidized organic intermediates which, under certain circumstances, can be more toxic than the initial materials [11].

Methods of controlling ozone production include the following.

Manual Operation–Manual Sampling

The change in water characteristics brought about by ozonation is determined by periodic visual inspection or chemical analyses. The required ozone production rate is controlled by manually changing either the voltage or the frequency of the electrical power supply, or by changing the number of ozone generators in operation (Figure 7).

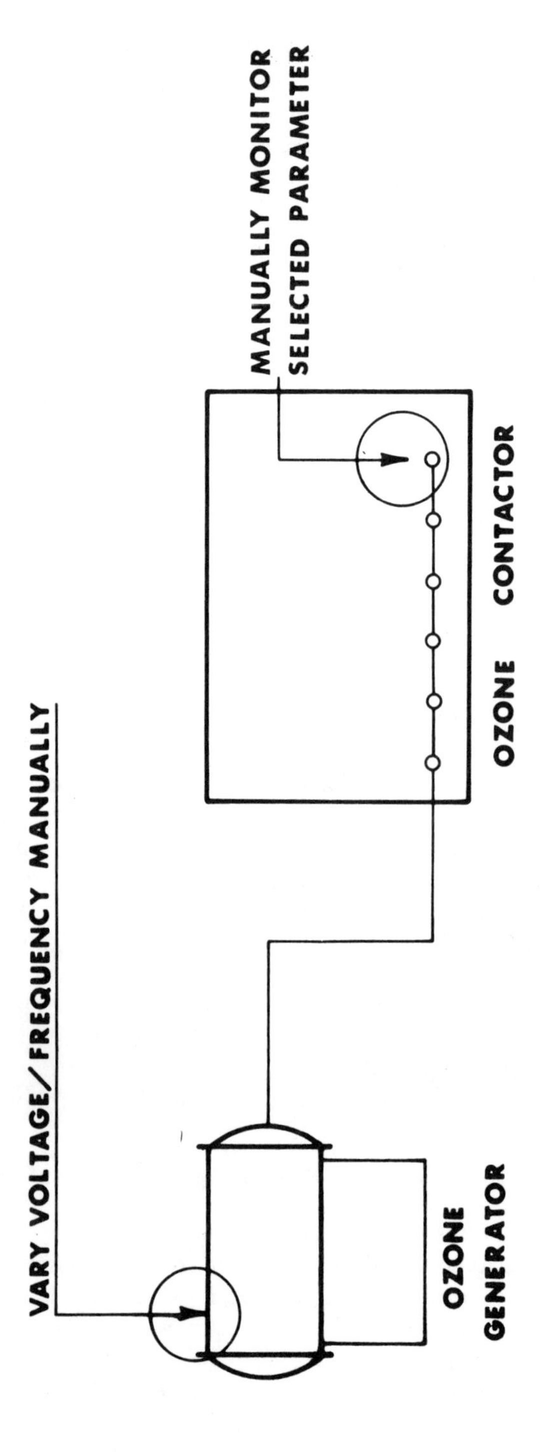

Figure 7. Ozonation control: manual operation–manual sampling.

Manual Operation–Automatic Sampling (Figure 8)

The ozone production rate is controlled manually through continuous monitoring of a selected parameter such as:

- residual dissolved ozone level in the water being treated;
- residual ozone level in the contactor offgases; or
- liquid characteristics such as color or turbidity.

Closed Loop Control–Automatic Sampling (Figure 9)

The ozone production rate is controlled automatically (voltage or frequency) by a signal generated by the continuous monitoring of a selected parameter, such as those listed above.

Closed Loop Control of Voltage/Frequency and Gasflow–Automatic Sampling (Figure 10)

The system is fully automatic and is designed to operate at peak electrical efficiency with ozone-containing gas concentration controlled for optimum dissolution. Many system parameters are monitored continuously and a preprogrammed controller adjusts both gas flow and voltage or frequency.

Full Computer Control–Automatic Sampling (Figure 11)

The computer receives the same information as in the immediately preceding system and more from the ozonation control system, including diagnostics, prefailure shutdowns and optimum energy balance.

Selection of a specific control strategy is based on the ozonation application as well as the size and sophistication of the overall treatment facility.

Larocque [6] divides instrumentation/control into the following categories:

1. necessary,
2. desirable, and
3. redundant

He lists the following ozone generator monitoring requirements as essential:

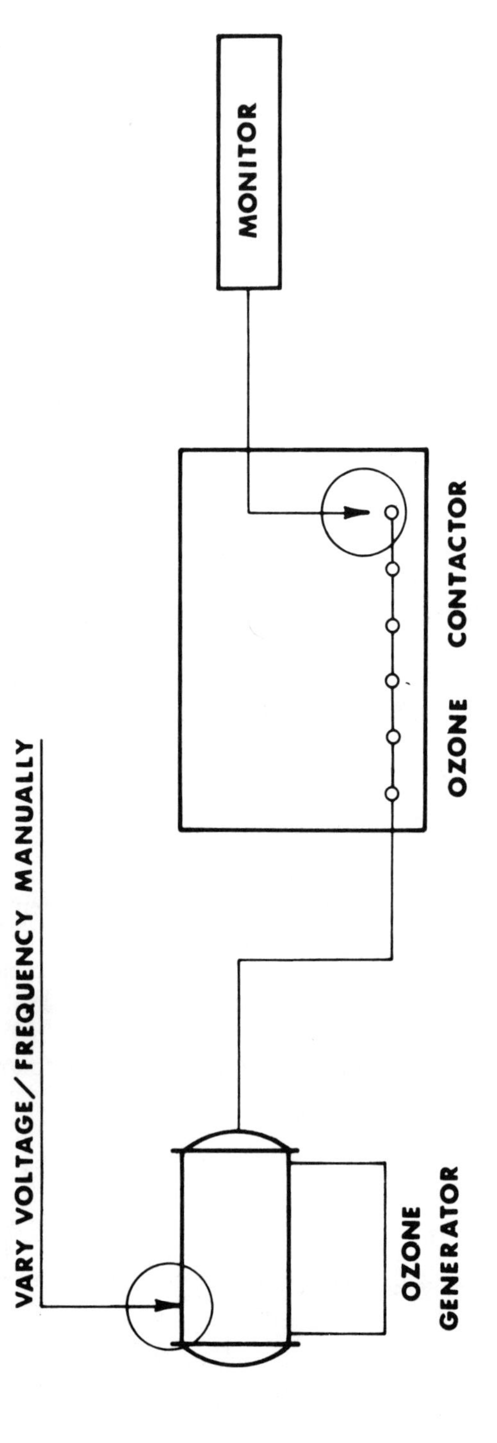

Figure 8. Ozonation control: manual operation–automatic sampling.

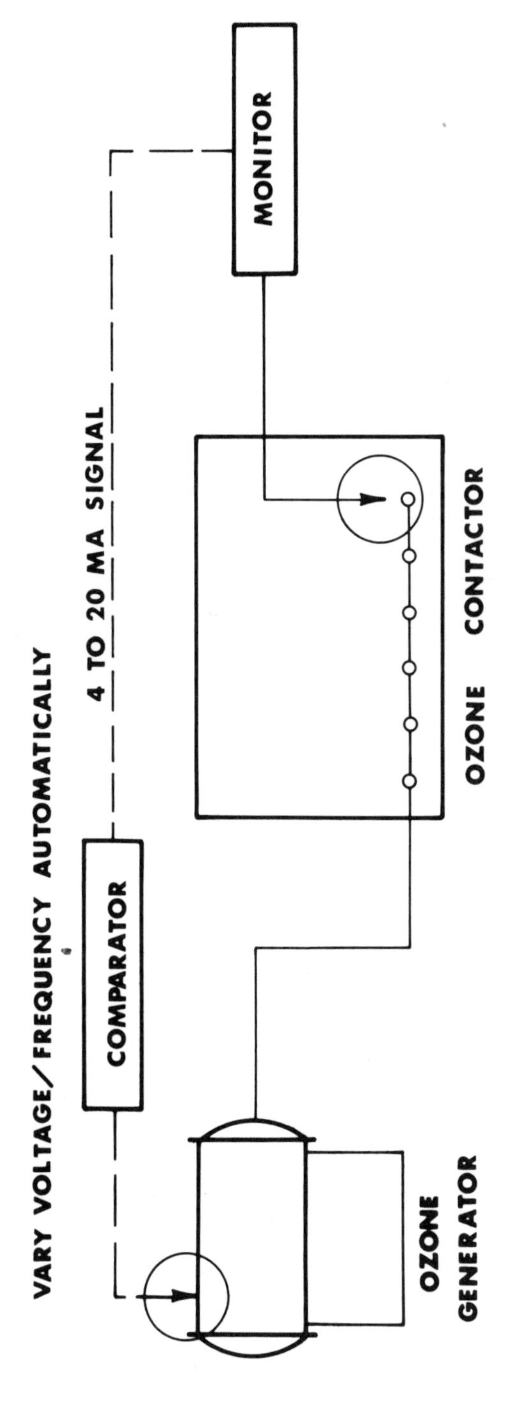

Figure 9. Ozonation control: closed-loop control–automatic sampling.

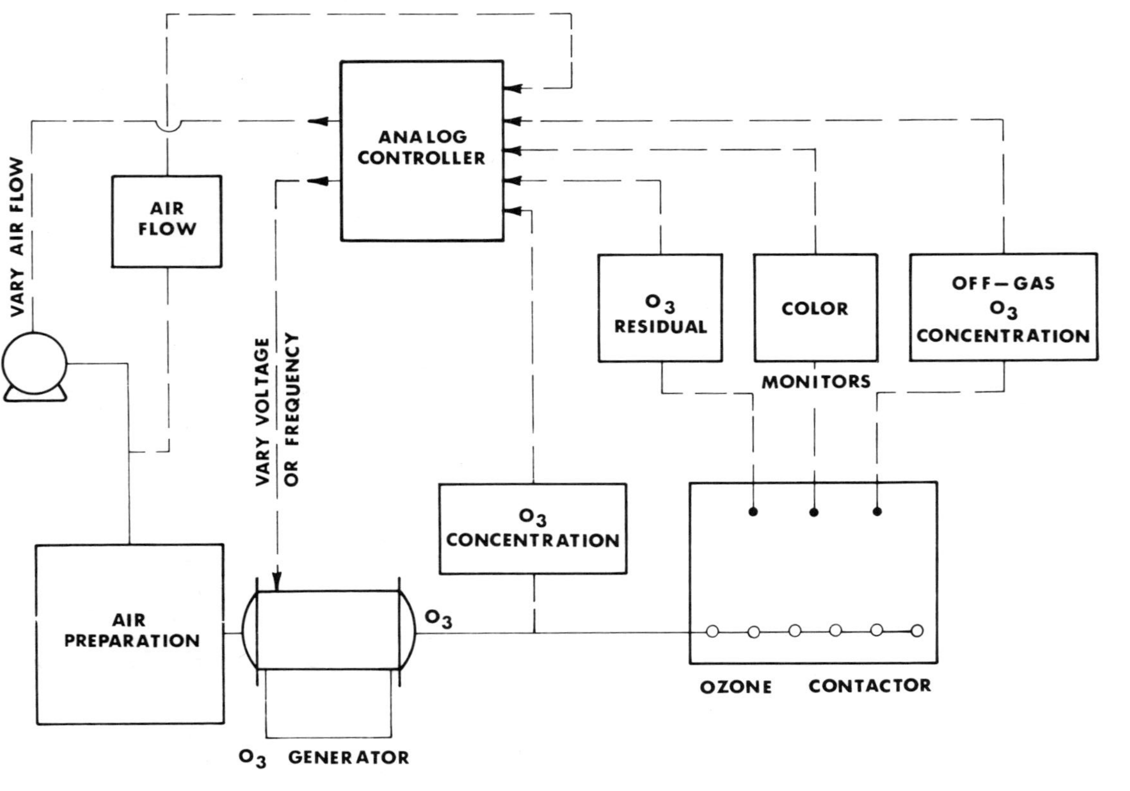

Figure 10. Ozonation control: automatic closed-loop control of voltage/frequency and gasflow.

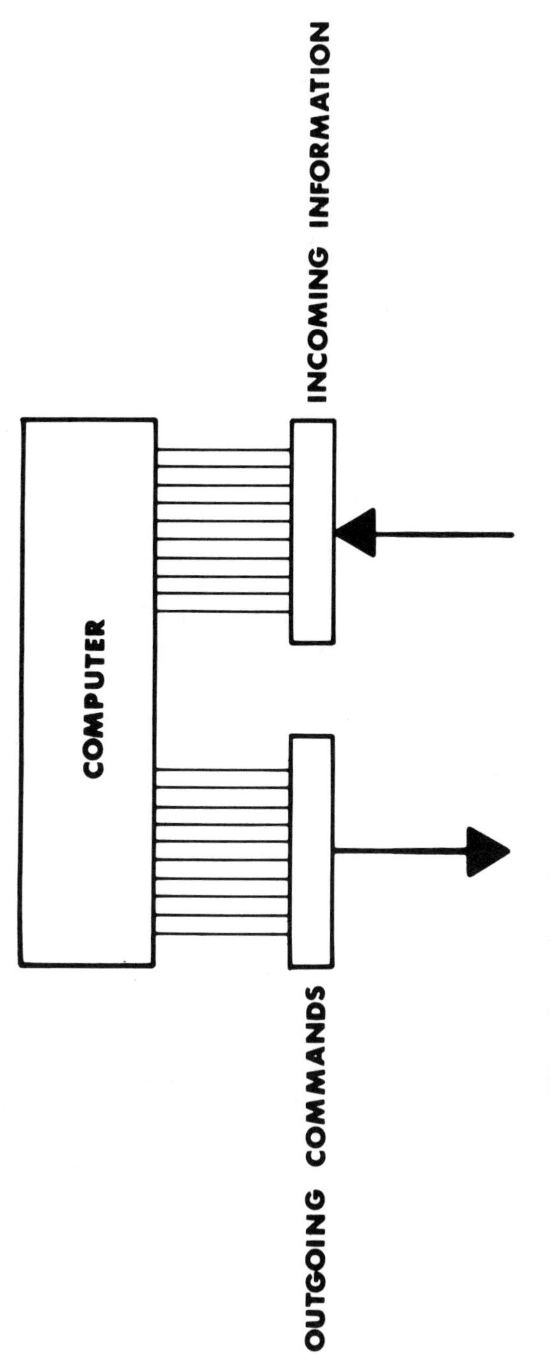

Figure 11. Ozonation control: full computer control–automatic sampling.

1. feed gas flow;
2. feed gas pressure;
3. feed gas temperature entering and leaving the ozone generator;
4. cooling medium inlet and outlet temperature; and
5. power consumption.

Experience elsewhere [11] indicates that feed gas dew point monitoring and associated calibration equipment should be considered as essential or very desirable. Another desirable control attribute is an equipment startup sequence that purges the gas preparation system and ozone generator of moisture prior to system startup.

Materials of Construction

Zawierucha and Charleson [12] give a more complete discussion of materials of construction of ozonation systems. However, design guidance can be derived from experiences of recent U.S. ozonation installations [13–15]. The following list of materials, considered appropriate for ozone applications, was prepared after a review of published literature and interviews with operators of U.S. ozonation facilities:

- "dry" ozonated gas piping: (1) flanged or screwed—304 and 316 stainless steel; (2) welded—304L and 316L stainless steel;
- "wet" ozonated gas piping: same as for dry service;
- valves: stainless steel face and body;
- gasket material: Viton, Teflon, Hypalon in compression (silicone under certain circumstances);
- flexible couplings: stainless steel; and
- concrete joint materials: Sikaflex-1A.

Welded stainless steel piping systems are preferred using tungsten-arc inert gas (TIG) welding procedures. Polyvinyl chloride (PVC) or unplasticized polyvinyl chloride (UPVC) piping systems are not recommended, based on a review of the experience at the previously referenced installations [13–15]. Although conventionally designed reinforced concrete appears appropriate for ozone contactor structures, the use of galvanized steel reinforcing bars for the gas space portions of the structures should be considered. This consideration is based on unpublished test work at the Springfield, MO, wastewater treatment plant ozone contactors. These contactors utilize surface aerators to diffuse ozonated gas into the nitrified and filtered wastewater. It may be argued that ozone concentrations in the headspace of the Springfield contactors were atypically high in contrast to those encountered with the more conventional porous tube diffuser type contactors. This question requires further study.

Environmental Considerations

Designs must take into account environmental considerations, including sound levels and ambient air quality. These factors must be considered in the absence of well established industry guidelines in areas of the world new to ozone technology.

Sound Levels

Noise is generated within the ozonation system by the gas pressurization and refrigerant gas drying subunits. Noise also may result from high-frequency electronic systems associated with certain types of ozone generators. Isolation of the gas pressurization equipment in a specially treated room is common practice. Careful consideration and specification of acceptable noise levels normally will allow the refrigerant cooling and other equipment to be housed in conjunction with other equipment of the ozonation system.

Air Quality

Ventilation requirements of enclosed spaces housing ozone generation facilities are currently ill-defined. Many operational Canadian and French drinking water ozonation facilities appear to be designed to take advantage of the fact that physically detectable ozone levels are far below hazardous levels as defined by current environmental standards [16]. Other European facilities incorporate more formal ozonation facility ventilation systems. Maintaining a pressure slightly below atmospheric in areas where ozone leakage could occur is practiced. A means of initiating an emergency, high rate of room ventilation frequently is provided at the exterior doorways of ozonation areas. Some German plants incorporate room ambient ozone level monitors which, when ozone levels exceed 0.1 ppm, automatically: (1) sound an alarm; (2) interrupt power to the generators; and (3) start exhaust fans. Current French design philosophy appears to be similar.

In the absence of any formal North American standard, the author suggests a two-level ventilation system; one level to meet normal room heating and cooling needs and a second, emergency level to provide an air change every two minutes for enclosed spaces that could be contaminated by ozone leakage. Separate areas, complete with separate heating, ventilation and air conditioning, as well as separate exterior access, are suggested for the ozonation facility. Comfort can be taken in regard to ozonation room ventilation in that ozone generation can be terminated by interrupting power to the ozone generator. This is in contrast to the hazards associated with handling a leak of chlorine gas from a pressurized container.

Treatment of the offgas from the ozone diffusion unit or other areas of high ozonated gas concentration is important with respect to the ventilation of other work spaces of the treatment facility. Several recently constructed U.S. ozonation facilities manifest the problem of ozonated offgas being short-circuited to the ventilation system intakes of other plant work areas when an offgas treatment system was not provided or was not being operated. The provision of an offgas treatment system and careful consideration of building ventilation intakes are two important aspects of maintaining outdoor and indoor plant ambient air quality.

General Considerations

Evaluation of operational ozonation facilities indicates that conventional building construction materials and design are adequate if noise and air quality problems are properly addressed. A minimum level of heating would be required in cold climates for worker comfort and to prevent equipment and piping damage due to cold temperatures. Since the ozone generation is an exothermic reaction, it has been suggested that the waste heat generated could be used for building heat. Equipment layout should provide for easy access to the equipment. The space requirements and piping arrangements of the various types of ozone generators should be taken into account. Here again, although perhaps bound by the low bid requirements, the design must establish the basic measure of quality. These would include the following:

1. architectural treatment, including exterior and interior finishes;
2. doors and windows, including: size, number and location;
3. vertical and horizontal clearances required for equipment, including aisleway considerations and the possible provision of an overhead crane; and
4. space and other provisions required for associated items, such as washrooms, lockers, sample points, tool storage and janitorial needs.

Plans and Specifications

Plans and specifications must be specific to guide general contractors and their suppliers bidding the project and to protect the owner from unsuitable materials and equipment.

Ozonation system supplier recommendations must be reviewed carefully, especially in North America, in view of the low bid supply concept. This is particularly true in view of the entry of promoters of European systems into the U.S. marketplace. Advice may be given in good faith by European-trained sales engineers who do not realize that many of the desirable, and even essential, features of a system that would be supplied under a negotiated contract would not be supplied under a low-bid contract unless they are clearly specified. An example of such an expensive, but essential operational tool, which would not be supplied if not specified, is a feed gas dew point meter and accompanying manual dew point calibration unit.

The ozonation system may not, in itself, be the reason for selection of the low bidder for the construction of an overall gas or liquid treatment facility. Since operational costs may vary from one ozone generator to another, an additional qualification frequently is used in evaluating the ozonation system. By this procedure, the bidder is required to submit the power requirements to produce the specified quantity of ozone as part of his bid. Liquidated damages would be assessed for actual power requirements in excess of those listed in the bid.

Preparation of ozonation facility bid documents which both protect the owner's interests and allow free competition between the wide range of available ozonation systems is extremely difficult, if not impossible. Therefore, the concept of prebidding, and thereby preselecting, key elements of the ozone generation system should be considered. Once the specific system has been determined, the designer is better able to design a facility to serve the owner's needs, in terms of operation and maintenance. The designer is also better able to work with the selected equipment supplier to optimize the application of the equipment to be supplied. The bid documents must carefully specify the procedures and conditions of the ozone generator tests. Industry standards are being developed slowly, but currently specifications should err on the side of being too detailed. A recent EPA document [11] includes such a test procedure. A review of comments of Naimie and Okey [17] regarding specific details of such testing is recommended before writing the specification.

Considerable initial startup and training assistance by the ozonation system vendor should be specified. Three or four visits by operationally oriented vendor personnel during the warranty period should be required.

CONSTRUCTION

In this chapter, construction is defined to include all activities conducted during the period between advertisement for bid and the owner's acceptance of the facility. There is little, if any, difference in constructing an ozonation system than any other treatment system. However, the general unfamiliarity of construction trades with installation of ozonation facilities requires extreme attention to assure that the specified materials and procedures are used in the work. The attack of wet and dry ozone will quickly reveal any deficiencies after commencement of operation, but this involves added delays and costs, which otherwise can be avoided.

Testing of the constructed ozone diffusion device is necessary to determine ozone transfer efficiency as part of a program to minimize operational costs. Procedures for testing the generator itself are covered in the preceding section. Effective use of startup and operator training are of great importance since ozonation systems are still comparatively uncommon in many parts of the world. Manufacturers' operation and maintenance manuals should be reviewed closely for clarity and adequacy.

OPERATION AND MAINTENANCE

The responsibility for operation and maintenance of the ozonation facility ultimately comes to rest on the plant operational staff. Their efforts now will determine whether the ozonation system will reliably meet the demands placed on it. Operation and maintenance procedures will be major factors in determining the uninterrupted operational life of the facility. Although the majority of any overall ozonation system consists of conventional equipment, the ozone generator and the electrical supply subunits are new to the average operator. Therefore, the plant management may choose to enter into a maintenance contract for the initial years of operation of the ozonation system. Full use should be made of the vendor services required by the contract.

Good maintenance procedures are as important for successful operation of the ozonation system as for any others. Oil changes, air filter media replacement and dielectric cleaning are obvious requirements. Record keeping procedures to provide baseline information for future troubleshooting evaluation must be established as well. Performance data, such as kilograms of ozone produced per kilowatt-hour, generator feed gas dew point, and other data will be important in determining maintenance cycles and solutions to operational problems.

The frequency with which the air preparation subunit filter medium is changed is a function of ambient air quality. Filter medium cleaning or replacement is commonly scheduled twice per year. Experience indicates a minimum absorber medium life in excess of 10 years for the desiccant dryers [11].

Two factors that affect ozone generator operation and maintenance include the effectiveness of the air preparation system and the period over which the ozone generator is required to operate at maximum capacity. Maintenance of the ozone generators commonly is scheduled once per year. However, many plants perform this maintenance every six months. Dielectric replacement, due to failure as well as to breakage during maintenance, can be held as low as 1–2% by adhering to the manufacturer's recommended maintenance program. However, it appears reasonable to predict an average tube life of ten years if a maximum feed gas dew point no higher than −60°C is maintained and if the ozone generator is not required to operate for prolonged periods at its rated capacity.

Ozone generators normally are operated at 50–75% of their maximum rated capacities. This allows more efficient use of electrical energy to produce a given amount of ozone, as well as to reduce the extent of dielectric wear and more frequent dielectric failure. Excess ozone generating capacity thus is required, but also becomes available for sudden surges in ozone demand and to allow for downtime during scheduled maintenance.

Operation and maintenance of the ozone contactor also must be considered. Turbine diffusers require electricity to power the drive motors. In comparison, porous diffusers require little or no added energy, but regular inspection and maintenance is required to ensure uniform distribution of ozonated gas in the contact chamber. Experience with periodic maintenance of the diffuser type contact chambers in the Morsang-sur-Seine plant in France indicates that even after purging of the contact chambers with air, maintenance personnel entering the chambers should be equipped with a self-contained breathing apparatus, since the density of ozone is heavier than air and is difficult to remove completely by air purging [11].

CONCLUSION

Studies and research are revealing the broad applications of ozone in treating gases and liquids. However, there is little nonproprietary information in scaling up laboratory scale or pilot-plant-scale processes into full-sized production facilities. This chapter addresses some of the frequently tedious details which should be considered by the designer to provide the owner with an operable installation with a minimum of unanticipated (and unappreciated) operational and maintenance problems.

REFERENCES

1. Nebel, C. "Pilot Plant Operations for Potable Water Plant Design," in *Proceedings of the Seminar on the Design and Operation of Drinking Water Facilities Using Ozone or Chlorine Dioxide,* R. G. Rice, Ed. (Dedham, MA: New England Water Works Association, 1979), pp. 241–248.
2. Légeron, J. P. "Design of Experimental Procedures for Determining Ozone Requirements," paper presented at the American Institute of Chemical Engineers 86th National Meeting, Houston, TX, April 1979.
3. Carlins, J. J., and R. G. Clark. "Ozone Generation by Corona Discharge," Chapter 2, this volume.
4. Weiner, A. L. "Drying Gases and Liquids," *Chem. Eng.* (September 16, 1974).
5. Higgins, J. "Air Preparation for Ozone Production," paper presented at the Ozone Technology Symposium, International Ozone Association, Los Angeles, CA, May 23–25, 1978.
6. Larocque, R. L. "An Overview of Design and Construction of Ozone Systems," paper presented at the Ozone Technology Symposium, International Ozone Association, Los Angeles, CA, May 23–25, 1978.
7. Masschelein, W. J. "Contacting of Ozone with Water and Contactor Offgas Treatment," Chapter 6, this volume.
8. Mignot, J. "Practique de la Mise en Contact de l'Ozone Avec le Liquide à Traiter," in *Proceedings of the Second International Symposium on Ozone Technology,* R. G. Rice, P. Pichet and M.-A. Vincent, Eds. (Vienna, VA: International Ozone Association, 1977), pp. 15–46.
9. Tritschler, F. C. "Ozone Contactor Off-Gas Treatment," in *Proceedings of the Seminar on the Design and Operation of Drinking Water Facilities Using Ozone or Chlorine Dioxide,* R. G. Rice, Ed. (Dedham, MA: New England Water Works Association, 1979), pp. 89–98.
10. Schalekamp, M. "Destruction of Ozone Off-Gas by Thermal Catalytic Method," *Ozone Sci. Eng.* 2(4):367–376 (1980).
11. Miller, G. W., R. G. Rice, C. M. Robson, R. L. Scullin, W. Kühn and H. Wolf. "An Assessment of Ozone and Chlorine Dioxide Technologies for Treatment of Municipal Water Supplies," U.S. EPA Report EPA-600-2-78-147 (Washington, DC: U.S. Government Printing Office, 1978).
12. Zawierucha, R., and H. Charleson. "Corrosion and Degradation of Materials in Ozone-Containing Environments," Chapter 7, this volume.
13. Jain, J. S., N. L. Presecan and M. Fitas. "Field Scale Evaluation of Wastewater Disinfection by Ozone Generated from Oxygen," in *Progress in Wastewater Disinfection Technology,* A. D. Venosa, Ed., U.S. EPA Report 600/9-79-018 (Cincinnati, OH: U.S. Environmental Protection Agency, 1979), pp. 198–209.

14. Hegg, B. A., K. L. Rakness, L. D. DeMers and R. H. Cheney. "Evaluation of Pollution Control Processes—Upper Thompson Sanitation District," U.S. EPA Report 600/2-80-016, Cincinnati, OH (1980).
15. Reid, Quebe, Allison, Wilcox & Associates. "Use of UPVC Piping in Ozone Diffuser Header Piping: Indianapolis Advanced Wastewater Treatment Project," unpublished report (1980).
16. Bollyky, L. J. "Ozone, Safety and Health Considerations," in *Proceedings of the Seminar on the Design and Operation of Drinking Water Facilities Using Ozone or Chlorine Dioxide,* R. G. Rice, Ed. (Dedham, MA: New England Water Works Association, 1979), pp. 281–290.
17. Naimie, H., and R. W. Okey. "An Interference-Free Procedure for the Ozone Generator Performance Evaluation," in *Proceedings of the Seminar on the Design and Operation of Drinking Water Facilities Using Ozone or Chlorine Dioxide,* R. G. Rice, Ed. (Dedham, MA: New England Water Works Association, 1979), pp. 131–150.

SECTION 6

MECHANISMS OF ORGANIC OXIDATIONS IN AQUEOUS MEDIA

CHAPTER 12

MECHANISMS, RATES AND SELECTIVITIES OF OXIDATIONS OF ORGANIC COMPOUNDS INITIATED BY OZONATION OF WATER

J. Hoigné

Swiss Federal Institute of Technology, Zürich
Institute for Aquatic Sciences and Water Pollution Control
Dübendorf, Switzerland

OBJECTIVES

The final products accumulating in water during ozonation reflect the mechanisms and kinetics with which the different oxidation reactions of the parent molecules and transient intermediates proceed. The goal of this chapter is to present kinetic data and to establish a framework within which observations of the chemical effects of ozonation can be localized and rationalized. This type of information is needed when one must consider ozonation of waters of varying composition; it can further be applied to decisions concerning reactor design. It also helps in the comparison of results of different research programs.

When describing the kinetics of ozonation processes in water, we are confronted with the following problem: in nonaqueous solvents, the direct reaction of dissolved ozone is generally the relevant reaction. In water, however, this direct reaction can hardly be separated from other reactions that arise from decomposition of aqueous ozone to secondary oxidants that are much more reactive than ozone itself. Here, the choice of pathways followed by the reaction is sensitive to changes in composition of the aqueous solution. Therefore, a comparison of results is rather complex unless sufficient information is presented to allow discrimination between different effects. Consequently, I have preferred to restrict my attention to those papers from which kinetic results can be evaluated.

Officials of waterworks and health departments are understandably eager to learn about possible treatment techniques to remove occasional contaminants that might appear in raw waters; one of the methods to be considered is ozonation. In Switzerland, for instance, ozonation processes are already installed in more than 150 drinking water treatment plants to improve disinfection, taste and odor. However, to learn about the efficiency of such processes for individual micropollutants and to optimize the processes involved, we must know the parameters that regulate the mechanisms and kinetics of the reactions. Such information could also be applied to advanced wastewater treatment processes, and treatment of air-scrubber water or treatment of cooling water systems. Moreover, kinetic data should give a basis for quantification of ozonation reactions that can occur in rain droplets exposed to ozone in the atmosphere.

The types of products expected to be formed during ozonation of aqueous solutions of organic material already have been described [1–5]. From such information we conclude that organic materials dissolved in water (solutes) may selectively become oxidized by ozonation, and that a parent molecule (M) yields some daughter products, which can be oxidized further to form additional daughter products if ozonation is continued (Figure 1).

In most cases, only extended ozonation would lead, by a series of consecutive oxidations, to "full mineralization" of the solutes, i.e., to carbon dioxide, water, nitrate, sulfate, etc. In practice, ozonation processes are stopped long before such a mineralization is achieved, e.g., in water treatment, the conventional processes are generally restricted, for physical, technical and economic reasons, to the conditions given in Figure 2. The values of these parameters are much lower than those normally used by organic chemists in their investigations and organic synthesis.

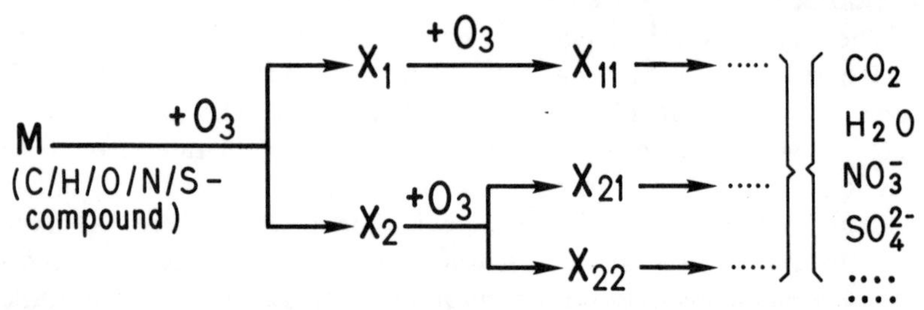

Figure 1. Scheme of consecutive reactions of a micropollutant and its daughter products with ozone.

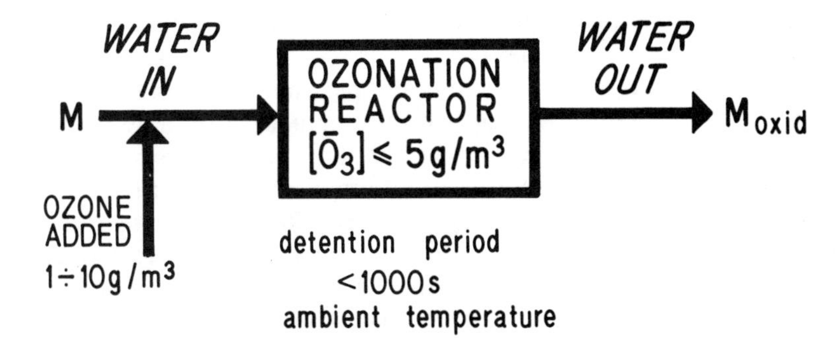

Figure 2. Reaction conditions typically encountered in drinking water ozonation. M = aqueous micropollutant.

PATHWAYS OF REACTIONS AND OXIDANTS INVOLVED

The primary reactions initiated by ozone in water can be described by the reaction sequences given in Figure 3. During ozonation, part of the dissolved ozone reacts directly with dissolved impurities. Such direct O_3 reactions are often rather slow. This makes them highly selective and often the most important reactions to be considered in water ozonation (disinfection, decoloration, etc.). The rates of these reactions are discussed below.

On the other hand, part of the added O_3 decomposes before it reacts

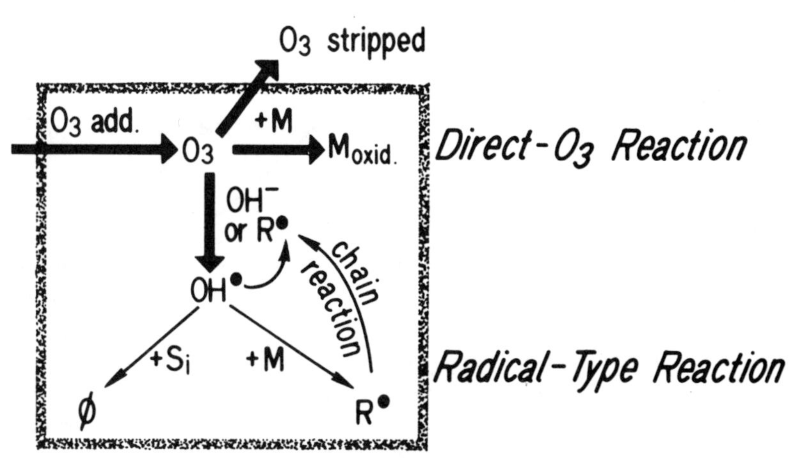

Figure 3. Scheme of primary reactions of ozone [6,7].

with solutes and before it is stripped off. Decomposition of O_3 is catalyzed by OH^- ions, and proceeds more rapidly with increasing pH. It is additionally accelerated by a radical-type chain reaction in which free radicals, such as $OH^\bullet$ and $HO_2^\bullet$ (which are produced from decomposed O_3), can act as chain carriers. Furthermore, benzene derivatives, for instance, react with $OH^\bullet$ radicals and form secondary radicals ($R^\bullet$) that also act as O_3-decomposing catalysts. Other solutes, such as bicarbonate ions [8] and methylmercury hydroxide [7] or chloride (at low pH), consume thc $OH^\bullet$ radicals by forming species that do not act as chain carriers. Such solutes thereby extend the lifetime of O_3 in water. Consequently, this lifespan depends on pH as well as on the types of solutes present. It is still uncertain as to which type of reaction dominates for decomposing O_3 in given systems [9]. We therefore base our approach on measured lifetimes of ozone in different types of waters (Figure 4).

As indicated above, secondary oxidants are produced by the decomposition of aqueous O_3. The most significant species formed are compiled

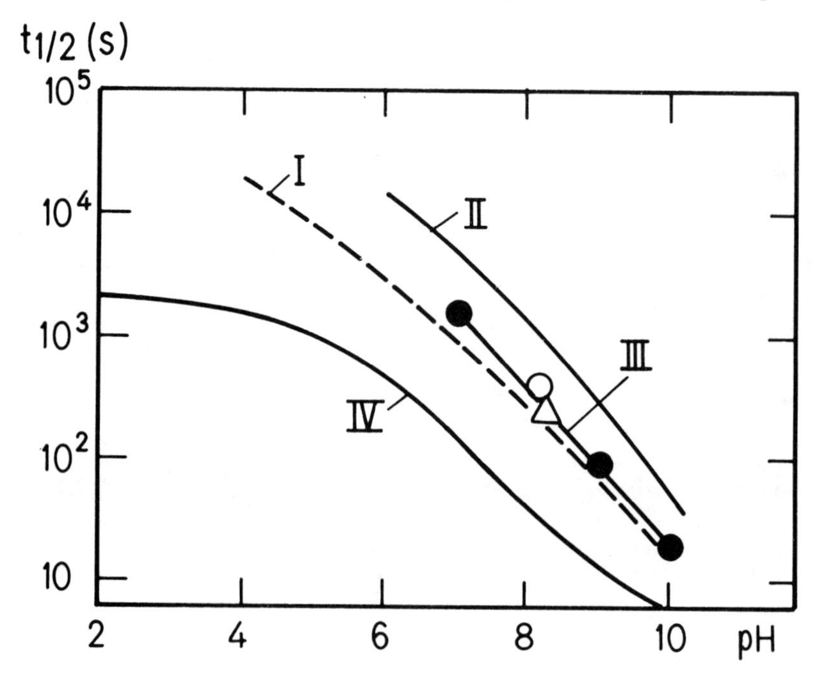

Figure 4. Half-life of ozone ($t_{1/2}$) in different types of waters [7,8,10]: I = distilled water; phosphate buffer; II = distilled water; 2 m*M* carbonate; phosphate buffer; III = 70% Lake Zürich water (DOC = 1.2/m³); ● 0.05 *M* borate buffer, Δ without buffer, O Lac de Bret water (DOC = 3.2 g/m³); IV = 70% municipal wastewater; secondary effluent; borate buffer (DOC = 7 g/m³).

in Figure 5. Many of these are identical to those which are produced by high-energy irradiation of (aerated) water [11]. Their reactions and products formed are thus well documented in the literature dealing with radiation chemistry, biology and food preservation technology [12].

Of all the secondary oxidants formed, the hydroxyl radicals ($OH^{\cdot}$) seem to play a key role. Their occurrence in ozonated water was reported by Weiss in 1935 [13]. Critical experimental identification of these species was later performed by Taube and Bray within an excellent series of kinetic studies [14–16]. These $OH^{\cdot}$ radicals are among the most reactive oxidants known to occur in water. They can easily oxidize all types of organic contaminants and many inorganic solutes by a radical-type reaction. They are therefore immediately consumed on their formation (within microseconds) and exhibit little substrate selectivity. Nonethe-

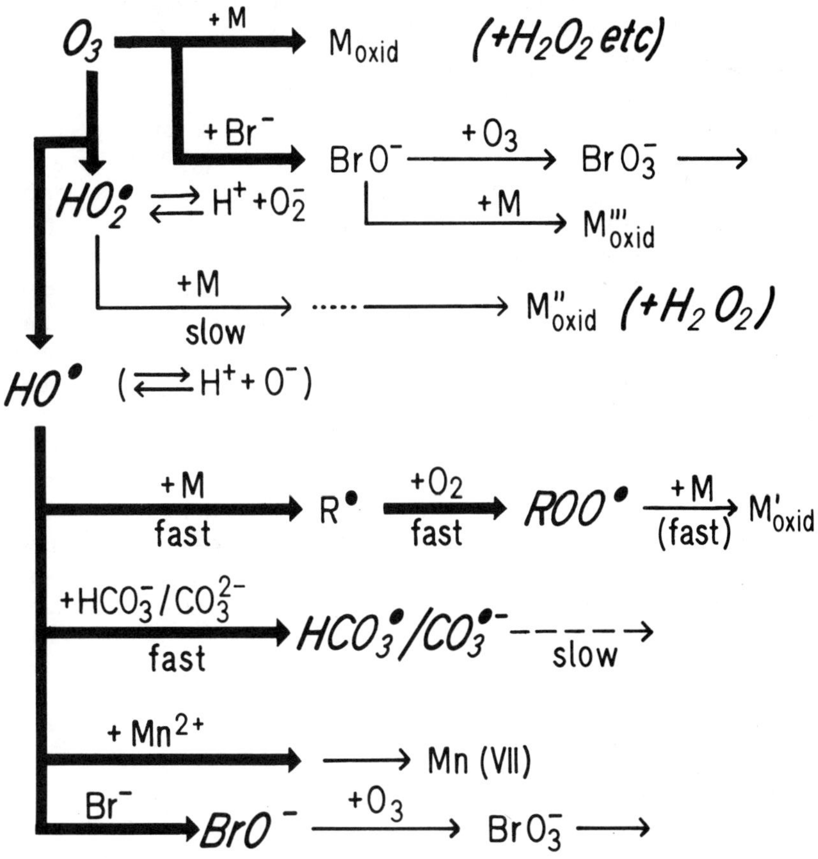

Figure 5. Secondary oxidants formed in ozonation processes.

less, if one considers the oxidation of relatively stable compounds, the reactions of these secondary oxidants may become of major importance. These radical-type reactions are therefore discussed explicitly below.

The occurrence of perhydroxyl radicals, which may dissociate to superoxide ions ($HO_2^{\cdot} \rightleftarrows O_2^{\cdot-}$; $pK_a = 4.9$), is also expected from the decomposition of ozone [9,13,14,16]. Generally, these species are of rather low reactivity in initiating oxidations [17]. Those types of chemical compounds which might be easily attacked by such radicals will already have been oxidized earlier by the direct action of ozone. However, these oxidants could still be important contributors to the overall oxidation following the primary atack of $OH^{\cdot}$ radicals on stable molecules.

In the presence of carbonate or bicarbonate ions, the $OH^{\cdot}$ radicals can oxidize these ions to $CO_3^{\cdot-}$ or $HCO_3^{\cdot}$, a carbonate or bicarbonate radical, which might act as a tertiary oxidant. These subsequent reactions are also expected to be rather slow and highly selective [8]. To date, their effects have not been observed in ozonated water.

In the presence of organic solutes, oxidations may lead to organic hydroperoxides or hydrogen peroxide. These species can also act as secondary oxidants or, in some cases, as reducing agents. The formation of permanganate on large doses of ozone is reported for some waters that contain dissolved manganese(II). In waters that contain bromide ions, a relatively rapid formation of bromine (hypobromous acid/hypobromite ion) is observed. These oxidants are selective and are more persistent than ozone. They are of importance when seawater is ozonated [16,18–22]. Addition of bromide to swimming pool water is proposed to convert ozone into a more persistent disinfectant [23]. (e.g., Hydrozone Process®, Hydro-Elektrik GmbH). Continued ozonation, however, may convert these oxidants to less reactive bromate ions [18–22,24].

All of these reagents formed during ozonation must be considered as potential secondary or tertiary oxidants, which contribute to the oxidation processes initiated by ozone in water. In the present survey, however, we will focus only on those two types of reactions that are of primary importance for drinking water: direct O_3 reactions and $OH^{\cdot}$ radical–type reactions.

DIRECT O_3 REACTION WITH SOLUTES

Reaction-Rate Parameters

Some solutes are oxidized directly by molecular O_3. If η denotes the stoichiometric factor, i.e., molecules of M oxidized per molecule of O_3

consumed by the sequence of reactions initiated by a primary attack of O_3 on M, we may write:

$$M + O_3 \xrightarrow{\eta \times k_{O_3}} X_1 \xrightarrow{+O_3} X_2 \xrightarrow{+O_3} \cdots \longrightarrow M_{oxid}. \qquad (1)$$

k_{O_3} is the apparent reaction-rate constant for the overall consumption of O_3, by means of all the reaction steps leading from M to a quasifinal end product, M_{oxid}. By definition, the oxidation of the transient intermediates, X_i, proceeds sufficiently fast (as compared with the rate of the initiating step), so as not to limit the overall kinetics. It follows from Equation 1 that k_M, the rate constant for the consumption of M, is given by:

$$k_M = \eta\, k_{O_3} \qquad (2)$$

Case A

Some substrates, if present at sufficiently high concentrations, such as nitrite and a few organic compounds, consume O_3 so quickly that the overall rate of the reaction, $-d[M]/dt$, is regulated only by the rate of O_3 supply:

$$-d[M]/dt = \eta \times (O_3 \text{ supplied/time})/\text{volume} \qquad (3)$$

If no buildup of intermediate daughter products takes place, the integration of this equation yields:

$$[M]_o - [M]_t = \eta \times (O_3 \text{ supplied/volume}) \qquad (4)$$

Here $[M]_o$ and $[M]_t$ are the concentration of the solute M at reaction times 0 and t. In this case, the absolute solute concentration decreases linearly with the amount of O_3 added (Figure 6A) until the concentration of the substrate becomes so small that the system becomes reaction rate–limited (Figure 6B).

Case B

On the other hand, if one considers ozonation of drinking water, most dissolved impurities are present in low concentrations and react so slowly with O_3 that the rate of the overall process is limited instead by the rate of the chemical reaction step. In these cases, experiments show that the rate

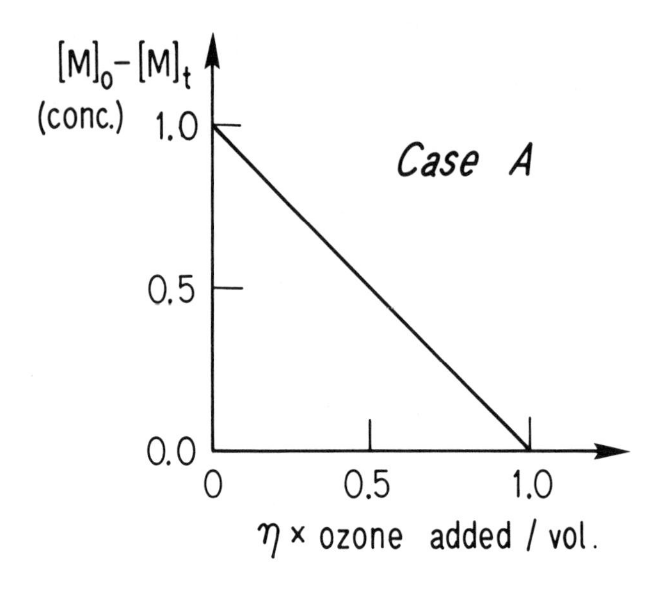

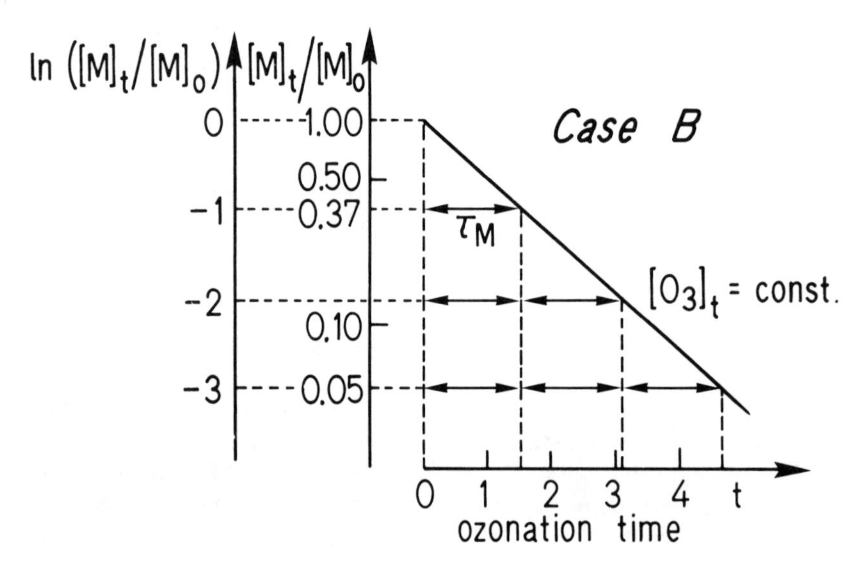

Figure 6. Oxidation rate of solute M by direct O_3-reaction. (A) Reaction rate is limited by the rate at which ozone is added to the system. Assumptions: absence of a significant buildup of secondary ozone-consuming intermediates (see Equation 4). (B) The reaction rate is limited by the rate of the chemical reaction. Assumptions: constant ozone concentration during process; M is an individual type of molecule; batch-type or ideal plugflow reactor (see Equation 6).

of the overall consumption of O_3 is proportional to the concentration of O_3 and the concentration of the solute. Therefore, the rates of oxidation of M and consumption of O_3 are described by:

$$-d[M]/dt = -\eta\, d[O_3]/dt = \eta\, k_{O_3} [O_3][M] = k_M [O_3][M] \quad (5)$$

Integrating Equation 5 over time, results in, for a batch or ideal plugflow reactor:

$$\ln([M]_t/[M]_o) = -\eta\, k_{O_3} [\overline{O_3}]\, t = -k_M [\overline{O_3}]\, t \quad (6)$$

or

$$[M]_t/[M]_o = \exp(-\eta\, k_{O_3} [\overline{O_3}]\, t) = \exp(-k_M [\overline{O_3}]\, t) \quad (7)$$

where $[\overline{O_3}]$ is the mean ozone concentration as averaged over the reaction time t:

$$[\overline{O_3}]\, t = \int_0^t [O_3]_t\, dt \quad (8)$$

(for mixed reactors see Equations 27 and 28).

Equation 6 shows that it is now the logarithm of the relative residual concentration that declines directly with ozone concentration and ozonation time, $[\overline{O_3}] \times t$. A straight-line relationship is found for systems where $k_M [\overline{O_3}]$ does not change during the reaction (see Figure 6B). This is the case if M represents a single type of compound. If M stands for a mixed system, then the apparent k_M value would change (decrease) as ozonation proceeds. This is the case if M is based on collective parameters of mixed systems, e.g., total dissolved organic carbons (DOC), ultraviolet (UV) absorbance, or the permanganate value.

The stoichiometric factor η is about 1.0 for cases in which M is an olefinic compound. It is somewhat lower for many other types of solutes, e.g., 0.25–0.5 for aliphatic or aromatic molecules [25].

The reaction rate can be characterized by the lifetime τ_M, during which the concentration of a specified individual solute M decreases to 37% (1/e) of its original value in the presence of a constant amount of ozone. This time is achieved when the exponent in Equation 7 becomes -1:

$$\tau_M = 1/(\eta\, k_{O_3} [\overline{O_3}]) \quad (9)$$

The τ_M value is indicated in Figure 6B.

The relationship between reaction-rate constants and the lifetime of a specified solute M is presented in Figure 7 for various O_3 concentrations. In this graph, the line for an O_3 concentration of 50 g/m^3 represents an

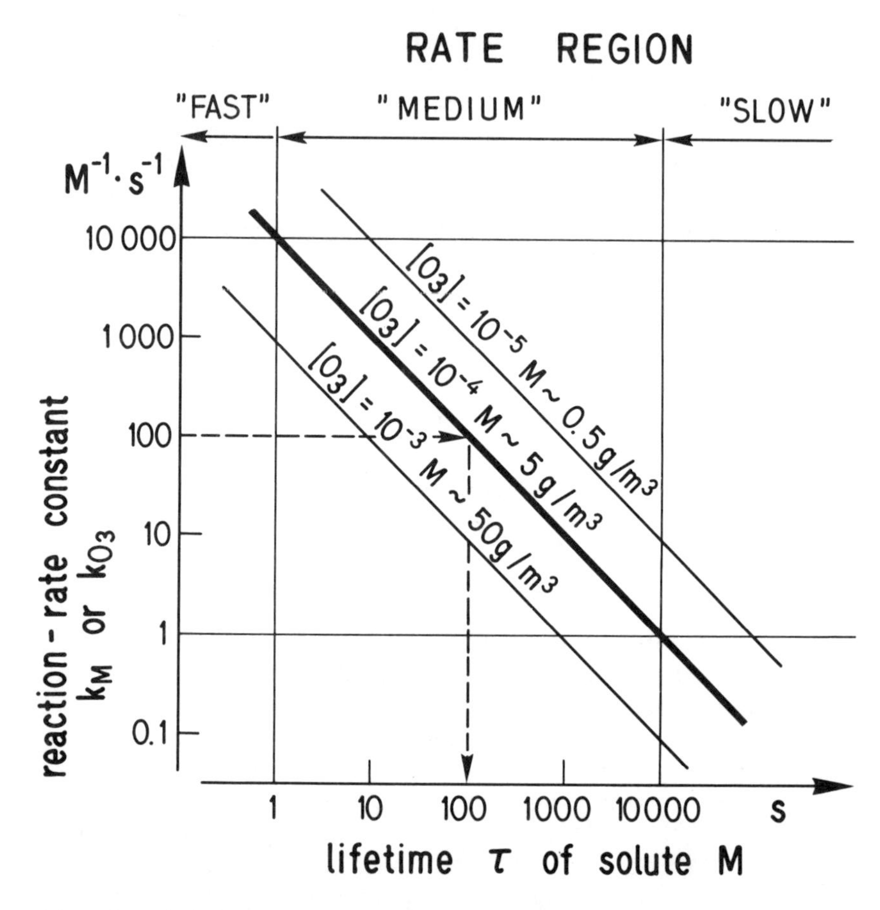

Figure 7. Relation between reaction-rate constant k and lifetime τ of solute M for different concentrations of ozone. Assumptions: $\eta = 1.0$.

upper theoretical limit in water treatment applications. Typical O_3 concentrations achieved in practical processes (atmospheric pressure) are much lower: the line for 5 g/m^3 represents a value achieved with commercial ozonators operating with dry air. However, 0.5 g/m^3 is more often used for drinking water plants. In Figure 7, the "slow" region denotes the region in which solutes have reaction-rate constants $<1\ M^{-1}\text{sec}^{-1}$. Their lifetimes τ_M are larger than 10,000 sec even if a concentration of O_3 of 5 g/m^3 should be applied. Solutes located in this "slow" region will not be oxidized within a practical process time. In the "fast" region, solutes have reaction-rate constants that exceed 10,000 $M^{-1}\text{sec}^{-1}$. Here, ozonation leads to eliminations within seconds, and the rate of the ozonation process is limited instead by physical transfer and distribution of ozone into the aqueous solution (Case A, Equation 2). The "medium" region represents solutes with reaction-rate constants between 1 and 10,000 $M^{-1}sec^{-1}$. Practical ozonation processes are relevant for these solutes and the rate of the chemical reaction determines the overall reaction rate (Case B, Equation 6).

For ozone application, it is of interest to learn which types of solutes (including intermediate ozonation products) have to be classified as fast, medium or slow. We need to know the value of the reaction-rate constants accurately for solutes in the "medium" category. It enables calculation of the rate of the chemical reaction, which here determines the overall rate of the process. The magnitudes of the reaction-rate constants of solutes classified as "fast," may also be of some interest as they give information on the selectivity with which different solutes react relative to each other (see Equation 12).

The parameters that control ozonation of a dissolved compound are:

Case A: rate of process limited by rate of O_3 supply

- $O_{3\,add.}$: total amount of O_3 added during ozonation
- η: stoichiometric factor (molecules M oxidized per molecule of O_3 consumed)

Case B: rate of process limited by the chemical reaction rate

- [M]: concentration of the solute
- $[\overline{O_3}]$: concentration of O_3 (mean over reaction time t)
- t: duration of ozonation
- k_M (or $\eta\, k_{O_3}$): second-order reaction-rate constants as defined by Equation 1 or 5
- reactor design: see below

Reaction-Rate Constants

Reaction-rate constants have been reported for many organic solutes in nonaqueous solvents. Preliminary comparisons with measurements may be rather different for aqueous solutions. For instance, aromatic hydrocarbons react 30 times faster in water than in chloroform at comparable temperature. Moreover, in water, acids and bases may dissociate to form ions, which react very differently than the nondissociated species [25].

We therefore determined the k values for many aqueous solutes by measuring the rate of ozone consumption. This was achieved in the presence of different concentrations of solutes in closed containers. [6,25,26]. For these measurements, the concentration of solutes was several times higher than the initial concentration of ozone. This made the reaction pseudo first-order, i.e.,

$$-d[O_3]/dt = k_{O_3}[M]_o[O_3]_t \tag{10}$$

The results of these measurements were supported by conclusions of further experiments, in which we measured the relative rates at which different organic substances are eliminated from a solution, i.e., when they compete with each other for reaction with ozone. The main problem with these experiments performed in water was to avoid interactions by radical reactions (see Figure 3). This could generally be achieved by lowering the pH and/or adding free radical scavengers. Criteria for the occurrence of pure direct ozone reactions were:

1. The kinetics remained first-order in ozone and solute concentrations over the entire concentration range measured.
2. Within the pH range employed, small changes in pH did not affect the apparent reaction-rate constant. (In case of acids or bases the reaction rates were based on the concentration of the appropriate species present at the given pH.)
3. The addition of further radical scavengers (bicarbonate, propanol, *tert*-butanol, methylmercury) had no effect on the results.

Figure 8 gives examples of measured reaction-rate constants. Diagonal lines for particular elimination factors are drawn for illustration. They indicate reductions of concentrations to 37% (A), 1% (B) and 0.01% (C) of the original value when a plugflow or batch-type reactor is used. The required ozonation time leading to such eliminations may be read off the horizontal time scale. This scale is calibrated for a mean ozone concentration of about 5 g/m^3 (10^{-4} *M*). It increases inversely with the ozone concentration. As shown in Figure 7, a rate constant of at least 10 $M^{-1}sec^{-1}$ is required for an appreciable reaction within 1000 sec detention time, even in the presence of 5 g/m^3 of ozone.

A more extended list of reaction-rate constants is presented in Table I. A comparison of the rate constants listed shows that direct reactions of ozone with dissolved compounds are highly selective. Even small modifications of a chemical structure can have a large effect on these values. (To visualize the lifetime of a solute M in the presence of a given ozone concentration, the rate constants given in Table I should be used in conjunction with Figures 7 or 8.)

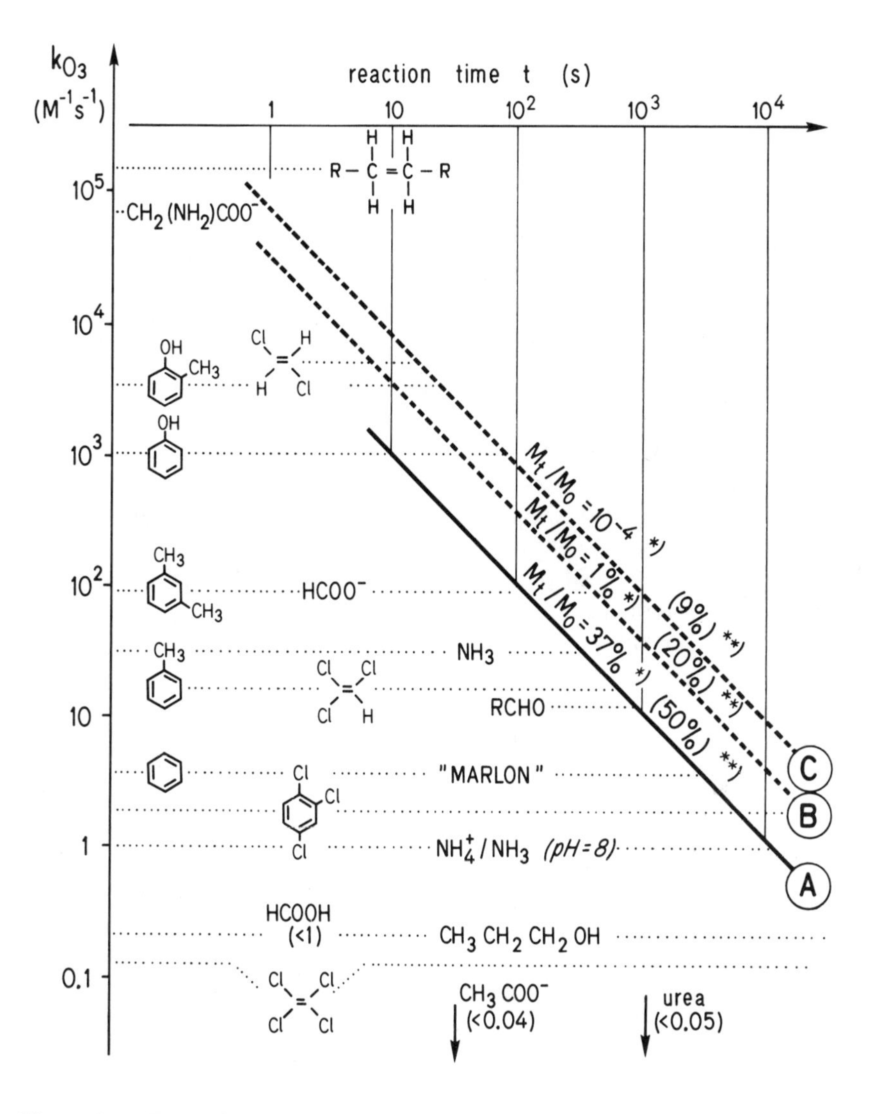

Figure 8. Examples of direct reactions of ozone with solutes. Vertical axis: reaction-rate constants k. Horizontal axis: reaction time required to reduce the solute concentration to the relative value indicated on the diagonal elimination lines. Assumptions: $\eta = 1.0$; $[O_3] = 10^{-4}\ M$ (about 5 g/m³). (*) values for a batch-type or ideal plugflow reactor (see Equation 22); (**) values for a completely stirred reactor (see Equation 28)

Discussion of Reaction-Rate Constants k_{O_3} [25]

The C=C double bond of olefinic compounds reacts so rapidly with ozone that its rate of ozonation is generally not limited by the rate of the chemical reaction, i.e., it belongs into the "fast" category of Figure 7 (see Table I). However, if hydrogen atoms adjacent to the double bond are replaced by electron-withdrawing groups, e.g., chlorine atoms, the reactivity of the bond becomes markedly depressed; tetra- and trichloroethylene are hardly attacked by the ozone molecule, even during extended ozonation (see Table I).

The reaction rate of benzene is rather low. Several hours are needed for oxidation even at elevated O_3 concentrations. The rate increases if benzene contains substituents that elevate the electron density on the ring. As few as three methyl groups on the ring are sufficient to increase the rate to such an extent that the reaction becomes appreciable within 10 min in the presence of 0.5 g/m^3 ozone. Phenol, nitrophenol, chlorophenols and cresols can even be oxidized within minutes in the presence of this low ozone concentration. In the pH region where phenolic compounds dissociate to phenolate ions, the rate constants increase to very high values. In the case of nitrophenol or polychlorinated phenols, which have relatively high acidity constants (e.g., on the order of $pK_a = 7$), the effect of the high reactivity of the phenolate species already becomes apparent at very low pH values ($pH > 2$). At $pH > 4$, the apparent rate will be mainly due to the phenolate-type species, even in the case of such phenols which have low acidity constants ($pK_a > 9$–10) (see Table I).

The electrophilicity of ozone in reactions with aromatic hydrocarbons can be quantified and correlated with rates of other types of reactions by comparing reaction rates within series of related compounds. From such correlations, we can extrapolate that dissolved polynuclear aromatic hydrocarbons (PAH) would also be relatively easily oxidized by molecular O_3 [25].

Based on measured data, aliphatic aldehydes and glyoxylic acid, which occur as ozonation products in lake water [6,27] or wastewater [28], are expected to accumulate as intermediates and only become oxidized during extended ozonation or if another type of oxidation mechanism becomes effective.

Ions of aliphatic saturated carboxylic acids (except formate and malonate), formed as oxidation products of larger organic molecules [29–37], also accumulate in water when only the direct O_3 reaction occurs. It is important to note that formate ions may still react directly with ozone; this is in contrast to the reactivity of oxalate or acetate ions, which are refractory. Therefore, if the pathway of oxidation proceeds via formate ions as intermediates, extended ozonation can result in a loss of

Table I. Reaction-Rate Constants (k_{O_3}, $M^{-1}sec^{-1}$) for Different Aqueous Solutes at 20°C[a]

Compound	k
Olefins	
1-Hexene-4-ol	$\cong 2 \times 10^5$
1-Hexene-3-ol	$\cong 1 \times 10^5$
Styrene	$>10^5$
1,1-Dichloroethylene	110
trans-Dichloroethylene	4×10^3
cis-Dichloroethylene	<800
1,1,2-Trichloroethylene	17
Tetrachloroethylene	<0.1
Maleic Acid	$\sim 1.0 \times 10^3$
Maleate Anion	$\sim 5 \times 10^3$
Fumaric Acid	$\sim 6 \times 10^3$
Fumarate Anion	$\sim 1 \times 10^5$
Substituted Benzenes	
Benzene	2
Naphthalene	3,000
Toluene	14
o-Xylene	90
m-Xylene	90
p-Xylene	140
1,2,3-Trimethylbenzene	400
1,2,5-Trimethylbenzene	700
Nitrobenzene	0.1
Benzenesulfonate	0.2
Benzoic Acid	1
Chlorobenzene	0.8
1,4-Dichlorobenzene	$\leqslant 3$
1,2,4-Dichlorobenzene	$\leqslant 2$
Anisole	300
Phenols	$\cong 1{,}000$ (see below)
Phenolates	10^7–10^9 (see below)
Aniline	$\cong 10^8$ (see below)

Compound	pK_a	$K_{\phi\text{-OH}}$	$k_{\phi\text{-O}^-}$[b]	$k_{app.\ (pH\ 8)}$[c]
Substituted Phenols[a]				
Phenol	9.9	1,300	1.4×10^9	2×10^7
o-Cresol	10.2	1.2×10^4	Very high[d]	Very high[d]
m-Cresol	10.0	1.3×10^4	Very high[d]	Very high[d]
p-Cresol	10.0	3.0×10^4	Very high[d]	Very high[d]
2-Chlorophenol	8.3	$\cong 1{,}000$	$\cong 10^9$	$\cong 3 \times 10^8$
4-Chlorophenol	9.2	$\cong 600$	5×10^8	$\cong 3 \times 10^7$
2,4-Dichlorophenol	7.80[i]	$\cong 1{,}500$	$\cong 8 \times 10^9$	$\cong 5 \times 10^9$
2,4,5-Trichlorophenol	6.94[i]	$<3{,}000$	$>10^9$	$>10^9$
2,4,6-Trichlorophenol	6.10[i]	$<10^4$	$(\cong 8 \times 10^9)$	$(\cong 6 \times 10^9)$

Table I, continued

Compound	pK_a	$K_{\phi\text{-}OH}$	$k_{\phi\text{-}O^-}$ [b]	$k_{app.\,(pH\,8)}$ [c]
Pentachlorophenol	4.70[i]		$\gg 3 \times 10^5$	$\gg 3 \times 10^5$
4-Nitrophenol	7.1	<50	$\cong 2 \times 10^7$	$\cong 2 \times 10^7$
Salicylic Acid		300		
Salicylic Acid Anion	13.4	2.8×10^4	Very high[d]	d
1,3-Dihydroxybenzene	9.3	$>3 \times 10^5$		

Compound	k
Aliphatic Alcohols	
Methanol	0.02
Ethanol	0.4
1-Propanol	0.4
2-Propanol	2
1-Butanol	0.6
tert-Butanol	≅0.003
Octanol	≤0.8
Cyclopentanol	2

Compound	pK_a	K_{AcOH}	k_{AcO^-} ($\cong k_{app.(pH\,8)}$)[b]
Carboxylic Acids[a]			
Formic Acid	3.8	<3	120
Acetic Acid	4.8	$<10^{-5}$	$\cong 3 \times 10^{-5}$
Propionic Acid	4.9	$<10^{-3}$	$\cong 10^{-3}$
Butyric Acid	4.8	$<10^{-2}$	$<10^{-2}$
Oxalic Acid	(1.0)/3.9	$\ll 10^{-2}$	0.04
Malonic Acid	(2.6)/5.2	<4	7
Succinic Acid	(4.2)/5.6		0.03
Glutaric Acid	(4.3)/5.4		≅0.008
Glyoxylic Acid	3.2	0.18	1.9
Benzoic Acid	4.0	1	2
Salicylic Acid	2.8	300	2.8×10^4

Compound	pK_a [f]	$k_{-NH_3^+}$ [g]	k_{-NH_2} [h]	$k_{app.(pH\,8)}$ [c]
Nitrogen Compounds[e]				
Ammonia	9.3	0.00	20	1
Methylamine	10.7	<1	14×10^4	280
Dimethylamine	11.0	<1	1.9×10^7	19,000
Trimethylamine	9.9	<1	4×10^6	50,000
Butylamine	10.7	<1	1.7×10^5	340
Glycine	9.9	$<10^{-2}$	1.3×10^5	1,600
α-Alanine	10.0	$<10^{-2}$	6.4×10^4	640
β-Alanine	10.3		6.2×10^4	310
Aniline	4.6		9×10^6	9×10^6
Urea			<0.05	<0.05

Table I, continued

Compound	k
Miscellaneous	
Hydrogen Cyanide	$\leq 1 \times 10^{-3}$
Cyanide Anion	$\sim 10^{5}$[b]
Acetone	0.03
2-Butanone	0.07
Dimedone	$>10^{5}$
Saccharose	0.1
Glucose	0.5
n-Propyl Acetate	0.03
Diethyl Malonate	0.06
Cystine, pH 2/3	600/1,000
Cysteine, pH 2	3×10^{4}
Creatine, pH 2/6	0.5/1.8
Creatinin, pH 2/5	$\cong 2/\cong 11$
Dimethylnitrosamine, pH 2/8	3/500
Ethylmercaptan, pH 2	$>10^{5}$
Dipropylsulfide, pH 2	$>10^{5}$
Methylmercury, pH 2/4	$<10^{-3}$
Trimethyltin Chloride, pH 3/4	$\cong 5 \times 10^{-3}$
Tributyltin Chloride, pH 3/4	$\cong 0.8$

[a]The values given here are only approximate. The original literature should be consulted for detailed information or discussion of errors involved [23].
[b]Extrapolated value for dissociated compound (values from data measured below pH 5).
[c]Apparent reaction-rate constant at pH 8 (extrapolation from data measured at varied pH values).
[d]Cannot be calculated by extrapolation (spontaneous O_3 decomposition initiated by solutes).
[e]All measurements were made at varied pH values (1.7 to <8).
[f]pK_a values of protonated amine.
[g]Reaction-rate constant of protonated compounds (such as present at low pH).
[h]Reaction-rate constant based on the amount of free amine present at pH value used for measurement.
[i]pK_a values from our measurements performed with the same buffer concentrations and pH calibrations as used for kinetic experiments. Method: concentrations of acid and base measured vs pH by UV absorption (multicomponent method).

total DOC. However, reduction in DOC levels cannot be expected if oxalate and acetate ions are intermediate degradation products, and if only direct O_3 reactions are operative.

Based on the electrophilicity of O_3, it is not surprising that amines show an appreciable reaction rate when they are present in the non-protonated form (see Table I). As only the free amino groups show high reactivity, the apparent rate with which the sum of protonated and non-protonated compounds react corresponds to only a fraction of the rate constant of the free compound given by the factor α:

$$\alpha = \frac{\text{free amine}}{\text{total amine}} = \frac{\text{R-NH}_2}{\text{R-NH}_2 + \text{R-NH}_3^+} = \frac{K_a}{K_a + [H^+]} \quad (11)$$

Hence, α depends on the difference between the pH of the aqueous solution and the pK_a of the protonated amine. For pH <9, α and, therefore, the apparent reaction rate of aliphatic amines, amino acids or ammonia, is reduced by a factor of 10 when the pH is lowered by one unit.

Free methylamines or amino acids react about 1000 times faster than NH_3 at the same pH [26,38]. The reaction-rate constant of free aniline is comparable to that of other amines, but free aniline exists even at fairly low pH values [25].

The accumulation and eventual depletion of intermediate oxidation products, observed during the continued ozonation of aqueous organic compounds, is in good agreement with the reaction-rate constants given in Table I. Examples: phenols [30,31,33,34,37,39–42] and even malonic acid [29,31,32] are always observed to be easily oxidized. Formic acid or glyoxylic acid often arise as dominant temporary intermediates and disappear on further ozonation [31,32,35,37]. In contrast, oxalic and acetic acids accumulate as final daughter products even during extended ozonation [29,32,35,37,41]. [A temporary accumulation of compounds such as oxalic acids is even observed in processes in which the $OH^{\cdot}$ radical-type reactions prevail; even $OH^{\cdot}$ radicals attack oxalic acid significantly slower than they do most other organic compounds (see Figure 10)].

All reported oxidations of aqueous amines also proceed as expected from the rate constants given in Table I [44–46]. For many of the systems studied at elevated pH, however, a preceding decomposition of ozone to $OH^{\cdot}$ radicals must also be accounted for when analyzing the experimental results [44,46]. Also, the inert character of urea can now be explained by its low reaction-rate constant.

Reaction rates of ozone often parallel those of other types of electrophilic reactions. One of the many implications of this result is that preozonation treatment is expected to oxidize functional groups that otherwise might react with chlorine or bromine in a subsequent chlorination process and produce undesirable products [24,41,42,47–49]. Rook et al. [19] and Maier and Mäckle [50], however, did not observe a real inhibition of haloform formation, but rather a shift in the type of halogenated products formed. An explanation of the different results has been submitted for discussion [41,51].

Selectivity of Direct O_3 Reactions

For the elimination of two solutes (M_1, M_2) exposed to O_3 in the same system, we may deduce from Equation 4, for a batch-type reactor:

$$\frac{\ln([M_1]_t/[M_1]_o)}{\ln([M_2]_t/[M_2]_o)} = k_1/k_2 \tag{12}$$

In Equation 12, it is the logarithm of the ratios of the residual concentrations to the concentration at time zero, $\ln[M]_t/[M]_o$, that is proportional to the relative rate constants with which the two solutes react. The high selectivity of the direct O_3 reactions can be visualized easily in Figure 8: different degrees of solute removal, $[M]_t/[M]_o$, are reached within a given reaction time, for two compounds of different rate constants, e.g., the elimination factor of xylene $k \sim 100$ $M^{-1}sec^{-1}$ approaches 0.37 after a reaction time of 100 sec, where the concentration of phenol ($k \sim 1000$) is already reduced to 0.01% of its original value.

Since the direct reaction of ozone is so highly selective, a clear distinction is possible between micropollutants that are oxidized by the direct O_3 reaction before a significant amount of ozone is decomposed and those that react so slowly with ozone that only the indirect, less selective, $OH^{\cdot}$-radical–type reaction is significant. The latter is easily depressed by the addition of scavengers such as bicarbonate or *tert*-butanol.

The disinfection power of ozone is largely due to the extreme selectivity with which O_3 reacts by direct reaction and thereby allows it to persist until it encounters a microorganism (a particle). It is this selectivity, rather than only its overall oxidation potential, that makes it a good disinfectant. The latter criterion is of secondary importance for the kinetics of such reactions which are primarily regulated by the activation energies required.

OXIDATIONS INITIATED BY $OH^{\cdot}$-RADICAL REACTIONS

Half of the O_3 introduced into water from Lake Zürich (pH 8; 1.5 meq/l HCO_3^-) is decomposed within 10 min (c.f. Figure 4). As indicated in Figure 2, this time corresponds approximately to the duration of a typical ozonation process. In other types of surface waters or in the event of elevated pH, an even higher percentage of added ozone decomposes during ozonation (see Figure 4). From experimental experience we deduce that about 0.5 mol of $OH^{\cdot}$ radicals is formed per mol of ozone decomposed in drinking water [8,11]. These are the predominant species that initiate further oxidations. Even organic substances that are known to be nonreactive with ozone or other oxidants can be attacked by $OH^{\cdot}$.

Speciation of the $OH^{\cdot}$ radicals does not change with pH as long as the pH remains well below 11 [$pK_a(OH^{\cdot}) = 11.8$]. However, at higher pH values, the reactivity of the O^- ion would also have to be considered.

Types of Reactions Initiated by $OH^{\cdot}$ Radicals

$OH^{\cdot}$ radicals initiate oxidations by three main types of reactions:

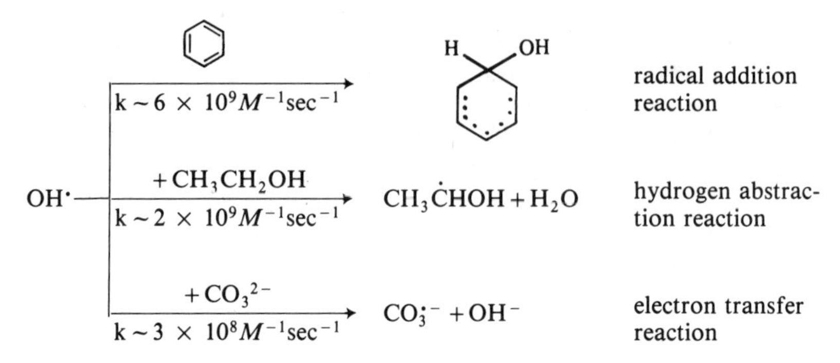

The secondary radicals ($R^{\cdot}$) formed by these three types of reactions may react with each other. In an ozonated system, however, there is always an ample amount of dissolved oxygen (O_2), so that the radicals may add an O_2 molecule before they encounter each other:

$$R^{\cdot} + O_2 \rightarrow ROO^{\cdot} \tag{13}$$

The peroxy radicals ($ROO^{\cdot}$) formed in this manner act as strong oxidants which can abstract many types of hydrogen atoms from organic compounds ($R'H$):

$$ROO^{\cdot} + R'H \rightarrow ROOH^{\cdot} + R'^{\cdot} \tag{14}$$

A new radical, $R'^{\cdot}$, is formed which again may add O_2. In principle, chain reactions that lead to autoxidations may become operative. In natural waters, however, such chain reactions are quenched by many types of dissolved impurities, which act as chain-terminating reagents. In most aqueous systems studied so far, only 0.4–0.5 organic molecules were oxidized per ozone molecule decomposed, even if M was the dominant solute.

Because $OH^{\cdot}$ radicals do not react very selectively with the dissolved materials, their reactions always lead to a great variety of daughter products.

Overall Efficiency of $OH^{\cdot}$-Radical Reactions

The reaction sequence can be represented as follows:

$$\Delta O_3 \rightarrow \eta'(OH^{\cdot}) \tag{15}$$

where ΔO_3 = amount of decomposed ozone
η' = yield of $OH^{\cdot}$ radicals produced per mole of ΔO_3 (~0.5)

$OH^{\cdot}$ radicals are so reactive that they are even reduced by carbonate or bicarbonate ions [8], humic material [52], or ammonia [38]. Oxidation of a specified trace impurity (M) following the decomposition of ozone will, therefore, only occur in competition with the reaction of all $OH^{\cdot}$-radical scavenger material present, $\Sigma[S_i]$:

$$OH^{\cdot} \xrightarrow[+\Sigma S_i]{+M} \begin{cases} \xrightarrow{k'_M [M]} & M'_{oxid} \\ \xrightarrow[\Sigma(k'_i [S_i])]{} & \text{consumption of } OH^{\cdot} \text{ by all scavengers } S_i \end{cases} \tag{16}$$

It follows that the fraction f of $OH^{\cdot}$ radicals that can oxidize solute M is given by the ratio of the rate with which $OH^{\cdot}$ radicals react with M to the rate with which $OH^{\cdot}$ radicals are concurrently consumed by the sum of all solutes present, ΣS_i, (ΣS_i includes M and O_3):

$$f = \frac{k'_M [M]}{\Sigma(k'_i [S_i])} \tag{17}$$

Taking Equations 15 and 17 into consideration, the rate at which M is oxidized by the $OH^{\cdot}$ radical mechanism is:

$$-\frac{d[M]}{dt} = \eta' \frac{d(\Delta O_3)}{dt} f \tag{18}$$

It is possible to estimate the factor f, defined by Equation 17, from the chemical composition of a water and from the rate constants known for $OH^{\cdot}$-radical reactions [8]. For practical use, however, it is more convenient to determine experimentally the $OH^{\cdot}$-radical consumption rate of a water relative to the rate with which a known reference solute becomes oxidized [7,10,52]. Therefore, we have defined a parameter Ω_M, i.e., the oxidation-competition value [10,52]:

$$\Omega_M = \frac{\Sigma(k'_i [S_i])}{\eta' k'_M} \tag{19}$$

This parameter now includes the stoichiometric yield factor η' which is difficult to determine by direct measurements.

Comparing Equations 17, 18 and 19, we find:

$$-\frac{dM}{dt} = -\frac{d(\Delta O_3)}{dt} \times \frac{[M]}{\Omega_M} \tag{20}$$

Integrating this equation we get:

$$\ln([M]_\infty/[M]_o) = -\Delta O_3/\Omega_M \tag{21}$$

or

$$[M]_\infty/[M]_o = \exp(-\Delta O_3/\Omega_M) \tag{22}$$

Assumptions made for Equations 21 and 22 are:

1. M is an individual, chemically uniform type of micropollutant.
2. $k'_M[M] \ll \Sigma k'_i[S_i]$.
3. Batch or ideal plugflow reactor is used (equations for completely stirred reactors, see below).

$[M]_\infty/[M]_o$ is the relative concentration of M remaining after a given amount of O_3 has been decomposed in water. According to Equation 22, Ω_M is the amount of decomposed ozone (ΔO_3) required to reduce the concentration of M to 37% (Figure 9) of the initial value in a given water. It does not depend on the solute concentration when $[M] \ll \Sigma[S_i]$.

The introduction of this empirical Ω parameter allows for a practical characterization of the effectiveness of the $OH^\cdot$-radical–type reaction with few parameters:

- $[M]_o$: initial concentration of micropollutant M
- ΔO_3: amount of O_3 decomposed during ozonation
- Ω_M: oxidation-competition value of the water based on the rate with which the solute M is eliminated

Ω_M itself is a collective parameter that comprises several subparameters, each of which depends on the type of water and on the ozonation process. It would be difficult to separate each of these subparameters and to evaluate them individually.

In principle, we expect that the Ω_M value of a water varies during progressive ozonation. However, in all natural waters we have examined, the Ω_M values remain rather constant if M denotes a uniform type of compound and after the region of immediate ozone consumption is overcome. This behavior can be explained as follows:

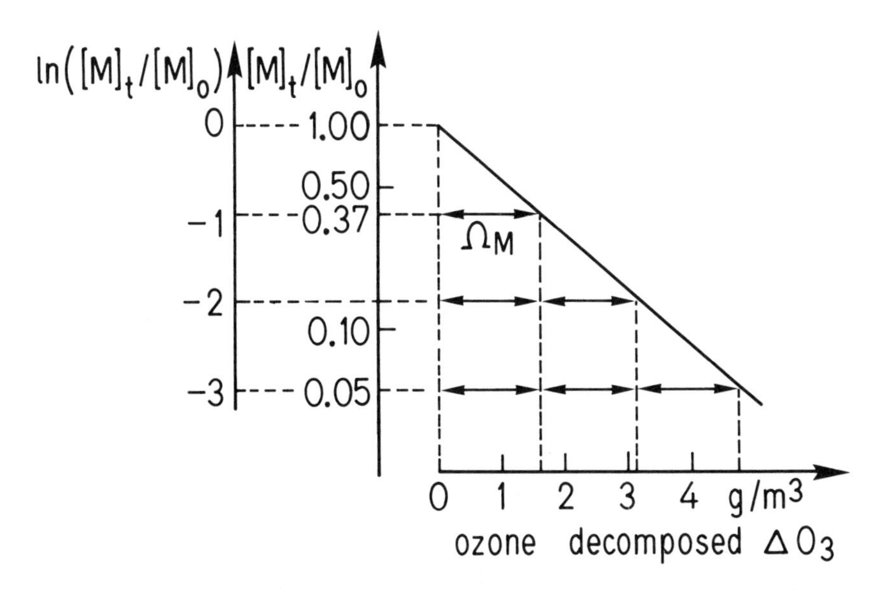

Figure 9. Oxidation of solute M by the radical-type reaction. Assumptions: Ω_M is not changed during ozonation; batch or ideal plugflow reactor (compare Equation 21).

1. The concentration of carbonate or bicarbonate ions which contributes to the oxidation competition is not changed significantly during ozonation.
2. The NH_3 consumed during ozonation becomes replenished by dissociation of further NH_4^+.
3. The reactivities of many of the organic materials present in a natural water or in a secondary effluent (DOC) are not changed by partial ozonation. Even octanol, which we have used as a model type of DOC compound, never changed its efficiency for scavenging the oxidants in ozonation experiments. In this case, the invariant competition value may be caused by the net effect of the sum of the oxidation products which scavenge $OH^{\bullet}$ at a rate which is comparable to that of the parent molecule.

Equation 19, as well as experimental experience, shows that the Ω_M value increases linearly with the concentration of $OH^{\bullet}$-radical–consuming solutes [52]; Ω_M is a composite value. It can therefore also be expressed by the sum of the contributions of all solutes S_i present, when the concentration of these solutes is multiplied by their appropriate oxidation-competition coefficient ω_i:

$$\Omega_M = \Sigma(\omega_{M,i}[S_i]) \tag{23}$$

Comparing Equations 23 and 19, we may express the individual coefficients by:

$$\omega_{M,i} = \frac{1}{\eta'} \frac{k_i'}{k_M'} \tag{24}$$

Ω Values and ω Coefficients for Different Types of Solutes

If a Ω_M or ω_M value is known for one solute A (Ω_A; ω_A), then the corresponding values for a solute B (Ω_B; ω_B) can easily be calculated since $\Omega_B/\Omega_A = (k_A'/k_B')$. The k_M' values can be found for many solutes in published tables, e.g., Dorfman and Adams have critically compiled hundreds of known reaction-rate constants of $OH^{\cdot}$ radicals with different types of organic solutes [53]. In addition, an extended list of data has been published by Farhataziz and Ross [54]. Some of the data are given in Figure 10. Most of the constants k′ lie within 10^9 to 10^{10} $M^{-1}sec^{-1}$, but the selection of entries in Figure 10 emphasizes also the existence of slowly reacting compounds, e.g., organic acids of low molecular weight.

Ω Values and ω Coefficients for Different Types of Waters

Examples of individually determined ω values are given in Table II. Where a comparison is possible, the ratios of these values, relative to

Table II. Examples of Oxidation-Competition Coefficients (ω) Exhibited by Different Types of Solutes (S)[a]

S	ω
HCO_3^-	0.1 g/mol
CO_3^{2-}	3.0 g/mol
NH_3	2.0 g/mol
NH_4^+	0.0 g/mol
Octanol	0.5 g/g
Glucose	0.13 g/g
DOC[b]	0.4 g/g (s = 0.1)
Dextran T-10	0.03 g/g
Fulvic Acid	0.2 g/g

[a]Measurements are based on the oxidation of benzene as a reference micropollutant [55,56].

[b]Dissolved organic carbon after filtration (0.45-μ filter). This ω is a mean value determined on 22 Swiss lake and river waters. s = standard deviation of single values.

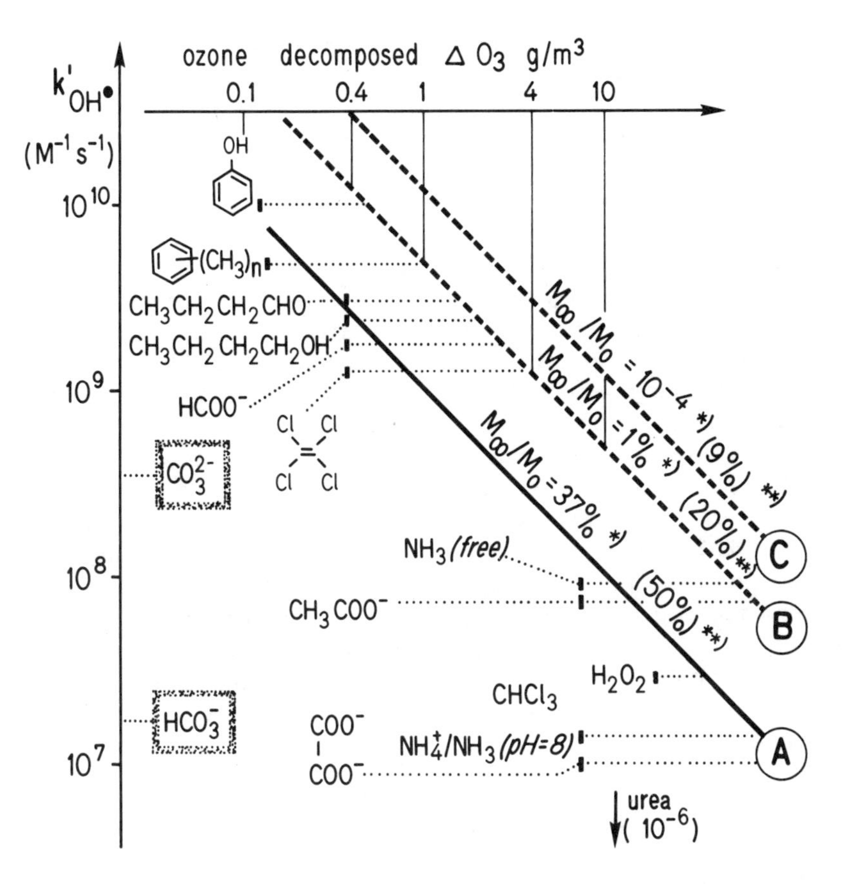

Figure 10. Examples of oxidations initiated by $OH^{\cdot}$-radical reactions. Vertical axis: reaction-rate constants k′. Horizontal axis: amount of ozone required to be decomposed to reduce the solute concentration to the relative value indicated on the diagonal elimination lines. The scale on the horizontal axis is calibrated for a water in which $\Omega_{benzene} = 1$ g/m³ (typical value for Swiss lake water, see Table 7). (*) values for a batch-type or ideal plugflow reactor (see Equation 22); (**) values for a completely stirred reactor (see Equation 28)

each other, range within a factor of two within those expected from the relative values deduced from published rate constants for $OH^{\cdot}$ consumption. These coefficients seem to be relatively independent of water type when only drinking waters are considered, and independent of pH in the range 7–10.5 (the pH effect on the ionic speciations of the dissolved compounds must naturally be accounted for) [55].

As can be seen from Table II and Figure 10, carbonate ions (CO_3^{2-}) are about 30 times more efficient than bicarbonate ions (HCO_3^-) in scavenging oxidants. This means that the Ω value of a water that contains bicarbonate increases if the pH is increased such that HCO_3^- dissociates to CO_3^{2-} (pH > 9). NH_3, typically present in wastewater, is also expected to act as a dominant scavenger for consuming the oxidants if the pH is increased to a value where ammonium ions begin to form appreciable amounts of free NH_3 (pH > 8) [10,38].

Some examples of Ω_M determinations are given in Figures 11 to 13. In certain natural waters or in secondary treated municipal effluents of communal wastewater, the Ω_M values only achieve a steady value (i.e., a constant slope in Figures 12 and 13) after a primary immediate ozone demand D is overcome. Easily oxidizable material, e.g., nitrite, contributes to such immediate demands.

Table III lists examples of Ω values for representative types of waters. In a recent compilation of Ω_M values of different types of waters in Switzerland, a good correlation was found between the Ω_M value of the water and the concentration of the dissolved organic material: the oxidation-competition value Ω increases by 0.4 g O_3/m^3 per mg DOC/l present [56]. Thus, the ozone required to achieve a certain removal of a specific micropollutant increases appreciably with the amount of total dissolved organic material. These Ω values do not change by prefiltration or prechlorination of the raw waters.

Selectivity of OH$^\cdot$-Radical Reactions

The relative yield with which different solutes present in the same water are simultaneously eliminated by OH$^\cdot$-radical–type reactions can be calculated from Equations 19 and 21, if the relative rate constants of the OH$^\cdot$ reactions are known. For solutes M_1 and M_2 present in a batch-type or plugflow reactor, we deduce:

$$\frac{\ln([M_1]_\infty/[M_1]_o)}{\ln([M_2]_\infty/[M_2]_o)} = \Omega_{M2}/\Omega_{M1} \sim k_1' \times k_2' = k_{rel}' \tag{25}$$

Relative eliminations of two solutes from the same solution can be estimated from Figure 10; vertical lines of given amounts of decomposed ozone meet the horizontal lines of given reaction-type constants at different elimination lines.

Equation 25 is formally the same expression as Equation 9, which

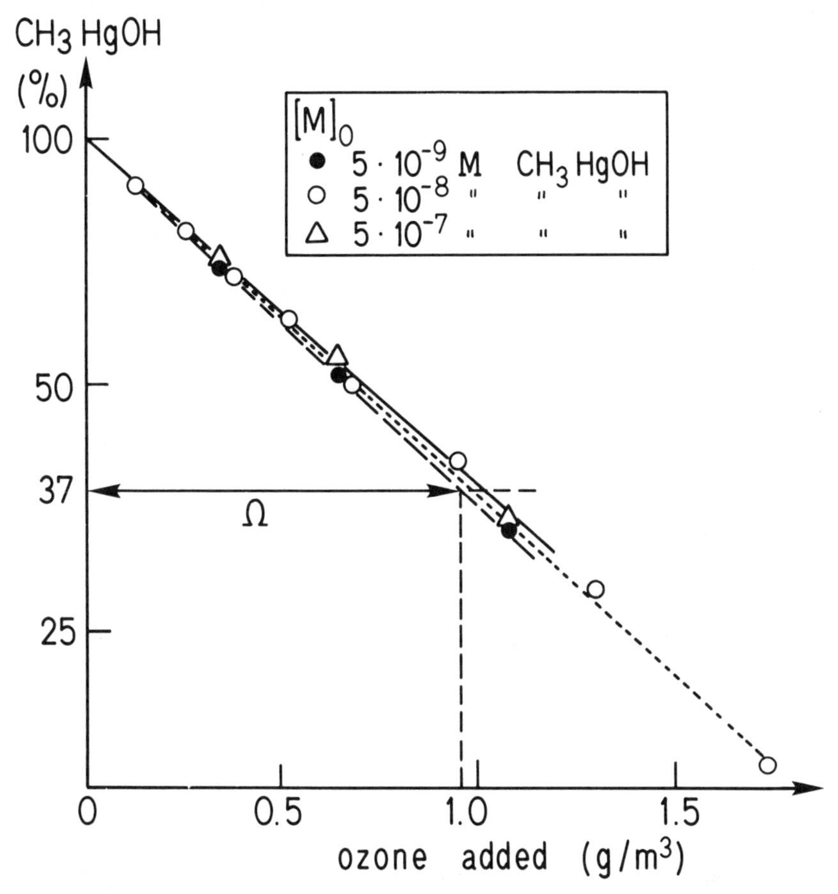

Figure 11. Mineralization of methylmercury hydroxide (CH_3HgOH) by the $OH^\cdot$-radical type reaction. The solutions containing 5 g/m³ octanol as a radical-scavenging-model-type impurity. pH = 10.5; 0.05 *M* sodium phosphate buffer; batch-type reactor [7]. This figure shows that Ω_{CH_3HgOH} is independent of initial concentration of methylmercury hydroxide and that it does not change with progressive ozonation.

describes the selectivity of the direct O_3 reaction. However, the relative rate constants for $OH^\cdot$-radical reactions are much smaller than the corresponding k'_{rel} values of direct O_3 reactions. The difference is evident if one compares the scales given in Figures 8 and 10. Due to this difference in selectivities, it is generally easy to determine if the direct O_3 reaction or the radical-type reaction is dominant.

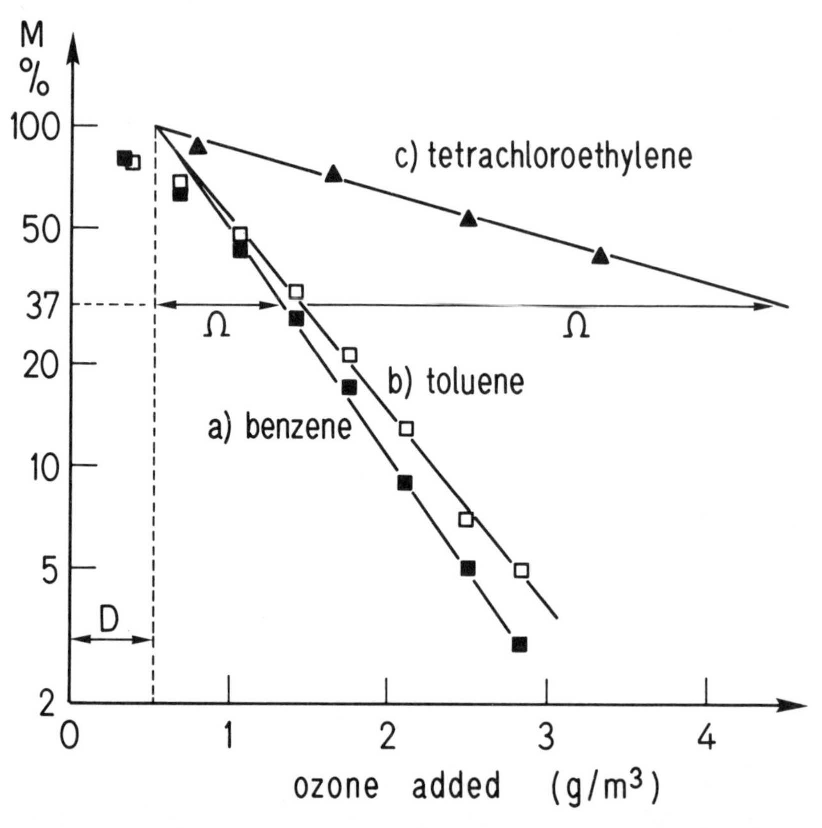

Figure 12. Oxidation by the OH•-radical–type reaction of different types of solutes spiked into water of Lac de Bret. (DOC = 4 g/m³; Σ[CO_2] = 1.6 mM; pH 8.3). (a) $[\text{benzene}]_o$ = 80 mg/m³; (b) $[\text{toluene}]_o$ = 100 mg/m³; (c) $[\text{tetrachloroethylene}]_o$ = 500 mg/m³. Batch-type reactor [7]. The slopes of the lines do not change with progressive ozonation. The relative Ω values are as expected from the different k′ values known for these solutes (compare Figure 10).

RATES AND SELECTIVITIES OF OXIDATIONS IN COMPLETELY STIRRED AND NONIDEAL PLUGFLOW REACTORS

We may summarize that if the rate of the oxidation of an individual solute in an ozonation process is regulated by the rate of the chemical reaction, and if the reaction proceeds in a batch- or plug-type reactor, the relative concentration of this solute decreases as:

$$[M]_t/[M]_o = \exp(-p_M) \tag{26}$$

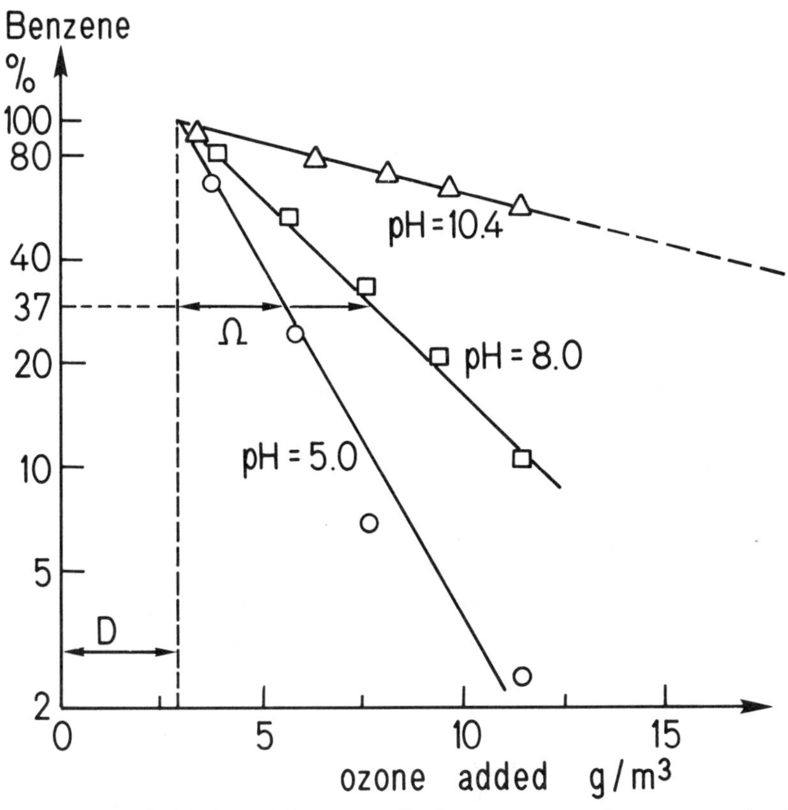

Figure 13. Oxidation of benzene spiked to a communal wastewater by the radical-type reaction (secondary effluent diluted to 0.6; DOC = 7 g/m^3; $NH_4^+ - N$ = 9 g/m^3; $\Sigma[CO_2]$ = 1.8 m*M*). pH varied by 0.05 *M* sodium borate buffer. Batch-type reactor. This figure illustrates that the Ω values increase with pH; at pH 8 part of the NH_4^+ and HCO_3^- present is dissociated to NH_3 and CO_3^{2-}, which act as more efficient $OH^{\cdot}$ scavengers (see Table II). At pH 10.4 this effect is even more pronounced.

where

$$\left.\begin{aligned} p_M &= p'_M + p''_M \\ p'_M &= k_M[\overline{O_3}]\bar{t} \\ p''_M &= \Delta O_3/\Omega_M \end{aligned}\right\} \qquad (27)$$

In Equation 27, p'_M corresponds to the elimination effected by the direct O_3 reaction (Equation 7) and p''_M corresponds to the radical-type reaction (Equation 22).

Table III. Examples of Oxidation-Competition Values in Different Types of Waters

Type of Water[a]	DOC (g/m^3)	Alkalinity (meq/1)	Ω (g/m^3)
Lake Constance, Switzerland[b]	1.1	2.4	0.6
Lake Zürich, Switzerland[b]	1.6	2.5	0.6
Baldeggersee, Switzerland	4.0	3.9	1.2
University Lake, Chapel Hill, NC[b]	5.7	0.4	2.4
Groundwater, Dübendorf, Switzerland[b]	0.6	6.0	1.0
Groundwater, Schaffhausen, Switzerland[b]	0.2	4.8	0.9
Groundwater, Samedan, Engadin, Switzerland[b]	0.1	0.6	0.25
Secondary Effluent, after Nitrification	13	2.5	3

[a] All pH values = 7.5–8.1. For detailed specifications of waters, see Hoigné and Bader [56].
[b] Raw waters from municipal waterworks.

Based on Equations 5 and 20, we may also describe the overall effect of ozonation processes for a completely stirred reactor for which the effluent concentration is equal to the concentration in the reactor. This becomes:

$$[M]_{\bar{t}}/[M]_o = 1/(1 + p_M) \qquad (28)$$

where p_M is the same as in Equation 27, and $\bar{t}$ is the hydraulic retention time in the stirred reactor. If we compare Equations 26 and 28, we may deduce that the elimination in a completely stirred reactor proceeds less effectively than in an ideal batch or plugflow reactor. A comparison of the efficiency of the different types of reactors is shown in Figures 8 and 10. The constant elimination lines drawn in these figures, indicated by A, B and C, for 37, 1 and 0.01% elimination in batch and ideal plugflow reactors, now correspond to eliminations to only 50, 20 and 9%, respectively, for completely stirred reactors. Note that the difference in performance between the two types of reactors becomes more significant as the design value of elimination efficiency is increased.

The ozonation reactors used in practice often operate as nonideal plugflow reactors: in many reactors, the air bubbles rising through the columns seem to induce mixing over the whole column. In a few installations, this nonideal flow is corrected somewhat by the installation of two or more reactors in series. An ideal plugflow, however, is still not achieved.

The selectivity with which a solute M_1, relative to a solute M_2, is oxidized in a completely mixed reactor is:

$$\frac{[M_1]_o/[M_1]_{\bar{t}} - 1}{[M_2]_o/[M_2]_{\bar{t}} - 1} = k_1/k_2 \quad \text{or} \quad k_1'/k_2' \qquad (29)$$

for a direct O_3 reaction or for a radical-type reaction.

This expression can be compared with Equations 12 and 25, which were derived for batch-type and ideal plugflow reactors. The selectivity with which a pair of solutes is eliminated in such a mixed reactor may again be estimated from Figures 8 and 10, but taking into account that the lines A, B and C now represent eliminations to only 50, 20 and 9% of the initial concentrations.

RELATIVE EXTENT OF DIFFERENT OXIDATION MECHANISMS

The foregoing section described how the two main oxidation pathways, the direct O_3 reaction and the $OH^{\bullet}$-radical–type reaction, lead to the oxidation of solutes in water. It is now advantageous to visualize the relative importance with which these different pathways contribute to the overall effect.

The extent of the direct O_3 reaction is a function of the ozonation time multiplied by the concentration of ozone present in the solution (see Equation 6). The $OH^{\bullet}$-radical–type reaction, however, increases with the amount of ozone decomposed and from which the $OH^{\bullet}$ radicals arise (see Equation 21). The situation is illustrated in Figure 14, where the horizontal axis represents $[\overline{O_3}]t$ and the vertical axis represents ΔO_3. For each type of water there will be a correlation between these two parameters: curve A traces the correlation estimated from measurements of the lifetime of ozone for a typical Swiss lakewater at pH 8 (Lake Zürich). If the pH is raised by one unit, the ozone would decompose about three times faster (line B): at pH 9, about three times more $OH^{\bullet}$ radicals are released within a given ozonation time than at pH 8, provided that the ozone concentration is kept the same in both cases. The relative importance of the $OH^{\bullet}$-radical–type reaction would therefore increase.

To compare the oxidation effects on a particular solute M, the horizontal scale in Figure 14 has to be calibrated with the rate constant k_M, and the vertical scale with $1/\Omega_M$, (compare p_M' and p_M'' parameters in Equations 26 and 27). This renders both scales dimensionless.

As an example, a fictitious straight line C is drawn into Figure 14, approximating the situation of ozone decomposition in a nonpolluted surface water of pH 8. This line is transformed from Figure 14 to Figure 15, taking into account different k_M and Ω_M values. The resulting line C′

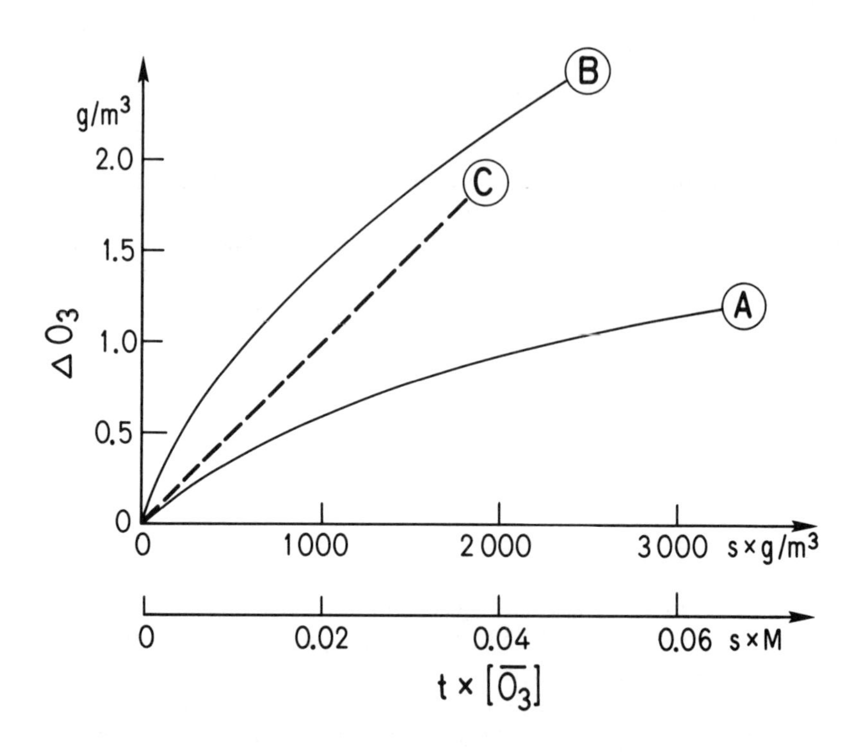

Figure 14. Amount of ozone decomposed to secondary oxidants (OH˙ radicals) vs ozonation time and ozone concentration. (A) correlation estimated from measurements on Lake Zürich water (pH 8; Σ [CO_2] ~ 1.3 meq/l; DOC = 1 g/m³). (B) correlation for similar water, after pH is raised to pH = 9; (C) correlation for ozone decomposition in a fictitious water. Approximate correlation value for a surface water (pH 8–9). (Line C is used for construction of lines C′, C″ and C‴ in Figure 15.)

would represent the effect of an ozonation process on a substance of $k_M = 250\ M^{-1}sec^{-1}$ and $\Omega_M = 1$ g/m³. Such reaction constants correspond to those of anisole in a good surface water (compare Table I for k_M value and Table III for typical Ω_M values, and assuming that Ω anisole $\sim \Omega$ benzene). For a compound of such a high (or higher) reaction-rate constant k_M, the effect of the direct reaction of ozone (value of p' parameter) clearly predominates. The line C″ is constructed for $k_M = 50\ M^{-1}sec^{-1}$ and $\Omega_M = 1$, i.e., for a compound such as xylene. For such a compound, the direct O_3 mechanism (p'_M parameter) and the OH˙-radical-type mechanism (p''_M parameter) are of comparable magnitude. The line C‴ is based on a $k_M = 10\ M^{-1}sec^{-1}$ and $\Omega_M = 1$ (such values correspond to

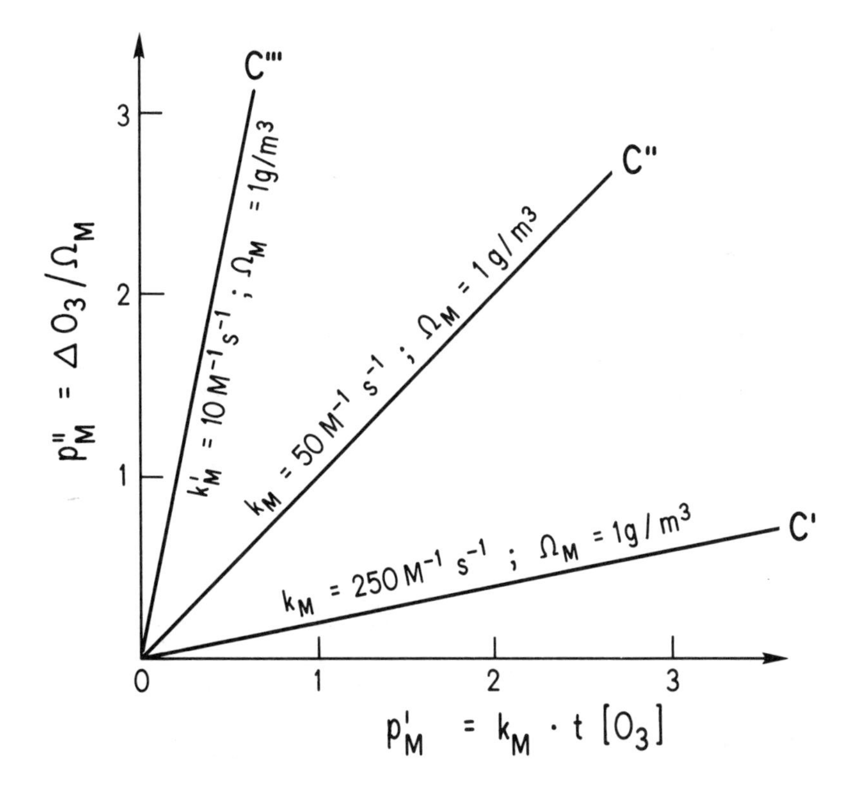

Figure 15. Extent of OH$^{\bullet}$-radical–type reaction (p''_M) vs extent of direct O_3 reaction (p'_M) in a fictitious water in which the ozone decomposition follows the ideal line C of Figure 10.

toluene). For solutes of such low (or lower) reaction-rate constants, k_M, the direct O_3 reaction is insignificant when compared with the extent to which the OH$^{\bullet}$-radical–type mechanism contributes to the oxidation in relatively clean waters.

When comparing the above examples with the list of rate constants presented in Table I, we may conclude that only a small fraction of all possible micropollutants in a water is eliminated simultaneously by the direct reaction of ozone and by the free radical type mechanism. For most compounds, either one or the other pathway of oxidation is clearly predominant. Therefore, situations encountered in practice are generally easier to handle than they might appear in the more general description given here.

The elimination of different types of organic micropollutants, which we observed in water of Lake Zürich, can well be quantified by the data shown in Figure 15 [6,57].

QUALIFICATIONS OF OZONATION PROCESSES

The direct O_3 reaction and the OH$^{\cdot}$-radical–type reaction, which is also initiated by ozonation of water, lead to two different branched trees of different alternative product formations. These trees will overlap with some of their branches as shown in Figure 16. From the foregoing sections we may conclude that the sizes of these two trees can both be quantified whenever the types of dissolved material, water and ozonation process are specified.

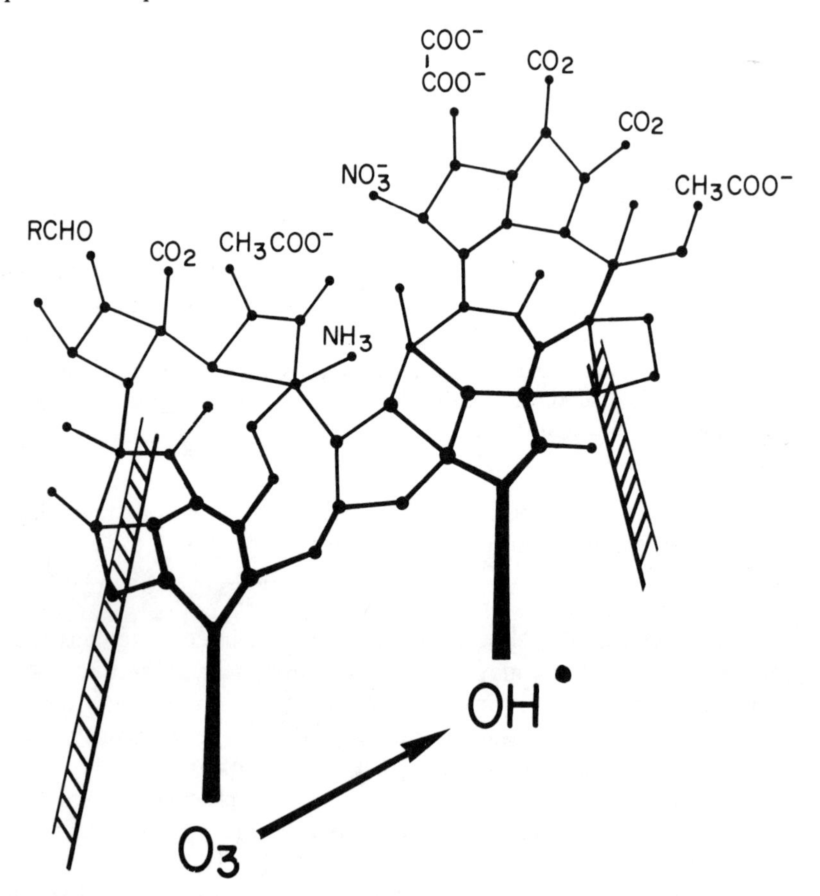

Figure 16. Entangled branched trees of alternative product formations initiated by the direct O_3 reactions and by the radical-type reactions [26].

It appears that most experiments in the hygienic domain have been performed under laboratory conditions favoring the O_3 tree. Organic chemists who work with nonaqueous solutions placed a ladder against this tree. With this ladder, at least the lowest branches of alternative product formations may easily be reached. Since the direct O_3 reactions are fairly selective, this O_3 tree is characterized by a few main branches of different daughter products.

In the discussion of the elimination of inert types of impurities, however, the branched tree of products grown by $OH^{\bullet}$-radical reactions becomes larger and more important. This tree also increases whenever the pH of the water is raised and more ozone is decomposed during the process. A ladder goes up even to this tree: the ladder built by radiation chemists who studied $OH^{\bullet}$-radical reactions. From the viewpoint of the classical ozone chemist, this $OH^{\bullet}$ tree of oxidized daughter products might often have been hidden by the more familiar O_3 tree. The tree of products formed by $OH^{\bullet}$-radical reactions generally will show more branches; the $OH^{\bullet}$-radical reactions are of low selectivity and may attack organic molecules in many different positions.

In special types of waters, even further trees of reaction products must be accounted for, e.g., in the ozonation of seawater and other waters containing bromide ions, a tree growing by secondarily produced hypobromite ions should also be considered (compare Figure 5). This tree would grow on account of the ozone and on account of the $OH^{\bullet}$ tree. Similar complications may also arise due to other circumstances. Therefore, to test the validity of the assumption made, experiments will still be necessary whenever special types of waters are considered. The basic information presented here should, however, help to plan experimental tests.

ACKNOWLEDGMENTS

The author would like to thank Heinz Bader for submitting many of the experimental data before a detailed publication was possible, as well as Philip Singer and Jim Davis for reviewing the manuscript and for their valuable suggestions. I thank Werner Stumm for his continuous interest in this project and for his engagement which made it possible to carry out such extended studies at our institute.

REFERENCES

1. Bailey, P. S. "The Reactions of Ozone with Organic Compounds," *Chem. Rev.* 58:925–1010 (1958).

2. Bailey, P. S. "Organic Groupings Reactive Toward Ozone: Mechanisms in Aqueous Media," in *Ozone in Water and Wastewater Treatment,* F. L. Evans III, Ed. (Ann Arbor, MI: Ann Arbor Science Publishers, Inc., 1972), pp. 29–59.
3. Bailey, P. S., Ed. *Ozonation in Organic Chemistry, Vol. 1, Olefinic Compounds* (London: Academic Press, 1978).
4. Oehlschlager, H. F. "Reactions of Ozone with Organic Compounds," in *Ozone/Chlorine Dioxide Oxidation Products of Organic Materials,* R. G. Rice and J. A. Cotruvo, Eds. (Vienna, VA: International Ozone Association, 1978), pp. 20–37.
5. Rice, R. G. "Reaction Products of Organic Materials with Ozone and with Chlorine Dioxide in Water," paper presented at the International Ozone Association Symposium on Advanced Ozone Technology, Toronto, Ontario, 1977.
6. Hoigné, J., H. Bader and F. Zürcher. "Kinetik and Selektivität der Ozonung organischer Stoff im Trinkwasser," in *Wasser Berlin—'77* (Berlin: Colloquium Verlag AMK Berlin, 1978), pp. 261–276.
7. Hoigné, J., and H. Bader. "Ozone and Hydroxyl Radical-Initiated Oxidations of Organic and Organometallic Trace Impurities in Water," in *Organometals and Organometalloids: Occurrence and Fate in the Environment,* F. E. Brinckman and J. M. Bellama, Eds., American Chemical Society Symposium Series No. 82 (Washington, DC: American Chemical Society, 1978), pp. 292–313.
8. Hoigné, J., and H. Bader. "Beeinflussung der Oxidationswirkung von Ozon und $OH^{\cdot}$-Radikalen durch Carbonat," *Vom Wasser* 48:283–304 (1977).
9. Peleg, M. "The Chemistry of Ozone in the Treatment of Water," *Water Res.* 10:51–55 (1976).
10. Hoigné, J., and H. Bader. "Ozone Initiated Oxidations of Solutes in Wastewater: A Reaction Kinetic Approach," *Prog. Water Technol.* 10: 657–671 (1978).
11. Hoigné, J., and H. Bader. "The Role of Hydroxyl Radical Reactions in Ozonation Processes in Aqueous Solutions," *Water Res.* 10:377–386 (1976).
12. Hoigné, J. "Aqueous Radiation Chemistry in Relation to Waste Treatment: An Introductory Review," in *Radiation for a Clean Environment* (Vienna, Austria, International Atomic Energy Agency, 1975).
13. Weiss, J. "Investigations of the Radical HO_2 in Solution," *Trans. Faraday Soc.* 31:668–681 (1935).
14. Taube, H., and W. C. Bray. "Chain Reactions in Aqueous Solutions Containing Ozone, Hydrogen Peroxide and Acid," *J. Am. Chem. Soc.* 62: 3357–3373 (1940).
15. Taube, H. "Chain Reactions of Ozone in Aqueous Solution. II. The Interaction of Ozone and Formic Acid in Aqueous Solution," *J. Am. Chem. Soc.* 63:2453–2458 (1941).
16. Taube, H. "Reactions in Solutions Containing O_3, H_2O_2, H^+ and Br^-. The Specific Rate of the Reaction $O_3 + Br^-$," *J. Am. Chem. Soc.* 64:2468–2474 (1942).
17. Bielski, B. H. J., and J. M. Gebicki. "Species in Irradiated Oxygenated Water," in *Advances in Radiation Chemistry, Vol. 2,* M. Burton and J. L. Magee, Eds. (New York: Wiley Interscience, 1970), pp. 177–279.

18. Williams, P. M., R. J. Baldwin and K. J. Robertson. "Ozonation of Seawater: Preliminary Observations in the Oxidation of Bromide, Chloride and Organic Carbon," *Water Res.* 12:385–388 (1978).
19. Rook, J. J., A. A. Gras, B. G. van der Heyden and J. de Wee. "Bromide Oxidation and Organic Substitution in Water Treatment," *J. Environ Sci. Health* A13(2):91–116 (1978).
20. Crecelius, E. A. "Measurement of Oxidants in Ozonized Seawater and Some Biological Reactions," *J. Fish. Res. Board Can.* 36:1006–1008 (1979).
21. Blogoslawski, W. J., L. Farrell, R. Garceau and P. Derrig. "Production of Oxidants in Ozonized Seawater," in *Proceedings of the Second International Symposium on Ozone Technology,* R. G. Rice, P. Pichet and M.-A. Vincent, Eds. (Vienna, VA: International Ozone Association, 1976), pp. 671–681.
22. Richardson, L. B., D. T. Burton, G. R. Helz and J. C. Rhoderick. "Residual Oxidant Decay and Bromate Formation in Chlorinated and Ozonated Seawater," *Water Res.* 15:1067–1074 (1981).
23. Meins, W. "Neuere Methoden bei der Langzeitdesinfektion im Schwimmbeckenwasser," *Arch. Badewesens* 26(2):88–92 (1973).
24. Umphries, M. D., R. R. Trussell, A. R. Trussell, L. Y. C. Leong and C. H. Tate. "The Effect of Preozonation on the Formation of Trihalomethanes," *Ozonews,* Tech. Paper Sec. 6(3):1–4 (1979).
25. Hoigné, J., and H. Bader. "Rate Constants of Direct Reaction of Ozone with Organic Compounds in Water (in preparation).
26. Hoigné, J., and H. Bader. "Ozonation of Water: Selectivity and Rate of Oxidation of Solutes," *Ozone Sci. Eng.* 1(1):73–86 (1979).
27. Schalekamp, M. "Die Erfahrungen mit Ozon in der Schweiz, speziell hinsichtlich der Veränderung von hygienisch bedenklichen Inhaltsstoffen," in *Wasser Berlin*—'77 (Berlin: Colloquium Verlag AMK Berlin, 1978), pp. 31–69.
28. Sievers, R. E., R. M. Barhley, G. A. Eiceman, R. H. Shapiro, H. F. Walton, K. J. Kolonko and L. R. Field. "Environmental Trace Analysis of Organics in Water by Glass Capillary Column Chromatography and Ancillary Techniques. Products of Ozonolysis," *J. Chromatog. Sci.* 142:745–754 (1977).
29. Dobinson, F. "Ozonisation of Malonic Acid in Aqueous Solution," *Chem. Ind.* (1959), pp. 853–854.
30. Gilbert, E. "Uber den Abbau von organischen Schadstoffen im Wasser durch Ozon," *Vom Wasser* 43:275–290 (1974).
31. Gilbert, E. "Reactions of Ozone with Organic Compounds in Dilute Aqueous Solution: Identification of Their Oxidation Products," in *Ozone/Chlorine Dioxide Oxidation Products of Organic Materials,* R. G. Rice and J. A. Cotruvo, Eds. (Vienna, VA: International Ozone Association, 1978), pp. 227–242.
32. Gilbert, E. "Chemische Vorgänge bei der Ozonänderung," in *Wasser Berlin—'77* (Berlin: Colloquium Verlag AMK Berlin, 1978), pp. 271–293.
33. Mallevialle, J. "Action de l'ozone dans la dégradation des composés phénoliques simples et polymerisés: Application des matières humiques contenues dans les eaux," *T.S.M-L'Eau* 75(3):107–113 (1975).
34. Gould, J. P., and W. J. Weber, Jr. "Oxidation of Phenols by Ozone," *J. Water Poll. Control Fed.* 48:47–60 (1976).

35. Struif, B., L. Weil and K. E. Quentin. "Verhalten herbizider Phenoxy-alkan-carbonsäuren bei der Wasseraufbereitung mit Ozon," *Z. Wasser-Abwasser-Forsch.* 11:118–127 (1978).
36. Kuo, P. P. K., E. S. K. Chian and B. J. Chang. "Identification of End Products Resulting from Ozonation of Compounds Commonly Found in Water," in *Ozone/Chlorine Dioxide Oxidation Products of Organic Materials,* R. G. Rice and J. A. Cotruvo, Eds. (Vienna, VA: International Ozone Association, 1978), pp. 153–168.
37. Yamamota, Y., E. Niki, H. Shiokawa and Y. Kamiya. "Ozonation of Organic Compounds. II. Ozonation of Phenol in Water," *J. Org. Chem.* 44:2137–2142 (1979).
38. Hoigné, J., and H. Bader. "Ozonation of Water: Kinetics of Oxidation of Ammonia by Ozone and Hydroxyl Radicals," *Environ. Sci. Technol.* 12: 79–84 (1978).
39. Eisenhauer, H. R. "The Ozonation of Phenolic Wastes," *J. Water Poll. Control Fed.* 40:1887–1889 (1968).
40. Doré, M., N. Merlet, T. Blanchard and B. Langlais. "Influence of Oxidizing Treatment on the Formation and the Degradation of Haloform Reaction Precursors," *Prog. Water Technol.* 10:853–865 (1978).
41. Hoigné, J. "Influence of Oxidizing Treatments on the Formation and Degradation of Haloform Reaction Precursors by M. Doré et al., Discussion by J. Hoigné," *Prog. Water Technol.* 10:1115–1117 (1978).
42. McGuire, M., I. H. Suffet, R. C. Patrick, B. Schultz, T. Gittelman and M. Shanahan. "The Effect of Oxidizing Agents on the Removal of Trace Organics from Drinking Water," paper presented at the Symposium of the International Ozone Association, Los Angeles, CA, 1978.
43. Elia, V. J., C. Clark, K. T. McGinnis, T. E. Cody and R. N. Kinman, "Ozonation in a Wastewater Reuse System: Examination of Products Formed," *J. Water Poll. Control Fed.* 50:1727–1732 (1978).
44. Hewes, C. G., and R. R. Davison. "The Kinetics of Ozone Decomposition and Reaction with Organics in Water," *Am. Inst. Chem. Eng. J.* 17: 141–147 (1971).
45. Somiya, I., H. Yamada and T. Goda. "The Ozonation of Nitrogenous Compounds in Water," paper presented at the International Ozone Association Symposium on Advanced Ozone Technology, Toronto, Ontario, 1977.
46. Neytzell-De Wilde, F. G. "Treatment of Effluents from Ammonia Plants. Part III. Ozonation of Amines in an Effluent from a Reforming Plant Serving an Ammonia Complex," *Water SA* 3:133–141 (1977).
47. Riley, T. L., K. H. Mancy and E. A. Boettner. "The Effect of Preozonation on Chloroform Production in the Chlorine Disinfection Process," in *Water Chlorination: Environmental Impact and Health Effects, Vol. 2,* R. L. Jolley, H. Gorchev and D. H. Hamilton, Jr., Eds. (Ann Arbor, MI: Ann Arbor Science Publishers, Inc., 1978), pp. 593–602.
48. Glaze, W. H., G. R. Peyton, R. Rawley, F. Huang and S. Lin. "A Comparison of Ozone and Ozone/UV for Destruction of Refractory Organic Compounds in Water," paper presented at the International Ozone Association Ozone Technology Symposium, Los Angeles, CA, 1978.
49. Lawrence, J. "The Oxidation of Some Haloform Precursors," paper presented at the Third Congress of the International Ozone Association, Paris (1977).

50. Maier, D., and H. Mäckle. "Wirkung von Chlor auf natürliche und ozonte organische Wasserinhaltsstoffe," *Vom Wasser* 47:379–397 (1976).
51. Hoigné, J. "Note on the Haloform Formation Potential of Pre-ozonized Water," in *Oxidation, Techniques in Drinking Water Treatment,* W. Kühn and H. Sontheimer, Eds., EPA Report No. 570/9-79-020 (Karlsruhe, FRG: Engler-Bunte-Institute, University of Karlsruhe, 1979), pp. 327–330.
52. Hoigné, J. "Ozone Requirement and Oxidation-Competition Value of Different Types of Waters for the Oxidation of Trace Impurities," in *Oxidation, Techniques in Drinking Water Treatment,* W. Kühn and H. Sontheimer, Eds., EPA Report No. 570/9–79–020 (Karlsruhe, FRG: Engler-Bunte-Institute, University of Karlsruhe, 1979), pp. 271–290.
53. Dorfman, L. M., and G. E. Adams. "Reactivity of Hydroxyl Radical in Aqueous Solutions," National Bureau of Standards, NSRDS-NBS 46 (Washington, DC: U.S. Government Printing Office, 1972).
54. Farhataziz and A. B. Ross. "Selected Specific Rates of Reactions of Transients from Water in Aqueous Solution. III. Hydroxyl Radical and Perhydroxyl Radical and Their Radical Ions," National Bureau of Standards, NSRDS-NBS 59 (Washington, DC: U.S. Government Printing Office, 1972).
55. Hoigné, J., and H. Bader. (in preparation).
56. Hoigné, J., and H. Bader. "Ozonation of Water: Oxidation-Competition Values of Different Types of Waters Used in Switzerland," *Ozone Sci. Eng.* 1(4):357–372 (1979).
57. Zürcher, F., H. Bader and J. Hoigné. "Verhalten organischer Spurenstoffe bei der Ozonung von Trinkwasser," in *Proceedings of the Symposium Cost 64B,* H. Otl, Ed. (Berlin, in press).

INDEX